Electronic Media
and Technoculture

Rutgers Depth of Field Series

Charles Affron, Mirella Jona Affron, Robert Lyons, Series Editors

Edited and with an introduction by
John Thornton Caldwell

Electronic
Media and
Technoculture

Rutgers
University
Press
New Brunswick,
New Jersey

Library of Congress Cataloging-in-Publication Data

Electronic media and technoculture / edited and with an introduction by John Thornton Caldwell.
    p. cm. — (Rutgers depth of field series)
   Includes bibliographical references (p.   ) and index.
   ISBN 0-8135-2733-3 (cloth : alk. paper) — ISBN 0-8135-2734-1 (paper : alk. paper)
   1. Digital media—Social aspects.   2. Technology and civilization.   I. Caldwell, John Thornton, 1954–   II. Series.
HM851 .E54 2000
306.4'6—dc21   99-038182

British Cataloging-in-Publication data for this book is available from the British Library

Manufactured in the United States of America

For Paul N. Caldwell

Who showed me at an early age
how electronics traced forces
far more profound than
their circuits

Contents

Electronic Media
and Technoculture

John Thornton Caldwell

Introduction: Theorizing the Digital Landrush

> *"The force of microelectronics will blow apart all the monopolies, hierarchies, pyramids, and power grids of established industrial society. It will undermine all totalitarian regimes. Police states cannot endure under the advance of the computer."*
> —*George Gilder,* Life After Television[1]

> *"(Bill) Gates promises not to rule TV Land."*
> —USA Today *headline, May 5, 1998*[2]

(Re)inventing the New

Disregard for a moment expectant tenth-century theologizing about the end of the first millennium. A very different *telos,* or final cause, has captured our own imaginations. Unlike the god of the first millennial apocalypse, ours is electronic, endlessly responsive, and digital. The future, in fact, has never been so systematically envisioned, aggressively analyzed, and grandly theorized as in the present rush to cyber and digital. In the mid-twentieth century, questions about media technologies-and-society first emerged as scholarly handwringing about the deleterious sweep of electronic media and information technologies in mass culture. Now, questions about new technologies and social and cultural impact are no longer limited to intellectual parapets in the academy but are pervasive parts of day-to-day discourses in newspapers, magazines, television, and film. Public consciousness about "contextual" social, aesthetic, and ideological issues, that is, is prevalent even in media organs that negotiate and promote the stuff destined to "move off the shelves": new hardware and software "applications." No longer a proprietary province restricted to enlightened, prophetic intellectuals like Lewis Mumford, Jacques Ellul, and Neil Postman, one is as likely to find critical anxieties, not just in the digerati's *Wired*, but in rather mundane places like the *L.A. Times*, the *Wall Street Journal*, the Sci-Fi Network, the Discovery Channel, and (at the top of the aesthetic pantheon) syndicated reality programs like *Access Hollywood.*[3]

This gush of workaday, real-time analysis of the future may seem to provide a bandwagon opportunity for media studies scholars, who have waited patiently for their place at the table of commercial, public discourse. But technospeak also issues an effusive density of conceptual clutter that may actually muddy the waters and meanings of digital technologies. In its haste to nail down useful angles and immediate interpretations of the new, cybergush tends to eclipse consideration of many nagging and problematic intellectual and social issues. Popular culture's articulations about the electronic future, for example, tend to look past an entire tradition of insights about electronic media in film and television studies. Although the digerati tend to presuppose novelty in framing fundamental critical questions, many of their chosen problems have churned historically through research and accounts of radio, television, cable, video art, and motion pictures. The master-paradigm of novelty does effectively drive editorial policy on the digital. Yet the historical amnesia that novelty presupposes serves few who care to adequately understand the social and cultural logics of new media. The problems and promises of computer and digital media, that is, can be usefully engaged and understood within a more extensive consideration of entertainment culture, and electronic and broadcasting media in particular. Among other things, approaching digital media and technoculture in this way—as historical formations animated by continuities as much as invention—serves to buffer several present orthodoxies: first, that digital is somehow inevitably emancipatory; and second, that high-technologies (idealized as "economic engines" by states and nations) emerge from a kind of immaculate conception of entrepreneurialism.

This book aims to anchor contemporary discussions of the digital future within a critical tradition of insights about the media arts, society, and culture. Rather than canonizing truisms from Silicon Valley trades or MIT's Media Lab, one can grasp the stakes involved in technological conversions by examining prototypical analyses of conversions to earlier electronic media technologies and their ideological and psychological dimensions. Historifying digital rhetoric and theory is essential for media studies scholars and students. Considering alternative critical explanations about technologies also ensures that truisms about electronic culture always remain in guarded dialogue. This collection of works addresses both dimensions.

Grounding theory about new media in this way means thinking broadly about symptomatic and recurrent issues. Doing so means that even when all of this year's cutting-edge technologies become obsolete next year (as they will), the critical capacity to engage new media technologies will not similarly be rendered to its own ash heap of past trends. Thinking critically about digital contexts and actual practices means forcing technologies to speak to questions and interests other than their own.

Attempting to collect representative, influential works in a single volume like this one helps chart a critical field, but the process also inevitably teases out several inherent contradictions. Even as anthologizing includes certain works, it writes other important works out. Even as anthologizing by nature demonstrates a range of different approaches, it may wrongly suggest (on this topic) that there is such a thing as a single, agreed-upon phenomenon or field called "digital culture," or "technoculture," or even "new media." In fact, the current, mediated age is neither singular, unified, nor stable. This very instability does not mitigate the need for conceptualization in a general sense, but places a premium on developing critical competencies that can look through transitional stages, technological conversions, and cultural and identity boundary-crossings. "Technoculture," as I am considering it here, is a shape-shifting phenomenon. Thinking critically about it means being able to step through those shifts with some sense of equilibrium and direction.

To give up notions of comparative, cross-institutional, and cross-cultural critical engagement because digital media is pitched commercially as unified and total is to acquiesce to the very cult of the new that fuels the consumer electronics industry. This book attempts to question such a unity, by highlighting a schema of resilient discursive and intellectual markers. By charting the diversity of the field in this way, I hope that the book provides readers with a more informed way to orienteer through digital culture. Electronic forms, after all, now define our cultural landscape, even as our history is commercially celebrated as a media spectacle in perpetual transition.

The pages that follow include a sequence of discussions about how the field has been construed and delineated in public discourses. I intend, through these introductory discussions, to raise a succession of issues and problems in four general areas: first, by describing recurrent framing "questions" (a foundational component of traditional media theorizing); second, by mapping the public "paradigming" ritual that epitomizes digital culture in the age of the "pitch" and "vaporware"; third, by delineating the different institutional cultures that originate and "author" the discourses of the future; and fourth, by discussing how theorizing new media also stands as a form of social and cultural positioning.

John Thornton Caldwell

Mapping Technoculture Questions

One way to start a book on the "theory" of any media form is to consider the kinds of basic questions that have defined the field. In classical aesthetic theory such questions attempted to define the arts and deliberate on issues of quality, value, and critical judgment. Classical film theory, in turn, waged ongoing theoretical dialogues about the nature of film form and film language, and the unique ways cinema structured experiential relationships with viewers. Traditional theorizing about broadcasting and electronic media (when it was not posed explicitly as a form of scientific, experimental method) tended toward macroscopic questions: about relationships between forms of transmission and social practices; between political-economic issues and cognitive or behavioral effects. No singular method or consensus about questions similarly defines the emerging fields of new media.

A cross-section of current digital talk, however, does provide a wealth of possibilities for understanding the stakes in electronic culture. Some framing questions about electronic culture are highly speculative and abstract. How is time and/or space altered within digital-mediated experience? How has "agency," or human activity, been altered by those who use digital media? How are notions of audience and pleasure reconfigured by the digital? These kinds of questions tend to reduce new media and the digital, as objects for analysis, to singularities. Such approaches allow—even in complex explanations of space, time, or agency—for implicitly universalized notions that there is such a thing as "a" digital technology. Philosophical questions of these sorts still frequently drive critical humanities studies of new media technologies. It is somewhat ironic that academic cultures tend to underestimate technical complexities and industrial specificities by essentializing "new media" in this way. Ironic, because the value of poststructuralist scholarship itself is showcased by the ability to shift framing categories; to marshal and "individuate" nonessentialist analysis; and to exploit "multiplicities" in the ways that theoretical problems are conceptualized. Technologies are seldom granted the same multiplicities by theorists.

An entirely different set of basic theoretical questions, however, ground industrial, technical, and trade discourses on new media. Industry tends to favor questions about how concepts, organization, and economics predispose practitioners for or against innovation, adaptation, and usage. How, for instance, does hierarchical organization work against creativity and technical innovation? How do high-technology synergies enable and reinforce adaptation? How can technical efficiencies be maximized? How do network externalities (the effect each new connection has on others in the network) add value to existing digital applications and intellectual properties? If there is, to take but one example, current industrial consensus on any issue, it comes from widely shared, public confidence about something called "convergence"; the idea that image, sound, and data will come together in a single delivery system. President and CEO of NBC Television Robert Wright responded to digital convergence this way: "You can be the guy who is still trying to make buggy whips or you can be the person who's always 10 years ahead and broke. Both are fraught with peril." By saying this, Wright was also summarizing television *management's* complicated, value-added, postnetwork, industrial equation; one that attempts to balance cutting-edge technology with consumer trends, and content provision with proven market share.[4] The folks at AOL (America Online) and broadcast.com—far less driven by program content issues and markets as yet unseen—flaunt convergence as the technical way that every imaginable form of digital content can cohere in a multiplexed stream through their proprietary gateway. Professional head-nods to the contrary, convergence means different things to different corporations. More so than managerial ones, *technical* questions in the converging industries tend to presuppose determinisms of one sort or another that are even more causal. How does networking and web activity exacerbate security and surveillance possibilities? How can increased bandwidth (a bigger digital "pipeline") provide leverage for audio/video/textual content "streaming" (onto the Net) or for revenue streaming (back to the provider)? Industrial discourses are seldom, as in academic cultures, untethered from social and economic contexts.

This book aims to lay out a series of influential statements and positions that problematize the critical-scholastic penchant for conceptual complication and technical reductivism, and the very different industrial penchant for technical determinism and marketing as natural law. To ask, as a reader of this collection might, for example, how aesthetic issues are informed by forces of political economy is to refuse to provide the neat interpretive package that either academic or industrial cultures presuppose. To ask how digital technologies structure experience or how users reorganize everyday life with new media, is to upend both cyberuser fantasies (of autonomy and radical authorship) and corporate fantasies of consumer responsiveness (buyer needs driving hardware design). Although new media and technoculture analyses in the context of film studies may suggest otherwise, digital screen and narrative forms are not unrelated to either political-economic determinations or domestic and familial relations. "Boutique" and visionary cyber practices, in addition, are not usefully disconnected from the mundane social and cultural ground of the vast majority of users and consumers. To continue to favor boutique cyberpractices over the realities of wired domesticity, for example—in academic high-theory—simply plays into the hand of hardware manufacturers. Those corporations, obviously, move more product by stroking niche-market fantasies of autonomy and individuation. By insisting, in this book, that questions about digital and new media experience be asked alongside institutional questions, I hope to encourage critical competencies in readers able to rise above the discursive roller coaster of a digital history theorized but not yet written.

Paradigming: Conceptual and Technical Survival in the Age of Pitch and Vaporware

"Hard drives on steroids."
—Sales pitch on storage capacity
for nonlinear at NAB[5]

"FireWire gets fast and fat."
—Cover story in Digital Video[6]

"Upgrade your head."
—NewMedia Invision 98 Festival theme[7]

Thomas Kuhn's work on the history of science, and Hayden White's on the narrativization of history, stimulated a slightly older structural truism: that history can no longer be credibly viewed as either a natural development or as a universally discernible, phenomenonological reality.[8] And while historiographic work since has allowed academics to continue "discovering" the play of framing paradigms (Kuhn) and discursive tropes (White) that lurk behind and construct culture, few analysts have acknowledged the extent to which a facility with inventing and "spinning" paradigms—both self-conscious competencies—now defines mass and electronic culture. Nicholas Negroponte alluded to the centrality of the process when he marveled at the "repurposing" of Madonna—"a thirty-four year old former cheerleader"—who for $1.2 billion was signed by Time-Warner as a "multimedia" enterprise in 1992.[9] Madonna's repurposing worked because Time-Warner effectively overhauled (re-troped) its figurative Madonna-model and extensively promoted (pitched) her as a cross-industry, "multimedia" phenom.

Digital culture, in particular, is fueled by a frantic economy of public paradigming and troping. The success of initial stock offerings for high-tech start-ups, for example, depends upon the process. Corporations rely not simply upon proprietary technical paradigms, but on marketing ideals tied to public consciousness of their product's individuated strengths and market niche. The high-stakes game of promoting vaporware (one corporation's next-generation product, still in development or pending release) is another example. Hyping vaporware can sabotage the high sales of a competitor's current product; but it can also send every buyer to the sidelines to wait, thereby hurting those selling the vapor as well. Theorizing a product in advance can cause markets to seize up, anxiously. Failing to actually deliver a theorized product in a timely fashion can lead to a corporation's demise. Selling vaporware, therefore, stands as both a corporate theoretical exercise and a marketing high-wire act.

My construal of high-technology troping and corporate niche-ing as forms of "theory"-in-practice may elicit unease in the critical, academic establishment. But corporate paradigming of this sort does provide the kinds of explanatory aspirations that academics presuppose in general theory. Media paradigming also attempts to provide general explanations for a broad range of conceptual and technical phenomena. The practice also frequently provides heuristic or interpretive models of the sort that have driven analysis in what has come be termed "critical studies" or "critical theory" in the humanities. The foundational value of technical paradigming and conceptual spinning is far from novel for MBA-types trained in marketing and public relations. Film and television theory, on the other hand, has more patiently looked past the marketing clutter of media culture to excavate deeper master codes: the gaze, glance theory, the mirror stage, suturing, gatekeeping, narrowcasting, and so on. But castigating marketing tropes as manipulative "bumpf" is shortsighted for scholars attempting to understand new media technologies. Given the frantic pace by which paradigms are invented, and because corporate troping does involve real-world technical practice, scholars would benefit by considering many of the symbolic, and occasionally problematic, relations between academic master-paradigms and industrial master-paradigms. Consider the parallels and tensions between the two discursive practices in the table that follows.

Academic Tropes (figure)	(promise)	**Industrial Tropes** (figure)	(incentive)
PULL TECH	Agency Choice	PUSH TECH	Target Delivery
INTERACTIVITY (Agency)	Responsiveness	VALUE-ADDED ENTERTAINMENT	Any interactivity increases value of linked properties and entire brand
CONNECTEDNESS (Space)	Community Identity	GLOBALISM STATUS CUES TASTE CULTURES	Postnational trade, digital habitus,[10] market segmentation
VIRTUALITY (Travel)	Simulation Virtual Reality	SPECTACLE HDTV	Digital FX, hi-res, alongside multiple low-res carriage and data
PRESENCE (Time)	Simultaneity Nowness Sense of Shared Time	BRANDING	Ensuring a shared, global, "Disney experience"
VELOCITY (Travel, Space)	Travel Speed	SERVER PROXIMITY GRAZING	Proximity and speed to digital vaults, versus PC databases

John Thornton Caldwell

BANDWIDTH (Time)	Real-time Processing Multitasking	PIPELINE	Broadband delivery, eliminates download, "always-on" features
NARRATIVITY (Agency)	Non-linearity Branching	MULTIPLEXING	Bundled streaming, piggybacked e-commerce
CONVERGENCE (Space, Agency)	Artisanal Localism Multimedia Desk-top Leverage[11]	CONVERGENCE (Corporate)	Merger synergies Horizontal integration, diversification
INDIVIDUATION (Agency)	Antidote to Information Clutter	NICHE MARKETING VIDEO-ON-DEMAND	Smart-agents, "bots" Personal profiling
PROSTHESIS (Agency, Space)	Out-of-body Gender Slippage Cyborg Extensions	CONSUMER-USER INTERFACE	Information appliance, Customized consumer

For each academic digital paradigm (pull technology, interactivity, individuation), the industry has a corresponding, and not unrelated ideal (push programming, value-added entertainment, niche marketing). For each theoretical promise (agency, community, collapsed space, and hyperspeed), there is a corporate incentive (targeted delivery, demographic segmentation within the network, and branded sites, portals, and servers). It may be that the value of academic theory lies in its resolute penchant for inculcating—and conceptually practicing—technocultural possibilities. Industrial theory, on the other hand, makes habitual the notion that each new technology and trope fulfills the value-free logic of responsive delivery. The notion is couched less as a consumerist task than as an obsessive and therapeutic responsiveness to the needs of the new media subject. As Harold Innis's pre-analog research suggested over a half-century ago, the speed of any new technology increases the centralization of power within any communication system.[12] Whether this accurately describes the hyperspeed of the digital era, the concept certainly characterizes a house fantasy for new digital corporations. Even in the diffused, global network, each start-up imagines creating a nodal center for its digital properties, guaranteed only by mastering a speed-inducing regimen.

Institutions Author the Future: Wired versus Screen

> "The fully wired world—the Metaverse, where people interact with the totality of gesture and facial nuance essential to good human communication. . . . I am interested in developing an infrastructure for telepresence."
>
> —James Crowe, CEO, Level 3, on internet telephony[13]

Introduction: Theorizing the Digital Landrush

*"My PC's having an identity crisis. It thinks it's my
stereo system."*
–Phillips Electronics advertisement, for PC, 1998[14]

*"Eventually, content will prevail on the internet just
like it's prevailing now on cable."*
–Michael Eisner, CEO, Disney,
on net programming[15]

On November 16, 1998, a grand coalition of corporate players, pushing America's transition to digital, webcast a multimedia event on the National Association of Broadcasters' (NAB) website that they touted as "The Dawn of Digital Television." Three different television and broadcasting trades—*Broadcasting and Cable, TWICE*, and *Digital Television*—simultaneously celebrated the initiative as the group "turned on the switch."[16] Their aim was to provide in-depth analyses of what the digital era would mean to broadcasters, programmers, manufacturers, advertisers, viewers, and consumers. The Dawn of Digital campaign led with stories like "The Medium They Couldn't Kill," which wrongly suggested that television had played a progressive visionary role in forging digital against all odds— which it had not. An article in the campaign entitled "The PC's Still Missing in PC/TV" betrayed television's smug sense of accomplished dominance over a computer industry still mired in technical incompatibilities, posturing, and speculation.[17] At the same time, trade press accounts in the information and computer trades, *PC World* and *Wired*, continued discussions of cultural and economic issues tied to the Web, browsers, mergers, and the anti-trust soap opera unfolding between the Justice Department, Netscape, and Microsoft. In the same month, Netscape allied with America Online (AOL), eliciting a wave of press that celebrated a future still tied to effective Net navigation and value-added, corporate, connectivity alliances. In the same month, PBS aired a three-hour comprehensive documentary entitled *Nerds 2.01: A History of the Internet*, alongside the publication of a book by the same name, to make sense of the flurry and front-page rhetoric about digital.[18]

Both the NAB's Dawn of the Digital campaign and the multimedia homage to Silicon Valley, *Nerds 2.01: A History of the Internet*, worked earnestly to sum up the digital age. Both were overdetermined efforts to historify and provide a master interpretative scheme for digital. Neither account, however, matched, mirrored, reflected, mimicked, or even overlapped with the other. Obviously, the digital era is either several very different things; or the same thing conceptualized in two different, and incompatible, ways, within parallel universes.

An archaeology of these divergent worlds reveals a fundamental difference not just in rhetoric but in foundational concepts as well. One discursive universe consistently construes technologies within the logic of the "screen." The other universe consistently construes new technologies within the logic of the "wire." When NAB broadcasters decided to web-cast programming that hypes the entertainment value of high-resolution digital sound and image, they perpetuate their long-standing bias toward content, delivery, and spectatorship. When Silicon Valley and cyberpundits boast about global reconfigurations due to digital, they are (like AT&T and telecom corporations before them) legitimizing an equally long-standing commitment to "connection" and wired connectivity. Consider in this re-gard that seventy years before television "went digital," film "went sound." When film finally talked, it also ignored a number of radical formal alter-natives in order to make the new sound technologies fit a proven cinematic deference to story and narrative. Bordwell, Thompson, and Staiger refer to the accommodation of sound technology as a form of "trended" rather than radical change.[19] Both classical Hollywood and the digital NAB work discursively and technically to ensure that cultural practices accommo-date, discipline, and mainstream technical change so that new technologies fit within their preeminent paradigm: the screen experience.

Computer culture, on the other hand, muses not on "trended change" but on "trigger apps"; or "killer applications" that will trigger shifts in both technology adaptation and cultural practice. Amazon.com has simply been one recent example of a new cultural practice (on-line merchandising) made possible by the killer app of the Web and http code. These divergent approaches—wired versus screened—suggest that digital technoculture incorporates alternative conceptual approaches that reflect institutional differences as well. To think and to theorize about digital cultures today means at least acknowledging the difference between the NAB and Comdex, Tektronix and Microsoft, entertainment brands and start-up/venture capital alliances.[20] Institutions and commercial com-munities that are clearly distinct and different from each other, that is, "author" the new digital discourses. Institutional authoring of the digital is no less significant than the critical discourse on digital.

Theorizing digital culture, then, does not mean simply examining how theorists have articulated and described cultural effects. It also means paying attention to how theorization itself is a culturally generated prac-tice, produced and circulated in and by specific professional communities. I have referred to but two of these communities: the wired world of com-puter culture and the screen-oriented world of digital broadcasters. Many other groups have weighed in on the digital debate as well, with different perspectives and interests, including Hollywood filmmakers. Slack has elaborated a useful three-part model of possible relations between tech-nology and society (e.g., traditions of technology assessment, alternative technology, and Luddite practices).[21] Examining current digital culture, however—especially entertainment forms—suggests an even more exten-sive taxonomy of positions and forms of theoretical engagment. This tax-onomy can be usefully examined, as it is in the pages that follow, in terms of commercial discourses, academic discourses, and counterpractices.[22]

Commercial industries publicly perform their digital-theoretical competencies, but do so in ways specific to the economies and professional cultures of each industry. One way to understand these differences is to consider the central object or focus that legitimizes and "authorizes" theoretical talk in each industry. Specific corporate cultures, that is, have centered their respective discourses around four different paradigms: the box, schedule, network or connectivity, and content. Computer industry (Intel, IBM, Apple, Compaq, Microsoft) trade publications, for example, have traditionally elaborated *box-centric* digital formulations. Multitasking, processing power and speed, database management, gaming, and multimedia production have all underscored the centrality and importance of the ever-higher performing PC or digital workstation. The broadcasting and cable industries (TCI, Time-Warner, NBC, QVC, the HDTV Grand Alliance, and the NAB), on the other hand, intervene in both digital talk and commercial alliances based on their economically proven *schedule-centric* precedents. Advertising, demographic targeting, network affiliation, syndication, event "webcasting," programming the Net (like TV), and merchandising the Net like (QVC on) cable television all make the now-cluttered Web look increasingly as much like "old media" as "new media." The networking, portal, and telecom industries (Oracle, Netscape, Yahoo, America Online, AT&T, MCI, snap.com), in turn, conceptualize and promote *network-centric* visions of digital culture. For these "authors," deregulation, connectivity, browsers, multiplexing, and broadband streaming all spur merchandising efficiencies on the World Wide Web. Finally, the Hollywood film and television industries (as distinct from broadcasters) provide their digital alliances with a *content-centric* discursive predilection. The buzzwords "branding," repurposing, entertainment value, syndication, ancillary marketing, and content provision drive promotional and marketing "theorizations" by studios (Sony/Columbia, Disney, and Warner Bros.).

Various players, however, make even the constellation of digital Hollywood far from monolithic. Cinematic production values, for example, animate the discourses that circulate around state-of-the-art digital imagers (Pixar, Digital Domain, PDI, Industrial Light and Magic, Silicon Graphics).[23] Tensions over piracy and artists' (and studio) rights in the age of digital, on the other hand, spur anxious critical (and promotional) practices on the part of the MPAA, the ASC, the DGA, and the WGA.[24] Reading the media trades alongside this anthology—especially for readers new to the field of electronic media—will highlight key differences among these institutional authori(zi)ng sites. *New Media, Daily Variety, Wired, Inter@ctive Week, Info World, Broadcasting and Cable,* and *Digital Video* all "act out" specific institutional alliances and economic interests in ways that make their analyses and generalizations about digital media and culture timely, acutely delimited, and problematic.

Academic scholarship has seldom flagged its own institutional bias as unabashedly as the media trades. Yet the classic tripartite sensibilities of theorizations on new media—as either utopian, utilitarian, or apocalyptic—make some sense given the disparate academic contexts and funding sources that produce such discourses. At least five genres of academic theory are worth considering in this regard. First, the field of technology assessment emerged only after Congress proposed a House Committee on Science and Astronautics in 1967, and the U.S. government established the Office of Technical Assessment in 1972. Research focused on economic feasibility, impact, new technology exploitation, productivity, and policy development all fit a utilitarian, government-generated agenda that saw new technologies, by definition, as benign, or at least inevitable. During the cold war, such technologies were sanctioned from the start as a necessary key to national progress. A second, very different and more apocalyptic tradition, however, had emerged in the far less grant-driven (and far more technologically disconnected) world of humanities departments. Lewis Mumford, Jacques Ellul, and Neil Postman all launched pessimistic condemnations of the threats new information technologies posed to humane culture and democratic values. Such broadsides tapped into fairly widescale cultural anxieties about the future.

Yet scholars in a third tradition—those that actually took film and television seriously as a starting point—provided far more complex and systematic explanations of the nature and significance of media technologies; explanations that were neither utopian nor apocalyptic. After film studies and communication were deemed legitimate as doctoral research disciplines in the academy, scholars like Bordwell, Thompson, and Staiger, Schatz, and Winston elaborated the complex interrelationships among mode of production, screen form, technological standardization, and professional practices.[25] Regulation and political economy also framed systematic media studies like Thomas Streeter's *Selling the Air* and Dan Schiller's *Digital Capitalism.*[26] A fourth tradition, what might be termed "grand theory," has had less patience for institutional specifics and industrial complications. Marshall McLuhan, J-L. Baudry, Jean Baudrillard, and, most recently, Pierre Levy and Paul Virilio have all favored speculative philosophical analysis, and a penchant for recurrent generalizations about total ideological and experiential effects.[27] Fifth, and finally, cultural studies has shifted media theorizations to notions of "consuming agency." This focus on the audience-user disregards the feasibility-centric bias of technology assessment, the industry-centric logic of film/television mode-of-production studies, and the utopian technotheologies of grand theory. The ethnographic methods utilized by David Morley, Ann Gray, and Ellen Seiter bracket key questions about new media within the lived social spaces of users. Sociologist/psychologists like Sherry Turkle help deindustrialize new media by highlighting cyberspace as an arena for gender-shifting and virtual sexual experimentation; a move that construes the digital as a fundamentally new force in the performance of personal "identity."

On the one hand, the fact that psychologists find identity, ethnographers find local culture, and political economists find industry in the digital seems fairly logical. On the other hand, few scholars have noted how commercial culture mirrors academic culture by similarly bracketing the personal, the cultural, and the industrial. The "multiplicities" and "performativities" of gender and sexuality, for example, that seem so progressive in Sherry Turkle's *Life on the Screen* and Judith Butler's *Gender Trouble* merit consideration in light of more recent popularizations of cyberculture as an opportunity for romance, binary coupling, and "virtual dating."[28] Advice- and rule-governed books like *Cyberflirt: How to Attract Anyone, Anywhere on the World Wide Web* makes cyberspace seem more like a fairly traditional singles bar or a personal column than a "transgressive" opportunity.[29] Recognizing, in this way, how institutional practices author(ize) divergent talk about new media may suggest to readers other opportunities for "comparative" analyses. Such comparisons include: (1) academic preoccupations with "identity" versus network-centric industrial preoccupations with "navigation" (both deal with user agency); (2) prescriptive theorizations about "girl culture" and consumerist empowerment in the age of "postfeminism" versus recurrent marketing and ethnographic findings that continue to show fairly rigid gendered relations in video gaming, personal homepages, and domestic media environments (both theorize gender, but one promotes, the other describes); (3) the branded digital properties of Time-Warner and Disney versus the packaging and scheduling practices of CNN-Interactive and ABC, their respective multinational cohorts that package content (different institutional authoring practices within the "same" corporations); (4) counterdiscourses that indict the lack of diversity and cyberaccess for marginalized groups (see Gómez-Peña's essay here) versus the target-marketing of those same groups by on-line services (the "success" of commercial on-line services for Hispanics launched in 1999). Doing comparative studies like these—with an eye toward cross-institutional logic and authoring—helps keep generalizations about digital and new media "effects" and "uses" tied to specific historical formations. Such an approach also suggests that one community's critical-theoretical intervention frequently becomes another institution's marketing opportunity.

The map of institutional authoring that I have sketched out here (even by excluding some communities such as computer science, artificial intelligence [AI], and what some consider the Web's killer application, the on-line sex industry) shows how extensive and variable accounts of electronic and digital culture will continue to be. Some of the theoretical positions, mostly in the industrial groups, work hard to maintain hardware or software as starting points—an order and protocol that ensures both technical causality and cultural effect as a consequence of technology. Counterpractices, on the other hand, presuppose, in general, the need for some form of intervention as a fundamental component of both conceptual and practical involvement. Academic theory, finally, ranges from the productivity bias and ideological disinterest of technology assessment, to the industrial analysis of mode-of-production historians, to the paranoid, surveillance, subjectivities of grand theory. But while some academic practices share an interventionist component with countertheory, others view worldly change as an outdated, even romantic aspiration, given the extensiveness of the present postmodern malaise.

As an institutional authoring site for theory, academic culture itself faces a digitally induced crisis of identity. As public funding dries up, claims that privatization, distant learning, and partnerships with industry will rescue higher education neatly downplay the challenge that the corporate digital culture brings to academia. Cherished notions about intellectual property, labor rights, and academic freedom pale in the shadow and draw of sponsored research and "real-world" development. Digital technologies, in some way, provoke or exacerbate each of these issues. High-technologies, after all, are now the bones offered universities by corporations to create what are publicly celebrated as "win-win" solutions. The divisive urgency—passed down by administrators—to secure "external support" will probably mean that the institutional authorship of technoculture by academia will become even more contentious and disparate rather than less. Even the nature of new media *theory*, therefore, is directly enmeshed in a contentious, socioeconomic collectivity. We speak cleanly, from seminar rooms and conferences, of a mediated cultural object. But it speaks us in return.

Digital and new media technologies are more than mere tools or machines. Technology and technoculture alike include the total means and systems that provide and enable digital machines and artifacts to circulate in culture. In this broader sense, then, surveying formative theoretical positions means considering as well who speaks and to what purpose.

Pre-Curve Theorizing: Historical Amnesia, Cultural Myopia, and Intellectual Orienteering

New technospeak is frequently as much a discourse about the future as it is an account of electronic machines. As such, digital theory also risks erasing history in problematic ways. George Lipsitz has characterized the emergence of postwar American television culture as being fueled by historical amnesia.[30] Selling modern consumerism meant that broadcasting, advertising, and electronic media had to reconstruct the American home and family as entities whose traditional values could be met with new electronic appliances, and whose Old World and ethnic identities could be replaced by a sense of common cause that was both contemporary and consumerist. In some ways academic theorizing about digital media risks a comparable form of amnesia, but with a different focus.

Many academic writers position their analysis by theorizing "ahead of the curve." That is, the value of a theoretical account is implicitly tied to the relative place the account has on the perceived cutting edge of a technological phenomenon. This relative position of radicality is generally determined by the nature of the chosen "object" upon which the theory is based. For example McLuhan's "global village" became a fashion in the very years that broadcast satellites were ushered into public consciousness. "Video art" had a bite in late 1960s artworld practice because video portapaks were in the hands of a very few male visionaries rather than homeowners or corporate types. Their very scarcity made them radical to art critics in the context of galleries and museums. Hackers became chichi to poststructuralist theorists in the 1980s because they fit a much earlier sociological role; one established in modernist culture for avant-gardists, anarchists, and dandies. By the late 1990s, however, all three of these cutting-edge "objects" were castigated as electronic antiquities. Broadcasting was couched as synonymous with Neanderthal by computer guru George Gilder. Art schools bailed on "video art" curricula in a rush to develop computer media and digital art studies programs. People who played disruptively on the Master's Net were no longer unique, either, but were slammed as flamers, lurkers, psychotics, or worse—the dark side of bulletin board activity and chatrooms. Players on the Net became the very consumers America Online was after.

By 1998 the trades were acknowledging the "failure" of the CD-ROM interactive revolution, even as universities were gearing up to develop interactive curricula and the production of CD-ROMs. Although producers had anticipated a fairly universal format for digital image and sound distribution, the DVD format as well was declared a financial disappointment less than two years into its far too protracted adoption. These recent "failures" evoke a litany of precursors: the "failure" of interactive cable in the 1970s, the "failure" of portapak activism in the 1970s, the "failure" of interactive television (by Time-Warner and others) in the years 1992–94. Books on the new are issued and critical prognoses offered during a new technology's pre-curve "moment," in a flurry of reflection on the promise of radical adoption. Such theorization tends to evaporate entirely when such technologies actually find their applications in society. The day-to-day is, apparently, far too mundane and useful for some theorists who gauge intellectual import on the conditions of early, technological scarcity.

This volume, unlike others, includes theorists who engage electronic media from *both* sides of the technology adoption curve: those that scan the horizon and research centers for cutting-edge promise and the potential for radical practice; and those that take seriously the important place that digital technologies play in lived experience and everyday life. In some ways, the pre-curve theorizers, even in their *postmodern* nonprescriptive cloaks, assume the traditional social position of *modernist* apologists. Discovery and radical practice have so informed academic culture that even the death of art cannot displace its hold. The truth is, post-curve theorizing is far less "sexy." For this reason much of the research on digital media culture, and audiences, takes place in social science, cognitive science, and communication departments and venues. Pre-curve theorizing, on the other hand, continues to stimulate scholarship in the arts and humanities, environments less burdened by practical, lived concerns than by the need for critical speculation and alternative practice.

Theories of contemporary technoculture are needlessly leashed if they do not at least acknowledge that the technical objects chosen for analysis and the theoretical points-of-view undertaken exist in relation to a concrete set of historical conditions. Technologies—whether cutting-edge and pre-curve or practical and post-curve—are fundamental parts of contemporary social formation. Pre-curve biases should be recognized as forms of intellectual promise and articulations of alterity. Post-curve bias can be engaged as meaningful evidence of socially symbolic practice. To denigrate either perspective is to produce a map of technoculture with vast uncharted areas, spaces lacking entire sets of important coordinates.

Contributors

This volume links key works in electronic media theory to enable critical dialogue under four different rubrics: first, theorizing technohistory and relations between "old" and "new media"; second, industrial considerations and producing technologies; third, audiences and consuming technologies; and, finally, identity processes and what used to be called "counterpractices." If this book charted the history of computer science, it might begin with Norbert Wiener's theories of feedback systems and cybernetics in 1948, or Claude Shannon and Warren Weaver's classical, mathematical model of information theory from 1949.[31] This volume, however, focuses on media and culture, broadly construed, rather than computer science. For this reason, arguments on electronic technologies by RCA's "general" David Sarnoff or Bertolt Brecht may be more useful in suggesting both the formative influences in the field and the conceptual stakes involved in how the field was defined. While Wiener and Shannon provided a kind of conceptual engineering that oriented technology developments, and Sarnoff and Brecht provided arguments for and against legitimacy, social control, and use, no one influenced the cultural proliferation of discourses on electronic media and the technological society more than Marshall McLuhan. I say this with some hesitation, for many scholars rightly point out the impressionistic, literary, and ultimately popular (therefore questionable) status of McLuhan's work.[32]

But while McLuhan's sloganeering made critical talk about television and media both popular and hip, academics generally refused to develop his insights in any acknowledged or systematic way. Two factors make the historic reaction to McLuhan's work worth reconsidering in this collection. First, the current wave of public attention given the digital does in fact privilege McLuhan's conceptualizations about the electronically mediated world. The inaugural issue of *Wired*, for example, ushered in the golden age of the Web and digital culture with an explicit celebration of McLuhan's influence. Second, many of the most recently celebrated concepts in academic media theory—interactivity, electronic presence, simultaneity and nowness, information as resource, glanced and distracted media perception, electronic technologies as extensions of consciousness, and electronic globalism—were all evoked, in sometimes problematic ways, by McLuhan. For these reasons, all of the essays that make up the first section of this book, "Theorizing Technohistory," can be seen—either explicitly or implicitly—as reactions or responses to the McLuhan legacy, a legacy built around utopian truisms about a global information society, mediated consciousness, and technological "revolution."

While celebrated in the corporate arena, McLuhan has frequently assumed the role of chief proponent and exemplar of "technological determinism." New media technologies emerge; society and human consciousness change as a result. Perhaps the most influential critique of this position came in Raymond Williams's 1974 book *Television, Technology and Cultural Form*. This work is most commonly cited for its insights on programming "flow" rather than individual programs as television's central organizing principle. The chapter included here, "The Technology and the Society," interrogates the very concept of determinism. Williams contrasts such views with "symptomatic" theories that treat technologies as essentially by-products of social processes that are in some ways already determined. To bridge the gap between determinism and symptomatism, Williams argues for the need, first, to examine the "social history of a technology," and second, to examine the "social history of the *uses*" of technology. This shift from McLuhan's perceptual-cognitive emphases in "mass" communications to Williams's view that technologies must be understood in terms of their specific social histories and privatized usage proved immensely influential in the field of contemporary television studies. Williams's focus in this chapter on institutional logic, ideology, and "mobile, privatized consumption" (arguably the basis for the commercial success of Net culture) would also, in the decades that followed, stimulate cultural studies work on electronic media in lived environments.

The utopian technological determinism of the 1960s was not simply questioned because of its historical and institutional glosses. An international tradition of activist theory also followed on its heels. Such work typically highlighted the fact that while technologies may be neutral, their uses and applications are persistently political in practice. Hans

Magnus Enzensberger, a German contemporary of Williams and influenced by the Frankfurt School, argued that even the technologies used by the "consciousness industry" to dominate society could be "appropriated" by activists and organizers in order to exploit their "emancipatory potential." Enzensberger's model of "mobilization"—one that rejected old-Left defeatism in the face of capital—linked media critique with a systematic plan for alternative production, together placed in the general service of cultural empowerment. Even as Enzensberger theorized how the media subjugated progressive potential through token, liberal forms—like public opinion forums, broadcast licensing, and fairness protocols—he laid out a call-to-arms for radical, alternative productions. These marching orders—decentralization versus centralized broadcasting, two-way transmitters versus reception-only receivers, mobility versus isolation, feedback and interactivity versus passivity, and collectivity versus professional specialization—may evoke the dated optimism of new-Left socialism, but they also prefigure digi-speak. Even though Enzensberger focused on analog electronic media—audio- and videotape production—his treatise provides a striking conceptual prototype for many current digital practices: from cyberpunk to current activist celebrations of the internet and the World Wide Web.

At the very same time that Williams in Britain and Enzensberger in Germany were writing about sociological, institutional, and political aspects of new media, French film theory had settled on a very different model of technological effect. Several influential, now canonized, essays formulated how dominant capitalist culture inscribed its regressive ideology into the very technical apparatus of the motion picture. Synthesizing the psychoanalytic theory of Jacques Lacan with Louis Althusser's revisionist Marxism, Jean-Louis Comolli provided an archaeology of the falsely "centered" representational urge that prefigured and determined the form that modern cinema took in Western culture.[33] Jean-Louis Baudry articulated how classical Hollywood cinema trapped its spectators in a prelinguistic developmental stage that produced a pliant spectator for capitalist culture.[34] These works were indeed revolutionary in that they linked media technologies to specific cultural ideologies and technical histories. By totalizing effects around a technological operation, however, the French psychoanalytic-ideological turn did in fact stimulate the emergence of the field of contemporary film studies. The work had little lasting effect, however, on thinking about new media technologies. No media technology or ideological effect, that is, proved to be as singular or as total as Comolli and Baudry proposed. Nevertheless, a wave of grand theory followed Comolli and Baudry under license of postmodernism. Jean Baudrillard totalized late capitalism around a unified theory of technical and cultural "simulation."[35] Paul Virilio's work started, like Comolli's, with some measure of historical specificity (systematic architecture studies) but evolved into a mystical and very problematic, Catholic cybertheology about technically induced, altered, space-time experience.[36] The ghost of McLuhan had strangely reappeared in French high-theory. Like their Canadian ancestor, Baudrillard and Virilio announced—without technical evidence or industrial grounding—total effects and unified global truisms. Unlike McLuhan's determined utopia, however, Baudrillard and Virilio now celebrated a technical dystopia that denigrated the material and the lived.

This cul-de-sac built of "ideological effect" was finally broken by a series of revisionist technical histories in the 1980s. Bordwell, Thompson, and Staiger's *Classical Hollywood Cinema* offered an exhaustive account of industrial organization and mode-of-production nuances that Comolli and Baudry had little apparent time for or interest in.[37] In the same year, Brian Winston's *Misunderstanding Media* took dead aim at the very "revolutionary" obsession that had fueled much of the work in media-and-society research. In the chapter included here, Winston offers instead a systematic model for how new technologies typically emerge. New media technologies cannot be theorized, according to Winston, without considering the interlocking forces of technical and scientific "competence," "ideation," and the presence of "supervening necessities." Winston's model is particularly useful in providing a method for giving technology studies the kind of historical specificity that Williams proposed but did not fully realize. Rather than predetermine an ideological agenda or psychological effect, Winston's model helps analysts sift through the complex historical and industrial factors that make some new media inventions viable while others—even superior alternatives—disappear. Certainly the start-up volatility and standards and format competition that define the present digital age demand a framework that can account for processes of technical innovation, adoptation, and diffusion.

The final essay in the history of theory section applies the tradition and arc of new media analysis directly to the growth of digital culture in the late 1980s. Bill Nichols's "The Work of Culture in the Age of Cybernetic Systems" makes direct reference to the revolutionary work of Walter Benjamin, "The Work of Art in the Age of Mechanical Reproduction." Whereas Benjamin argued that photographic reproduction destroyed the aura and conditions so necessary for the survival of artculture, Nichols suggests that the shift to cybernetic systems represents a change in mass culture that is no less fundamental. Nichols's essay is remarkable in its synthesis of disparate theoretical perspectives: Wiener's cybernetics with Benjamin's modernism; film theory with psychoanalytic modes of subjectivity; Bateson's communicative ecology with new legal issues in copyright and patent law. Nichols's essay provides a key turn in our survey "theorizing technohistory" as the first in the collection to grapple with the extensiveness of computer culture per se. Following Nichols, the "work of culture" includes not just computerized media technologies, but biological cybernetic systems in nature as well. Benjamin, Williams, and Enzensberger all worked from models of industrial capitalism. This essay introduces a schematic model for how cybernetic culture can be understood in relation to more recent developments as well: "entrepeneurial capitalism" and "multinational capitalism." Many of the essays that follow in the volume will revisit the significance of these economic formations, along with other issues highlighted in this account—corporate practices, categorical slippages, and computer-mediated subjectivities.

Most popular accounts of contemporary digital culture focus on either consumer usage or entrepeneurial origins. Many academic cultural studies, on the other hand, tend to presuppose that media technologies are best understood within lived or domestic environments. Of course, when academics emphasize the audience/consumer in this way, they may in fact only help to reinforce a basic industrial point-of-view. Corporate America, after all, justifies its R&D and business strategies on perceptions of "responsiveness" to the needs of the consumer. Examining where technoculture comes from, and how it is produced, is in many ways more difficult than asking the same questions about audiences, especially given the industry's penchant for proprietary isolation. As a social phenomenon, industry is arguably less transparent, and certainly less accessible to researchers, than lay cultures. The three accounts of "producing technoculture" included in the second section of this book take for analysis an "industry" far less singular or bounded than, say, the classical Hollywood cinema described by Bordwell, Thompson, and Staiger. Faced with the post-Fordist, multinational, corporate world of digital, these essays suggest a range of very different ways to think about the origins and forces that generate electronic culture.[38] Each account synthesizes disparate methodologies in some way. The first offers a political-economic critique of digital culture aimed at the "telematic ethics" of virtuality; the second, a "phenomenological" study of how cinematic and electronic representational practices envision and prefigure digital experience; and the third, a highly personal account of the emergence of VR technologies that is simultaneously autobiographical, ethnographic, and theoretical.

Arthur Kroker and Michael Weinstein's "The Theory of the Virtual Class" provides a pointed cultural critique of the digital world and the Net grounded in political-economic concerns. Kroker and Weinstein refuse to let the information industries dictate the symbolic terms used to legitimate the wiring of the world. Instead, they challenge celebrated notions of technological progress and digital emancipation by constantly forcing such terms to speak to issues of class, economics, and politics. Although the information superhighway is typically hyped as a triumph where individuals can "command information for whatever purposes they entertain," for the authors the information superhighway fuels the creation of a dystopia they term "technotopia." Kroker and Weinstein argue that the superhighway means the demise of human agency and the triumph of its substitute, the "expert program." While Kroker and Weinstein build their argument from a critique of transnational capitalism, they ultimately use it to make arguments about "ethics": first, that "virtuality" only succeeds by denigrating the human body and subjectivity; and second, that the "technical class" uses virtuality to exploit and abuse working classes and developing nations.[39] Kroker and Weinstein deconstruct what they consider to be the false promises of interactivity, the "hypertexted body" and "soft ideology," and then explain how the "will to virtuality" functions as an obsession for today's elite technological class. After providing a schema for how virtuality works to render ethics (economic justice, democratic discourse, social solidarity, and creativity) obsolete, the authors describe how the virtual class uses technical deregulation to exploit the demise of nation-states and the emergence of supranational trading blocs. Their analysis of virtual global strategies includes troubling notions that one would never read in the computer trades—including the argument that the virtual class continues to win by offering interactive, "creative participation" as a substitute for real employment and economic well-being.

Vivian Sobchack's article, "The Scene of the Screen," exposes a dimension in electronic media that Kroker's political economic perspective on culture by definition looks beyond. Utilizing a phenomenological framework that considers "lived experience" central, Sobchack interrogates what she considers to be fundamental experiential differences between cinema and electronic media. Like Nichols, she sets electronic media in opposition to cinematic modes, and links computer-mediated practices to the epochal shifts toward postmodernism and multinational capitalism. Cinema conjoins the "presentational and the representational" for viewers in order to provide a "thickened and concrete world" of experience, one that animates in the mind a sense of duration and memory. Electronic media, on the other hand, is "ultimately bound up in a centerless, network-like structure of instant simulation" and desire rather than either nostalgia or anticipation. The "lived-body" of the film spectator is thus replaced by the psychological sense of a "no-body" in electronic presence.

Sobchack's model makes links between the producing world and the viewing world, but it does so based not on social or industrial relationships, but on explications of subject-object relations. Sobchack's phenomenological account could serve, that is, as a bridge to the unit on "consuming technoculture" that follows this section. Yet I consider the exemplary use and analysis of motion picture and electronic "texts" in the "Scene of the Screen" to be of great value in understanding the intellectual and perceptual capital that fuels emerging technologies. Close textual analysis of the sort Sobchack integrates here provides substantive evidence that macroscopic sociological accounts of political economy and public relations never can. In this sense, Sobchack's analysis, from traditions of philosophy and film studies, suggests the important place that cultural icons and motion picture representations play in prefiguring popular consciousness about digital. *La jetée, Blade Runner, Repo Man,*

Terminator, and other industrial films prefigure experiential tastes and overdetermine the accepted symbolism of electronic culture and its technologies. While Winston might consider this textual evidence to play a role in technical "ideation," the viewing practices and relations Sobchack targets suggest both technical and cognitive reasons for a shift in the sense of electronic experience. Scholars would benefit by paying as much attention to how new technologies are textually and culturally "envisioned" as they do to how new technologies are industrially launched and commercially consumed.

The final essay in this section, Allucquère Rosanne Stone's remarkable "Sex, Death, and Machinery," weaves a story of the development of VR and cyberculture that is both autobiographical and ethnographic. Rejecting the grand abstractions of generalized social or ideological theories, Stone works through the emergence of the "virtual age" novelistically and inductively, by offering the reader what she terms a "set of provocations." Her thesis argues that electronic communication technologies, their discourses, and attendant social formations have created a trend toward increasing self-consciousness, isolation, and the displacement of shared physical space. These factors have placed a premium on the need for what she terms "textuality" and "prosthetic communication." Stone's work is significant in this collection: first, for its emphasis on play and pleasure (rather than work and tools), as determining paradigms in digital and VR development; second, for its application of poststructuralist perspectives (multivocality, situatedness, and "emergence" rather than cause, effect, or closure) in describing the virtual age; and finally, for the way Stone grounds her analysis in real social communities (a women's phone sex collective, The Atari Lab, and Xerox's PARC). This ethnographic dimension makes Stone's taxonomy of multiuser modes in VR and her accessible explications of digital "interactivity" even more compelling.

Media industries make huge investments in new technologies, new delivery systems, and new looks that together work to link producing cultures with consuming cultures. The next unit of this collection considers, in the context of reception, how new media technologies reproduce and/or make problematic issues of power, class, and gender. Many of the best recent writings on new media technologies mine their territories by showing how power and pleasures tend to be unequally distributed across the (now dizzying) array of delivery devices and available home entertainment electronic systems. Ien Ang's work has been most influential in clarifying the relationship between audience consumption and electronic media. The essay included here, "New Technologies, Audience Measurement, and the Tactics of Television Consumption," is taken from Ang's suggestively titled book *Living Room Wars.* In this account, Ang examines how the current multimediated home environment creates a crisis for an industry whose success depends upon objective audience measurement. While marketing research demands rationality—and a fixing of the audience/consumer as an "object" that acts for or against an entertainment product—users of electronic media evidence far more complex forms of behavior. New technologies like the remote, VCR, and DSB have encouraged complicated textual, perceptual, and cognitive actions—"zipping," "zapping," "time-shifting," and "churn." Attempting an objective measurement of these acts of audience control, however, finds more ambiguity rather than less. According to Ang, these audiences can never be "domesticated" or fixed within the Foucauldian "regime of truth" that governs marketing. Ang's thesis—that the "elusive tactics" of audience consumption subvert the industry's need to fix the audience as a commodity—holds as true for the internet as it did for television and cable. The current rush to advertising sponsorship on the Web continues to engineer this tension into many other digital developments as well.

Cynthia Cockburn's "The Circuit of Technology: Gender, Identity, and Power" moves past determinist and functionalist explanations to consider the "social shaping" of media technologies. Drawing on Bruno Latour's work on technoscience, Cockburn utilizes an "actor-network" model to help explain how new technologies emerge from "collectivities" comprised of various agents and interests. Rather than explain how technologies are "diffused" or "socially constructed"—tactics that presuppose a stable, localized "interest" behind the diffusion or construction—Cockburn describes how technology projects must continually "change shape and content as alliances are stitched together." Many "actors" must be "enrolled" for any new technology to be transformed into a "black box," or a standardized technical application. Cockburn's contribution is to interrogate this actor-network model over and against feminist theory. While the actor-network views "power" as a capacity, it shortchanges specific histories of domination. Feminism, on the other hand, assuming that power suppresses subjectivity, takes the liberation of subjectivity as its project. While the basic question for the actor-network model has concerned change, the basic question for feminism has been about continuity. Cockburn bridges these differences by suggesting that power deploys and utilizes—rather than represses—subjectivity. Feminism's insight that domination persists through successive technological changes means that analysts of technology must place renewed emphasis on how networks refigure these power relations with and as a result of each change. Cockburn's performative approach to actor-network study does not then simply locate "agency." It also underscores the importance of how power is played out in collectivities that elicit agency.

Helen Cunningham's "Moral Kombat and Computer Game Girls" provides a specific case study of a form of technology "gendering" theorized by Cockburn. Cunningham does this by utilizing what Ang would refer to as an "ethnographic" methodology. She takes as her project fallout from the overdetermined way that the press had constructed computer gaming as a violence-fixated, male enclave that promotes "addictive" social effects. Before considering how girls incorporate the computer game into their existing "bedroom culture," Cunningham summarizes how the Sega and Nintendo corporations had achieved and legitimized their market dominance in Britain. Promoted as "edutainment," computer games appeared to have some useful value without being strictly educational— and therefore "uncool." Cunningham rejects earlier research that argued that game culture is dominated by boys, by noting that the domestication of computer games actually gave girls increased access to technology. Cunningham makes genre distinctions among game types, describes how even violent games can be therapeutic, and cites evidence for how girls rationalize their use of computer games in telling ways. By showing how shifts from arcade to home altered media usage, how game-playing provided transgender identifications, and how denigrated consumer electronics helped develop technical skills for those left out of male-oriented computer culture, Cunningham demonstrates the power of close, ethnographic observation to challenge popular critical notions about new media.

The final contribution on technoculture consumption, Ellen Seiter's "Television and the Internet," provides an opportunity both for a consideration of computer and internet use, and for reflection on methodological issues involved in studying new media audiences. In challenging the many claims that the internet is going to revolutionize communications, Seiter calls for attention to the kinds of issues that most research in information science, management, education, and computer science tends to ignore. Public perceptions about computers and the cultural contexts in which they are used, and the images, sounds, and worlds found on computer screens, suggest that the internet may in fact reinforce cultural hierarchies of class, race, and gender. An examination of this process benefits by considering television audience research in relation to computer usage in the domestic sphere, and by considering the articulation of gender identities in popular genres and culture. "Television and the Internet" is taken from a larger study that develops a model for "ethnographies of media audiences." The selection included here describes the kinds of cultural connections that a systematic recording and analysis of this kind can elucidate. Seiter usefully elaborates a range of cultural issues and problems in three broad categories: (1) the relationship between television and computers (in terms of technology, economics, interactivity, programming, and home use); (2) gendered uses of computers at home (issues of control, space, leisure time, and age); and (3) computerized work (issues involving "pink-collar" employment, women working at home, and the role of education in technological-gendering). In some ways, Seiter's broader field research project suggests a corrective both to speculative theories about the *endless* performativity of gender possible in cyberspace, and to notions of *limitless* media empowerment possible in commercialized youth culture. Both notions are prescriptive scenarios that fail to describe the lived, social realities faced by many new media consumers. Both prescriptions also fail to highlight a key *generational* question operative in new media. Pink-collar, class, and familial constraints continue to co-exist extensively for many women—even if, and as, adolescent cyber-culture simultaneously offers its many lifestyle options and generational pleasures to those with enough economic and cultural capital to access and exploit them. Many theorists premise their explanations of new technologies on assumptions of radical discontinuity. Seiter reminds us that daily choices—about how work and leisure are organized around media—also ground the promises of technoculture in lived ways that can be either exclusionary or liberating, restrictive or enabling.

The fourth and final section of this book focuses on studies and proposals of new media practices that exist far from the marketed norms espoused by new media corporations and their publishing cohorts. Twenty years ago, when there was still vague consensus among intellectuals and artists about the split between commercial "insiders" and cultural "outsiders," this section might have been termed "alternative media." Against the dominant media industry these positions would be posed as "countermedia." But the very nature of "radical" has since become a pervasive and valued sign in the industrial world—even as it used to reside on the artculture fringes. One is as likely to find cutting-edge, "outsider" postures cultivated in Silicon Valley as in Soho. Venture capital looks for alliances that promise disruptive potential, even as major entertainment "brands" like Sony/Columbia and Murdoch/Fox scan the horizon for programming trend-fodder: signs of "unevenness" in cultural, fashion, and ethnic worlds. Institutions now complicate things by animating inside and outside through shifting and ever-permeable boundaries. Yet resistant media practices—especially ones that leverage and cohere around issues of identity—somehow proliferate.

Andrew Ross provides a compelling account in "Hacking Away at the Counterculture" of how popular and journalistic discourses on computer activities can dramatically invert widely held views on new media. Ross takes as his subject computer hacking and the cultural activities that surround, resist, and recuperate it. Ross traces the roots of hacking from 1960s counterculture to the hackers' bestselling notoriety as romantic technolibertarians in 1984. Their mantra: all users have a right to free access to all information. Their public spin: hackers are "rebels with a modem." To Ross this was "a principled attempt . . . to challenge the tendency to use technology to form information elites." What follows the golden age of hacking in the mid-1980s, however, is a detailed story of increasing governmental outrage, growing public scorn, and the transformation of corporate and journalistic opinion into a consensus "war-on-hacking." Ross's account stands as a valuable, historically grounded account of how delegitimation works to symbolically subvert—

and actually exploit for profit—threats from media activists and alternative producers.

Ravi Sundaram's "Beyond the Nationalist Panopticon: The Experience of Cyberpublics in India" is one of the few detailed analyses of how the shift to cyber and digital culture has impacted the postcolonial cultures of India. The electronic geography that Sundaram plots here is far less binary and monolithically resistant to the West than Third World media theory once would have supposed. What is perhaps most extraordinary about this account is how Sundaram synthesizes a range of frameworks not normally tethered in the same critical work: historical accounts of postcolonial development with Foucault; political analyses of leaders Nehru and Gandhi with discussions of postmodernists Jameson and Clifford; indigenous national development and communications policy with explications of grand digital-theorist Paul Virilio. Sundaram finds in this multimodal account three important "cyberpublics" locked in tension: those of the state, the transnational elite, and the activist movements. Only the state cyberpublic pursues virtuality to "reestablish the dream of nationalism in India." The transnational cyberpublic, by contrast, pursues accelerated access to the West as a solution to underdevelopment. The activist cyberpublic, finally, has set aside historic suspicions of Western technology in order to use digital to upset and transform technoculture. The topic of "Digital Third Worlds" is occasionally brought up—almost as a guilty afterthought—in computer conferences and publications. Anyone interested in generalizing about digital globalism, electronic media in international development, or "cybermarginality" would do well to consider the complicated historical struggle over digital that Sundaram has theorized in India.

The final entry, Guillermo Gómez-Peña's "The Virtual Barrio @ The Other Frontier" throws a different kind of wrench into the gospels of digital technoculture. Multimedia and performance artist Gómez-Peña rejects the easy, binary "us-them" distinctions that litter many academic discourses on digital. As a Latino experimental artist Gómez-Peña cruises through the "liminal"—or in-between—social ground that simultaneously connects and walls the cultures of the United States and Mexico. Even as he sarcastically muses on racial and ethnic barriers and ironies in net-culture, he rejects the easy role of cultural victim. Rather, this essay, like Gómez-Peña's performance art in general, is a toying provocation that raises all sorts of important but unresolved issues. Gómez-Peña's shifting registers of discourse, his juxtaposition of clashing identity symbols, and his pointed play with figures of speech all work to fuel broader goals: "to 'brownify' virtual space, to 'Spanglishize the Net,' and to 'infect' the linguas francas." Gómez-Peña refuses to let power create clean digital "insides" and "outsides," wholly separate Silicon Valleys and Tijuana macquiladoras. Gómez-Peña's electronic play of hybrid identities—a sensibility geared to urban life in cities like Los Angeles and elsewhere—makes grand theories of the digital suspect, and certainly impossible to impose.

Moore's Law of Digital Theorizing

Silicon Valley's most important hardware and software product cycles are governed by a truism set in motion by Intel Chairman Gordon Moore in 1965: "with price kept constant, the processing power of microchips doubles every 18 months."[40] The extreme and abrupt, upward geometric curve that illustrates this "law" of the memory chip arguably holds for writings about digital technologies as well: "explanations and theories about digital technologies double in quantity in increments tied to the diffusion of digital technologies in consumer culture." In the same way that PC buyers remorsefully (and inevitably) moan that RAM amounts and processing speed purchased last month are sure signs of outmoded obsolescence this week, students and analysts of electronic culture may recoil at the manic increase in books and essays on the digital as well. One thing is certain. Given the central place that electronic media and the digital continue to play in contemporary culture, and the stiff learning curve that many face when grappling with new hardware technologies, talk and texts explaining the surge will continue to have a premium. Even if society embraces a future involving "information appliances" and pervasive or "ubiquitous computing"[41]—where everything in the consumer's lifestyle (transportation and cooking and banking, not just entertainment and the PC) is connected to a single networked source—it is unlikely that diverse institutional theorizations will cease proliferating. Far from it. The singular matrix forecast may make variable conceptualizations on digital culture even more imperative and contentious, rather than less so.

   Dramatizing differences between new media and old media helps hype product, sell stock, and spin editorials. But doing so also shortchanges the odds of a successful Web launch, as many digital practitioners now discover. Anxious role-switching now defines the white-knuckles world of high-techs; where pre–buy-out corporation life spans are counted in weeks and months rather than decades. In speculative ventures to be the "first one in," traditionally branded studios and "content providers" recoil from their timid failures to own and profit from networked delivery as well. At the same time, conventional networking heavies get burned by underestimating the formidable obstacles of successful content creation. In a resolution-impoverished Net industry now obsessed with forcing the hand of broadband technical convergence, some now quietly concede that there may already be a workable "broadband network"; something called television. "Delivering eyeballs" has proven to be a daunting task in the digital era; one that mocks anyone with the compulsion to harden boundaries. With the millennium at hand, then, one nagging question returns. What becomes of critical theory when the showcased digital art, hacker, and counter practices that first set theory in motion fade; when boutique cyber aesthetics recede in the gush of the workaday e-commerce and e-business onslaught; when digital activity finally becomes as commonplace, pervasive, and ubiquitous as television?

The geometric public increase in digital-speak does not preempt the need for collections like the one at hand. Rather, the permeation of technodiscourses in commercial popular culture means that the need for a set of guiding coordinates and questions becomes even more paramount. The essays that follow—influential signposts in emergent electronic culture—suggest ways to conceptualize and scale the approaching curve of technoculture; suggest ways, that is, to orienteer a digital landscape that is volatile-by-design. Written at specific historical moments, these writers raise questions—about experience, cognition, pleasure, perception, and creativity—that have outlasted innumerable electronic products. Such issues will surely outlast the conveyer of new technologies that define us in a never-ending cycle, a cycle that takes as marching orders the claim "new/used/obsolete." In the face of this celebrated market mantra, reflective users, artists, and thinkers have much to offer popular culture. Critical talk about the digital helps keep everyone a bit more honest—and a bit more realistic as well—as the spectacle of a fully mediated cultural horizon, the digital promised land, approaches.

NOTES

1. George Gilder, *Life After Television: The Coming Transformation of Media and American Life* (New York: Norton, 1992).

2. David Lieberman, "Gates Promises Not to Rule TV Land," *USA Today*, May 5, 1998, p. B3.

3. Key works in the scholastic tradition that protest how and why technology threatens culture include: Lewis Mumford, *Technics and Civilization* (New York: Harcourt, Brace and World, 1934); Jacques Ellul, *The Technological Society* (New York: Alfred A. Knopf, 1964); Neil Postman, *Amusing Ourselves to Death: Public Discourse in the Age of Show Business* (New York: Penguin, 1985); Neil Postman, *Technopoly: The Surrender of Culture to Technology* (New York: Knopf, 1992).

4. Randall Rothenberg, "Go Ahead, Kill Your Television. NBC Is Ready," *Wired*, December 1998, p. 207.

5. NAB stands for National Association of Broadcasters, and also refers to the annual convention, trade show, and extravaganza in Las Vegas mounted to introduce and sell each year's new electronic technologies and practices.

6. "Fire Wire Gets Fast and Fat: Will It Challenge SDI and replace SCSI?," *Digital Video*, January 1999, cover page.

7. "Upgrade Your Head," advertisement in *New Media*, October 1998, pp. 26–27.

8. Thomas Kuhn, *The Structure of Scientific Revolutions* (Chicago: University of Chicago Press, 1962); and Hayden White, *The Content of Form: Narrative Discourse and Historical Representation* (Baltimore: Johns Hopkins University Press, 1987).

9. Nicholas Negroponte, *Being Digital* (New York: Alfred A. Knopf, 1995), p. 62.

10. "Habitus" is French sociologist Pierre Bourdieu's term for a person's specific cultural space, defined by lifestyle, consumption, and popular aesthetic. Digital lifestyles—acute "taste cultures"—are efficient ways of realizing cultural power through the display and performance of "distinction." New media companies compete to enter and lock down these distinctive lifestyle boundaries. What looks like a global network is actually a stratified constellation of higher- and lower-caste networks.

11. Multimedia artists have a very different view of convergence than corporate types. Such artisans praise the "local" possibilities and the creative efficiencies that follow from shrinking and combining multiple digital graphics and sound capabilities on a single desktop workstation. Convergence to Disney or Microsoft or Time-Warner means something very different—a kind of global, horizontal, market integration.

12. The major works of Innis synthesize an analysis of natural resources, industry, transportation, and communication, with speed proving a critical force in altering the nature of such systems. See his *Empire and Communications* (Toronto: University of Toronto Press, 1950), and *The Bias of Communication* (Toronto: University of Toronto Press, 1951) for accounts of communication technologies in historical context. I thank Peter Wollen for pointing out the importance of speed in centralization, and for the connection between Innis and McLuhan.

13. As quoted in Joshua Shapiro, "Split Visionary: Level 3 CEO James Crowe Has Seen the Future of Telephony and Its Metaversal," *Wired*, November 1998, p. 168.

14. Phillips Electronics advertisement for hi-fi sound on PC, *Wired*, November 1998, p. 67.

15. As quoted in Michael White, "Disney Will Work the Net," *Long Beach Press Telegraph*, April 30, 1998, p. A17.

16. "The Dawn of Digital Television" was simultaneously published on November 16, 1998, by *Broadcasting and Cable*, *TWICE*, and *Digital Television.*

17. "The Medium They Couldn't Kill," pp. S2–S3, and "The PC's Still Missing in PC/TV," pp. S55–S56.

18. *Nerds 2.01: A History of the Internet* aired on PBS on November 25, 1998.

19. David Bordwell, Janet Staiger, and Kristin Thompson, *The Classical Hollywood Cinema: Film Style and Mode of Production to 1960* (New York: Columbia University Press, 1985), pp. 247–48.

20. Comdex is one of many computer trade shows. While Microsoft is by far the dominant corporate force in software and operating system development, Tektronix is an established video engineering company that now develops, manufactures, and sells "digital" post-production workstations, the Lightworks and AVID. Hollywood's market muscle has been linked to the importance of its long-standing corporate "brands" (Disney, Fix, Warner Bros., Sony/Columbia), whereas the Silicon Valley phenomenon is tied to the persistent "successes" of alliances between technical start-ups and "venture capital" funds.

21. Jennifer D. Slack, *Communication Technologies and Society: Conceptions of Causality and the Politics of Technological Intervention* (Norwood, N.J.: Ablex, 1984).

22. A fuller discussion of what I am terming "counterpractices" is beyond the scope of this introduction, but readers might consider four frameworks when investigating this tradition: alternative technology, artculture, activism, and anarchism (or hacking). In books like Victor Papanek's *Design for the Real World* the alternative technology movement argued for "convivial tools," "soft-technology," and low-impact technologies. Artculture's theorizations on media include the "art and technology" movement in the 1960s, video art and installations in the 1970s, and interactive and digital arts in the 1980s and 1990s. Key texts go back at least to Andre Breton's *Surrealist Manifesto*, and would include Michael Shamberg's *Guerrilla Television*, John Hanhardt's (ed.) *Video Culture*, and Lynn Hershman's (ed.) *Clicking In*. New media activism has traditionally promoted "appropriations" of mass media, emancipatory and democratic media usage, and both popular front and identity politics. Key texts would include Brecht's treatise on "Radio as an Apparatus," the 1970s journal *Radical Software*, Hans Magnus Enzensberger's *The Consciousness Industry*, and

cooperative production initiatives like TVTV, Paper Tiger Television, and the Deep Dish Satellite Network. Finally, computer hackers carry on a neo-anarchist tradition of resistance that finds roots, even if unacknowledged, in Dadaism, situationism, and cyberpunk. The spirit of cyberanarchism plays out in texts like de Bord's *Society of the Spectacle,* Cornwall's *Hacker's Handbook,* Levy's *Hackers,* and Rushkoff's *Cyberia.* See Andrew Ross's entry in this collection, and the annotated bibliography, for a more in-depth discussion of hacking culture and these specific books in particular.

23. PDI is a commercial digital effects company on the order of Pixar and Digital Domain.

24. The MPAA protects the interests of the major studio producers and stands for the Motion Picture Producer's Association of America. The ASC (American Society of Cinematographers), WGA (Writer's Guild of America), and DGA (Director's Guild of America) are all professional associations that codify, promote, and publicly and legally guard the interests of their practitioners during moments of institutional configuration and technological change. The ASC, WGA, and DGA all made relatively concerted attempts to flag and negotiate the coming of digital for their members.

25. Similar work, systematic studies of mode of production and screen form, has yet to be undertaken in the area of digital media. An excerpt of Winston's work is included in this collection. A discussion of Bordwell, Thompson, and Staiger's work is included in the pages that follow. See also Thomas Schatz, *Old Hollywood, New Hollywood* (Ann Arbor: UMI Press, 1983).

26. Dan Schiller, *Digital Capitalism: Networking the Global Market System* (Cambridge, Mass.: MIT Press, 199), and Thomas Streeter, *Selling the Air: A Critique of the Policy of Commercial Broadcasting in the United States* (Chicago: University of Chicago Press, 1998).

27. See the annotated bibliography in this volume for further discussion of these authors and the others mentioned in the section that follows.

28. Sherry Turkle, *Life on the Screen: Identity in the Age of the Internet* (New York: Simon & Schuster, 1995). See also Judith Butler, *Gender Trouble: Feminism and the Subversion of Identity* (New York: Routledge, 1990), and *Bodies that Matter: On the Discursive Limits of "Sex"* (New York: Routledge, 1993).

29. Susan Rabin, *Cyberflirt: How to Attract Anyone, Anywhere on the World Wide Web* (New York: Plume, 1999).

30. George Lipsitz, *Time Passages: Collective Memory and American Popular Culture* (Minneapolis: University of Minnesota Press, 1990).

31. Norbert Wiener, *On Control and Communications in Animal and Machine* (Cambridge, Mass.: MIT Press, 1976). Originally published in 1948. Claude E. Shannon and Warren Weaver, *The Mathematical Theory of Communication* (Urbana: University of Illinois Press, 1949).

32. Lynn Spigel, "The Making of a TV Literate Elite," in Christing Geraghty and David Lusted, eds., *The Television Studies Book* (London: Arnold, 1998), p. 77.

33. Jean-Louis Comolli, "Technique and Ideology: Camera, Perspective, Depth of Field," in *Narrative, Apparatus, Ideology* ed. Philip Rosen, tr. the British Film Institute (New York: Columbia University Press, 1986), pp. 421–43.

34. Jean-Louis Baudry, "Ideological Effects of the Basic Cinematographic Apparatus," *Film Quarterly* 28, no. 2 (Winter 1974–75).

35. Jean Baudrillard, *Simulations* (New York: Semiotext(e), 1983).

36. Paul Virilio, *Open Sky* (New York: Verso, 1995).

37. David Bordwell, Kristin Thompson, and Janet Staiger, *Classical Hollywood Cinema* (New York: Columbia University Press, 1985).

38. One of the most effective models for describing this contemporary landscape is Roger Silverstone's chapter "The Tele-technological System" from *Television and Everyday Life* (New York: Routledge, 1994), pp. 78–103. Silverstone's research makes it impossible to see electronic media industries as instigators of technical change in any kind of linear, orig-

inating, or causal way. The "origins" of any new media technology can only be understood if one considers how a "socio-technical system" is also "tactically" manufactured alongside any new hardware development. Silverstone argues that technologies successfully emerge within these socio-technical systems "as a result of the potential space created within a network for the actions of individuals." That is, the industrially sanctioned role of user/buyer/viewer "agents" within the "consumption junction" of the home is a key to understanding how any new technologies succeed. A "double articulation" of "leading technologies" means that television and the computer function both as objects and as media that animate social and industrial relations. This double articulation gives them both specific technical functions and more general, symbolic ones that can only be sustained by technical and social relations. Silverstone's account suggests that we examine each new media technology as a center and "articulating principle" that animates a far broader network of social relations and operations. The "production" of technoculture for Silverstone, therefore, cannot be fully explained without reference to the "consumption" of technoculture.

39. Whereas traditional political economic studies of the digital might examine industrial relations in maquiladora factories, later in *Data Trash* (the book from which this essay comes) Kroker and Weinstein address the broader importance of technotopia's global reconfiguration of new, electronic "Third Worlds" in Haiti and Bangladesh.

40. John Browning and Spencer Rice, "Encyclopedia of the New Economy, Part II," *Wired*, April 1998, p. 102.

41. Gary Chapman, "Troubling Implications of Internet's Ubiquity," *Los Angeles Times*, July 5, 1999, pp. C1, C6.

Acknowledgment

Given the churn of ideas that circulate in contemporary media studies, many influences came to bear on this project. I would like to thank Leslie Mitchner for her editorial guidance, and express my particular appreciation for the contributions and work of Professor Ellen Seiter of the University of California at San Diego. Although I take sole responsibility for the form of this book, Professor Seiter challenged a number of my initial (production-oriented) assumptions about new media in productive ways. I thank her for her initial assistance and help in undertaking this project.

Theorizing Technohistory: Old Media/New Media

"Once the broadcasting company is established as a public service and the general public educated to the idea that the sole function is to provide the public with a service as good and extensive as its total income permits, I feel that with suitable publicity activities, such a company will ultimately be regarded as a public institution of great value, in the same sense that a library, for example, is regarded today."
—"General" David Sarnoff, 1922,
proposing national broadcasting network
to General Electric Company, 1922

"It must follow the prime objective of turning the audience not only into pupils but into teachers. It is the radio's formal task to give these educational operations an interesting turn, i.e., to ensure that these interests interest people. Such an attempt by radio to put its instruction into an artistic form would link up with the efforts of modern artists to give art an instructive character. . . . It is not at all our job to renovate ideological institutions on the basis of the existing social order by means of innovations. Instead our innovations must force them to surrender that basis."
—Bertolt Brecht, "The Radio as an Apparatus
of Communication," July 1932

"Synesthesia, or unified sense and imaginative life, had long seemed an unattainable dream to Western poets, painters, and artists in general. . . . Yet these massive extensions of our central nervous systems have enveloped Western man in a daily session of synesthesia."
"Ours is a brand new world of allatonceness. 'Time' has ceased, 'space' has vanished. We now live in a global village . . . a simultaneous happening."
—Marshall McLuhan, Understanding Media,
and The Medium is the Message, 1963, 1967

Raymond Williams

The Technology
and the Society

It is often said that television has altered our world. In the same way, people often speak of a new world, a new society, a new phase of history, being created—'brought about'—by this or that new technology: the steam engine, the automobile, the atomic bomb. Most of us know what is generally implied when such things are said. But this may be the central difficulty: that we have gotten so used to statements of this general kind, in our most ordinary discussions, that we can fail to realize their specific meanings.

For behind all such statements lie some of the most difficult and most unresolved historical and philosophical questions. Yet the questions are not posed by the statements; indeed they are ordinarily masked by them. Thus we often discuss, with animation, this or that 'effect' of television, or the kinds of social behavior, the cultural and psychological conditions, which television has 'led to,' without feeling ourselves obliged to ask whether it is reasonable to describe any technology as a cause, or, if we think of it as a cause, as what kind of cause, and in what relations with other kinds of causes. The most precise and discriminating local study of 'effects' can remain superficial if we have not looked into the notions of cause and effect, as between a technology and a society, a technology and a culture, a technology and a psychology, which underlie our questions and may often determine our answers.

It can of course be said that these fundamental questions are very much too difficult; and that they are indeed difficult is very soon obvious to anyone who tries to follow them through. We could spend our lives trying to answer them, whereas here and now, in a society in which television is important, there is immediate and practical work to be done: surveys to be made, research undertaken; surveys and research, moreover, which we know how to do. It is an appealing position, and it has the advantage, in our kind of society, that it is understood as practical, so that it can then

From *Television: Technology and Cultural Form* (New York: Schocken, 1974). Reprinted by permission of Pantheon Books/Schocken, a division of Random House, Inc., and Harper-Collins UK.

be supported and funded. By contrast, other kinds of questions seem merely theoretical and abstract.

Yet all questions about cause and effect, as between a technology and a society, are intensely practical. Until we have begun to answer them, we really do not know, in any particular case, whether, for example, we are talking about a technology or about the uses of a technology; about necessary institutions or about particular and changeable institutions; about a content or about a form. And this is not only a matter of intellectual uncertainty; it is a matter of social practice. If the technology is a cause, we can at best modify or seek to control its effects. Or if the technology, as used, is an effect, to what other kinds of cause, and other kinds of action, should we refer and relate our experience of its uses? These are not abstract questions. They form an increasingly important part of our social and cultural arguments, and they are being decided all the time in real practice, by real and effective decisions.

It is with these problems in mind that I want to try to analyze television as a particular cultural technology, and to look at its development, its institutions, its forms, and its effects, in this critical dimension. In the present chapter, I shall begin the analysis under three headings: (a) versions of cause and effect in technology and society; (b) the social history of television as a technology; (c) the social history of the uses of television technology.

Versions of Cause and Effect in Technology and Society

We can begin by looking again at the general statement that television has altered our world. It is worth setting down some of the different things this kind of statement has been taken to mean. For example:

(i)   Television was invented as a result of scientific and technical research. Its power as a medium of news and entertainment was then so great that it altered all preceding media of news and entertainment.

(ii)   Television was invented as a result of scientific and technical research. Its power as a medium of social communication was then so great that it altered many of our institutions and forms of social relationships.

(iii) Television was invented as a result of scientific and technical research. Its inherent properties as an electronic medium altered our basic perceptions of reality, and thence our relations with each other and with the world.

(iv)   Television was invented as a result of scientific and technical research. As a powerful medium of communication and entertainment it took its place with other factors—such as greatly increased physical mobility, itself the result of other newly invented technologies—in altering the scale and form of our societies.

(v) Television was invented as a result of scientific and technical research, and developed as a medium of entertainment and news. It then had unforeseen consequences, not only on other entertainment and news media, which it reduced in viability and importance, but on some of the central processes of family, cultural, and social life.

(vi) Television, discovered as a possibility by scientific and technical research, was selected for investment and development to meet the needs of a new kind of society, especially in the provision of centralized entertainment and in the centralized formation of opinions and styles of behavior.

(vii) Television, discovered as a possibility by scientific and technical research, was selected for investment and promotion as a new and profitable phase of a domestic consumer economy; it is then one of the characteristic 'machines for the home.'

(viii) Television became available as a result of scientific and technical research, and in its character and uses exploited and emphasized elements of a passivity, a cultural and psychological inadequacy, which had always been latent in people, but which television now organized and came to represent.

(ix) Television became available as a result of scientific and technical research, and in its character and uses both served and exploited the needs of a new kind of large-scale and complex but atomized society.

These are only some of the possible glosses on the ordinary bald statement that television has altered our world. Many people hold mixed versions of what are really alternative opinions, and in some cases there is some inevitable overlapping. But we can distinguish between two broad classes of opinion.

In the first—(i) to (v)—the technology is in effect accidental. Beyond the strictly internal development of the technology there is no reason why any particular invention should have come about. Similarly it then has consequences which are also in the true sense accidental, since they follow directly from the technology itself. If television had not been invented, this argument would run, certain definite social and cultural events would not have occurred.

In the second—(vi) to (ix)—television is again, in effect, a technological accident, but its significance lies in uses, which are held to be symptomatic of some order of society or some qualities of human nature which are otherwise determined. If television had not been invented, this argument runs, we would still be manipulated or mindlessly entertained, but in some other way and perhaps less powerfully.

For all the variations of local interpretation and emphasis, these two classes of opinion underlie the overwhelming majority of both professional and amateur views of the effects of television. What they have in common is the fundamental form of the statement: "television has altered our world."

It is then necessary to make a further theoretical distinction. The first class of opinion, described above, is that usually known, at least to its opponents, as *technological determinism*. It is an immensely powerful and now largely orthodox view of the nature of social change. New technologies are discovered, by an essentially internal process of research and development, which then sets the conditions for social change and progress. Progress, in particular, is the history of these inventions, which 'created the modern world.' The effects of the technologies, whether direct or indirect, foreseen or unforeseen, are as it were the rest of history. The steam engine, the automobile, television, the atomic bomb, have *made* modern man and the modern condition.

The second class of opinion appears less determinist. Television, like any other technology, becomes available as an element or a medium in a process of change that is in any case occurring or about to occur. By contrast with pure technological determinism, this view emphasizes other causal factors in social change. It then considers particular technologies, or a complex of technologies, as *symptoms* of change of some other kind. Any particular technology is then as it were a by-product of a social process that is otherwise determined. It only acquires effective status when it is used for purposes which are already contained in this known social process.

The debate between these two general positions occupies the greater part of our thinking about technology and society. It is a real debate, and each side makes important points. But it is in the end sterile, because each position, though in different ways, has abstracted technology from society. In *technological determinism*, research and development have been assumed as self-generating. The new technologies are invented as it were in an independent sphere, and then create new societies or new human conditions. The view of *symptomatic technology*, similarly, assumes that research and development are self-generating, but in a more marginal way. What is discovered in the margin is then taken up and used.

Each view can then be seen to depend on the isolation of technology. It is either a self-acting force which creates new ways of life, or it is a self-acting force which provides materials for new ways of life. These positions are so deeply established, in modern social thought, that it is very difficult to think beyond them. Most histories of technology, like most histories of scientific discovery, are written from their assumptions. An appeal to 'the facts,' against this or that interpretation, is made very difficult simply because the histories are usually written, consciously or unconsciously, to illustrate the assumptions. This is either explicit, with the consequential interpretation attached, or more often implicit, in that the history of technology or of scientific development is offered as a history on its own. This can be seen as a device of specialization or of emphasis, but it then necessarily implies merely internal intentions and criteria.

To change these emphases would require prolonged and cooperative intellectual effort. But in the particular case of television it may be possible to outline a different kind of interpretation, which would allow us to see not only its history but also its uses in a more radical way. Such an interpretation would differ from technological determinism in that it would restore *intention* to the process of research and development. The technology would be seen, that is to say, as being looked for and developed with certain purposes and practices already in mind. At the same time the interpretation would differ from symptomatic technology in that these purposes and practices would be seen as *direct:* as known social needs, purposes, and practices to which the technology is not marginal but central.

The Social History of Television as a Technology

The invention of television was no single event or series of events. It depended on a complex of inventions and developments in electricity, telegraphy, photography and motion pictures, and radio. It can be said to have separated out as a specific technological objective in the period 1875–90, and then, after a lag, to have developed as a specific technological enterprise from 1920 through to the first public television systems of the 1930s. Yet in each of these stages it depended for parts of its realization on inventions made with other ends primarily in view.

Until the early nineteenth century, investigations of electricity, which had long been known as a phenomenon, were primarily philosophical: investigations of a puzzling natural effect. The technology associated with these investigations was mainly directed toward isolation and concentration of the effect, for its clearer study. Toward the end of the eighteenth century there began to be applications, characteristically in relation to other known natural effects (lightning conductors). But there is then a key transitional period in a cluster of inventions between 1800 and 1831, ranging from Volta's battery to Faraday's demonstration of electromagnetic induction, leading quickly to the production of generators. This can be properly traced as a scientific history, but it is significant that the key period of advance coincides with an important stage of the development of industrial production. The advantages of electric power were closely related to new industrial needs: for mobility and transfer in the location of power sources, and for flexible and rapid controllable conversion. The steam engine had been well suited to textiles, and its industries had been based on local siting. A more extensive development, both physically and in the complexity of multiple-part processes, such as engineering, could be attempted with other power sources but could only be fully realized with electricity. There was a very complex interaction between new needs and new inventions, at the level of primary production, of new applied industries (plating) and of new social needs which were themselves related to industrial development (city and house lighting). From 1830 to large-scale generation in the 1880s there was this continuing complex of need and invention and application.

In telegraphy the development was simpler. The transmission of messages by beacons and similar primary devices had been long established. In the development of navigation and naval warfare the flag system had been standardized in the course of the sixteenth and seventeenth centuries. During the Napoleonic wars there was a marked development of land telegraphy, by semaphore stations, and some of this survived into peacetime. Electrical telegraphy had been suggested as a technical system as early as 1753, and was actually demonstrated in several places in the early nineteenth century. An English inventor in 1816 was told that the Admiralty was not interested. It is interesting that it was the development of the railways, themselves a response to the development of an industrial system and the related growth of cities, which clarified the need for improved telegraphy. A complex of technical possibilities was brought to a working system from 1837 onward. The development of international trade and transport brought rapid extensions of the system, including the transatlantic cable in the 1850s and the 1860s. A general system of electric telegraphy had been established by the 1870s, and in the same decade the telephone system began to be developed, in this case as a new and intended invention.

In photography, the idea of light-writing had been suggested by (among others) Wedgwood and Davy in 1802, and the camera obscura had already been developed. It was not the projection but the fixing of images which at first awaited technical solution, and from 1816 (Niepce) and through to 1839 (Daguerre) this was worked on, together with the improvement of camera devices. Professional and then amateur photography spread rapidly, and reproduction and then transmission, in the developing newspaper press, were achieved. By the 1880s the idea of a 'photographed reality'—still more for record than for observation—was familiar.

The idea of moving pictures had been similarly developing. The magic lantern (slide projection) had been known from the seventeenth century, and had acquired simple motion (one slide over another) by 1736. From at latest 1826 there was a development of mechanical motion picture devices, such as the wheel-of-life, and these came to be linked with the magic lantern. The effect of persistence in human vision—that is to say, our capacity to hold the 'memory' of an image through an interval to the next image, thus allowing the possibility of a sequence built from rapidly succeeding units—had been known since classical times. Series of cameras photographing stages of a sequence were followed (Marcy 1882) by multiple-shot cameras. Friese-Greene and Edison worked on techniques of filming and projection, and celluloid was substituted for paper reels. By the 1890s the first public motion picture shows were being given in France, America, and England.

Television, as an idea, was involved with many of these developments. It is difficult to separate it, in its earliest stages, from phototelegraphy. Bain proposed a device for transmitting pictures by electric wires in 1842; Bakewell in 1847 showed the copying telegraph; Caselli in 1862 transmitted pictures by wire over a considerable distance. In 1873, while working at a terminal of the Atlantic telegraph cable, May observed the light-sensitive properties of selenium (which had been isolated by Berzelius in 1817 and was in use for resistors). In a host of ways, following an already defined need, the means of transmitting still pictures and moving pictures were actively sought and to a considerable extent discovered. The list is long even when selective: Carey's electric eye in 1875; Nipkow's scanning system in 1884; Elster and Geitel's photoelectric cells in 1890; Braun's cathode-ray tube in 1897; Rosing's cathode-ray receiver in 1907; Campbell Swinton's electronic camera proposal in 1911. Through this whole period two facts are evident: that a system of television was foreseen, and its means were being actively sought; but also that, by comparison with electrical generation and electrical telegraphy and telephony, there was very little social investment to bring the scattered work together. It is true that there were technical blocks before 1914—the thermionic valve and the multistage amplifier can be seen to have been needed and were not yet invented. But the critical difference between the various spheres of applied technology can be stated in terms of a social dimension: the new systems of production and of business or transport communication were already organized, at an economic level; the new systems of social communication were not. Thus when motion pictures were developed, their application was characteristically in the margin of established social forms—the sideshows—until their success was capitalized in a version of an established form, the motion picture *theater*.

The development of radio, in its significant scientific and technical stages between 1885 and 1911, was at first conceived, within already effective social systems, as an advanced form of telegraphy. Its application as a significantly new social form belongs to the immediate postwar period, in a changed social situation. It is significant that the hiatus in technical television development then also ended. In 1923 Zworykin introduced the electronic television camera tube. Through the early 1920s Baird and Jenkins, separately and competitively, were working on systems using mechanical scanning. From 1925 the rate of progress was qualitatively changed, through important technical advances but also with the example of sound broadcasting systems as a model. The Bell System in 1927 demonstrated wire transmission through a radio link, and the prehistory of the form can be seen to be ending. There was great rivalry between systems—especially those of mechanical and electronic scanning—and there is still great controversy about contributions and priorities. But this is characteristic of the phase in which the development of a technology moves into the stage of a new social form.

What is interesting throughout is that in a number of complex and related fields, these systems of mobility and transfer in production and communication, whether in mechanical and electric transport, or in telegraphy, photography, motion pictures, radio, and television, were at once incentives and responses within a phase of general social transformation. Though some of the crucial scientific and technical discoveries were made by isolated and unsupported individuals, there was a crucial community of selected emphasis and intention, in a society characterized at its most general levels by a mobility and extension of the scale of organizations: forms of growth which brought with them immediate and longer-term problems of operative communication. In many different countries, and in apparently unconnected ways, such needs were at once isolated and technically defined. It is especially a characteristic of the communications system that *all were foreseen—not in utopian but in technical ways—before the crucial components of the developed systems had been discovered and refined.* In no way is this a history of communications systems creating a new society or new social conditions. The decisive and earlier transformation of industrial production, and its new social forms, which had grown out of a long history of capital accumulation and working technical improvements, created new needs but also new possibilities, and the communications systems, down to television, were their intrinsic outcome.

The Social History of the Uses of Television Technology

It is never quite true to say that in modern societies, when a social need has been demonstrated, its appropriate technology will be found. This is partly because some real needs, in any particular period, are beyond the scope of existing or foreseeable scientific and technical knowledge. It is even more because the key question, about technological response to a need, is less a question about the need itself than about its place in an existing social formation. A need which corresponds with the priorities of the real decision-making groups will, obviously, more quickly attract the investment of resources and the official permission, approval, or encouragement on which a working technology, as distinct from available technical devices, depends. We can see this clearly in the major developments of industrial production and, significantly, in military technology. The social history of communications technology is interestingly different from either of these, and it is important to try to discover what are the real factors of this variation.

The problem must be seen at several different levels. In the very broadest perspective, there is an operative relationship between a new kind of expanded, mobile, and complex society and the development of a modern communications technology. At one level this relationship can be reasonably seen as causal, in a direct way. The principal incentives to first-stage improvements in communications technology came from problems of communication and control in expanded military and commercial operations. This was both direct, arising from factors of greatly extending distance and scale, and indirect, as a factor within the development of transport technology, which was for obvious reasons the major direct response. Thus telegraphy and telephony, and in its early stages radio, were secondary factors within a primary communications system which was directly serving the needs of an established and developing military and commercial system. Through the nineteenth and into the twentieth centuries this was the decisive pattern.

But there were other social and political relationships and needs emerging from this complex of change. Indeed it is a consequence of the particular and dominant interpretation of these changes that the complex was at first seen as one requiring improvement in *operational* communication. The direct priorities of the expanding commercial system, and in certain periods of the military system, led to a definition of needs within the terms of these systems. The objectives and the consequent technologies were operational within the structures of these systems: passing necessary specific information, or maintaining contact and control. Modern electric technology, in this phase, was thus oriented to uses of person to person, operator and operative to operator and operative, within established specific structures. This quality can best be emphasized by contrast with the electric technology of the second phase, which was properly and significantly called *broadcasting.* A technology of specific messages to specific persons was complemented, but only relatively late, by a technology of varied messages to a general public.

Yet to understand this development we have to look at a wider communications system. The true basis of this system had preceded the developments in technology. Then as now there was a major, indeed dominant, area of social communication, by word of mouth, within every kind of social group. In addition, then as now, there were specific institutions of that kind of communication which involves or is predicated on social teaching and control: churches, schools, assemblies and proclamations, direction in places of work. All these interacted with forms of communication within the family.

What then were the new needs which led to the development of a new technology of social communication? The development of the press gives us the evidence for our first major instance. It was at once a response to the development of an extended social, economic, and political system and a response to crisis within that system. The centralization of political power led to a need for messages from that center along other than official lines. Early newspapers were a combination of that kind of message—political and social information—and the specific messages—classified advertising and general commercial news—of an expanding system of trade. In Britain the development of the press went through its major formative stages in periods of crisis: the Civil War and Commonwealth, when the newspaper form was defined; the Industrial Revolution, when new forms of popular journalism were successively established; the major wars of the twentieth century, when the newspaper became a universal social form. For the transmission of simple orders, a communications system already existed. For the transmission of an ideology, there were specific traditional institutions. But for the transmission of news and background—the whole orienting, predictive, and updating process which the fully developed press represented—there was an evident need for a new form, which the largely traditional institutions of church and school could not meet. And to the large extent that the crises of general change provoked both anxiety and controversy, this flexible and competitive form met social needs of a new kind. As the struggle for a share in decision and control became sharper, in campaigns for the vote and then in competition for the vote, the press became not only a new communications system but, centrally, a new social institution.

This can be interpreted as response to a political need and a political crisis, and it was certainly this. But a wider social need and social crisis can also be recognized. In a changing society, and especially after the Industrial Revolution, problems of social perspective and social orientation became more acute. New relations between men, and between men and things, were being intensely experienced, and in this area, especially, the traditional institutions of church and school, or of settled community and persisting family, had very little to say. A great deal was of course said, but from positions defined within an older kind of society. In a number of ways, and drawing on a range of impulses from curiosity to anxiety, new information and new kinds of orientation were deeply required: more deeply, indeed, than any specialization to political, military, or commercial information can account for. An increased awareness of mobility and change, not just as abstractions but as lived experiences, led to a major redefinition, in practice and then in theory, of the function and process of social communication.

What can be seen most evidently in the press can be seen also in the development of photography and the motion picture. The photograph is in one sense a popular extension of the portrait, for recognition and for record. But in a period of great mobility, with new separations of families and with internal and external migrations, it became more centrally necessary as a form of maintaining, over distance and through time, certain personal connections. Moreover, in altering relations to the physical world, the photograph as an object became a form of the photography of objects:

moments of isolation and stasis within an experienced rush of change; and then, in its technical extension to motion, a means of observing and analyzing motion itself, in new ways—a dynamic form in which new kinds of recognition were not only possible but necessary.

Now it is significant that until the period after the First World War, and in some ways until the period after the Second World War, these varying needs of a new kind of society and a new way of life were met by what were seen as specialized means: the press for political and economic information; the photograph for community and entertainment; telegraphy and telephony for business information and some important personal messages. It was within this complex of specialized forms that broadcasting arrived.

The consequent difficulty of defining its social uses, and the intense kind of controversy which has ever since surrounded it, can then be more broadly understood. Moreover, the first definitions of broadcasting were made for sound radio. It is significant and perhaps puzzling that the definitions and institutions then created were those within which television developed.

We have now become used to a situation in which broadcasting is a major social institution, about which there is always controversy but which, in its familiar form, seems to have been predestined by the technology. This predestination, however, when closely examined, proves to be no more than a set of particular social decisions, in particular circumstances, which were then so widely if imperfectly ratified that it is now difficult to see them as decisions rather than as (retrospectively) inevitable results.

Thus, if seen only in hindsight, broadcasting can be diagnosed as a new and powerful form of social integration and control. Many of its main uses can be seen as socially, commercially, and at times politically manipulative. Moreover, this viewpoint is rationalized by its description as 'mass communication,' a phrase used by almost all its agents and advisers as well, curiously, as by most of its radical critics. 'Masses' had been the new nineteenth-century term of contempt for what was formerly described as 'the mob.' The physical 'massing' of the urban and industrial revolution underwrote this. A new radical class-consciousness adopted the term to express the material of new social formations: 'mass organizations.' The 'mass meeting' was an observable physical effect. So pervasive was this description that in the twentieth century multiple serial production was called, falsely but significantly, 'mass production': mass now meant large numbers (but within certain assumed social relationships) rather than any physical or social aggregate. Sound radio and television, for reasons we shall look at, were developed for transmission to *individual* homes, though there was nothing in the technology to make this inevitable. But then this new form of social communication—broadcasting—was obscured by its definition as 'mass communication': an abstraction to its most general characteristic, that it went to many people, 'the masses,' which obscured the fact that the means chosen was the offer of individual sets, a method much better described by the earlier word 'broadcasting.' It is interesting that the only developed 'mass' use of radio was in Nazi Germany, where under Goebbels's orders the Party organized compulsory public listening groups and the receivers were in the streets. There has been some imitation of this by similar regimes, and Goebbels was deeply interested in television for the same kind of use. What was developed within most capitalist societies, though called 'mass communication,' was significantly different.

There was early official intervention in the development of broadcasting, but in form this was only at a technical level. In the earlier struggle against the development of the press, the State had licensed and taxed newspapers, but for a century before the coming of broadcasting the alternative idea of an independent press had been realized both in practice and in theory. State intervention in broadcasting had some real and some plausible technical grounds: the distribution of wavelengths. But to these were added, though always controversially, more general social directions or attempts at direction. This social history of broadcasting can be discussed on its own, at the levels of practice and principle. Yet it is unrealistic to extract it from another and perhaps more decisive process, through which, in particular economic situations, a set of scattered technical devices became an applied technology and then a social technology.

A Fascist regime might quickly see the use of broadcasting for direct political and social control. But that, in any case, was when the technology had already been developed elsewhere. In capitalist democracies, the thrust for conversion from scattered techniques to a technology was not political but economic. The characteristically isolated inventors, from Nipkow and Rosing to Baird and Jenkins and Zwyorkin, found their point of development, if at all, in the manufacturers and prospective manufacturers of the technical apparatus. The history at one level is of these isolated names, but at another level it is of EMI, RCA, and a score of similar companies and corporations. In the history of motion pictures, capitalist development was primarily in production; large-scale capitalist distribution came much later, as a way of controlling and organizing a market for given production. In broadcasting, both in sound radio and later in television, the major investment was in the means of distribution, and was devoted to production only so far as to make the distribution technically possible and then attractive. Unlike all previous communications technologies, radio and television were *systems primarily devised for transmission and reception as abstract processes, with little or no definition of preceding content.* When the question of content was raised, it was resolved, in the main, parasitically. There were state occasions, public sporting events, theaters, and so on, which would be communicatively distributed by these new technical means. *It is not only that the supply of broadcasting facilities preceded the demand; it is that the means of communication preceded their content.*

The period of decisive development in sound broadcasting was the 1920s. After the technical advances in sound telegraphy which had been made for military purposes during the war, there was at once an economic opportunity and the need for a new social definition. No nation or manufacturing group held a monopoly of the technical means of broadcasting, and there was a period of intensive litigation followed by cross-licensing of the scattered basic components of successful transmission and reception (the vacuum tube or valve, developed from 1904 to 1913; the feedback circuit, developed from 1912; the neutrodyne and heterodyne circuits, from 1923). Crucially, in the mid-1920s, there was a series of investment-guided technical solutions to the problem of building a small and simple domestic receiver, on which the whole qualitative transformation from wireless telegraphy to broadcasting depended. By the mid-1920s—1923 and 1924 are especially decisive years—this breakthrough had happened in the leading industrial societies: the United States, Britain, Germany, and France. By the end of the 1920s the radio industry had become a major sector of industrial production, within a rapid general expansion of the new kinds of machines which were eventually to be called 'consumer durables.' This complex of developments included the motorcycle and motorcar, the box camera and its successors, home electrical appliances, and radio sets. Socially, this complex is characterized by the two apparently paradoxical yet deeply connected tendencies of modern urban industrial living: on the one hand mobility, on the other hand the more apparently self-sufficient family home. The earlier period of public technology, best exemplified by the railways and city lighting, was being replaced by a kind of technology for which no satisfactory name has yet been found: that which served an at once mobile and home-centered way of living: a form of *mobile privatization*. Broadcasting in its applied form was a social product of this distinctive tendency.

The contradictory pressures of this phase of industrial capitalist society were indeed resolved, at a certain level, by the institution of broadcasting. For mobility was only in part the impulse of an independent curiosity: the wish to go out and see new places. It was essentially an impulse formed in the breakdown and dissolution of older and smaller kinds of settlement and productive labor. The new and larger settlements and industrial organizations required major internal mobility, at a primary level, and this was joined by secondary consequences in the dispersal of extended families and in the needs of new kinds of social organization. Social processes long implicit in the revolution of industrial capitalism were then greatly intensified: especially an increasing distance between immediate living areas and the directed places of work and government. No effective kinds of social control over these transformed industrial and political processes had come anywhere near being achieved or even foreseen.

Most people were living in the fallout area of processes determined beyond them. What had been gained, nevertheless, in intense social struggle, had been the improvement of immediate conditions, within the limits and pressures of these decisive large-scale processes. There was some relative improvement in wages and working conditions, and there was a qualitative change in the distribution of the day, the week, and the year between work and off-work periods. These two effects combined in a major emphasis on improvement of the small family home. Yet this privatization, which was at once an effective achievement and a defensive response, carried, as a consequence, an imperative need for new kinds of contact. The new homes might appear private and 'self-sufficient' but could be maintained only by regular funding and supply from external sources, and these, over a range from employment and prices to depressions and wars, had a decisive and often a disrupting influence on what was nevertheless seen as a separable 'family' project. This relationship created both the need and the form of a new kind of 'communication': news from 'outside,' from otherwise inaccessible sources. Already in the drama of the 1880s and 1890s (Ibsen, Chekhov) this structure had appeared: the center of dramatic interest was now for the first time the family home, but men and women stared from its windows, or waited anxiously for messages, to learn about forces, 'out there,' which would determine the conditions of their lives. The new 'consumer' technology which reached its first decisive stage in the 1920s served this complex of needs within just these limits and pressures. There were immediate improvements of the condition and efficiency of the privatized home; there were new facilities, in private transport, for expeditions from the home; and then, in radio, there was a facility for a new kind of social input—news and entertainment brought into the home. Some people spoke of the new machines as gadgets, but they were always much more than this. They were the applied technology of a set of emphases and responses within the determining limits and pressures of industrial capitalist society.

The cheap radio receiver is then a significant index of a general condition and response. It was especially welcomed by all those who had least social opportunities of other kinds; who lacked independent mobility or access to the previously diverse places of entertainment and information. Broadcasting could also come to serve, or seem to serve, as a form of *unified* social intake, at the most general levels. What had been intensively promoted by the radio manufacturing companies thus interlocked with this kind of social need, itself defined within general limits and pressures. In the early stages of radio manufacturing, transmission was conceived before content. By the end of the 1920s the network was there, but still at a low level of content-definition. It was in the 1930s, in the second phase of radio, that most of the significant advances in content were made. The transmission and reception networks created, *as a by-product*, the facilities of primary broadcasting production. But the general social definition of 'content' was already there.

This theoretical model of the general development of broadcasting is necessary to an understanding of the particular development of television. For there were, in the abstract, several different ways in which television as a technical means might have been developed. After a generation of universal domestic television it is not easy to realize this. But it remains true that, after a great deal of intensive research and development, the domestic television set is in a number of ways an inefficient medium of visual broadcasting. Its visual inefficiency by comparison with the cinema is especially striking, whereas in the case of radio there was by the 1930s a highly efficient sound broadcasting receiver, without any real competitors in its own line. Within the limits of the television home-set emphasis it has so far not been possible to make more than minor qualitative improvements. Higher-definition systems, and color, have still only brought the domestic television set, as a machine, to the standard of a very inferior kind of cinema. Yet most people have adapted to this inferior visual medium, in an unusual kind of preference for an inferior immediate technology, because of the social complex—and especially that of the privatized home—within which broadcasting, as a system, is operative. The cinema had remained at an earlier level of social definition; it was and remains a special kind of theater, offering specific and discrete works of one general kind. Broadcasting, by contrast, offered a whole social intake: music, news, entertainment, sport. The advantages of this general intake, within the home, much more than outweighed the technical advantages of visual transmission and reception in the cinema, confined as this was to specific and discrete works. While broadcasting was confined to sound, the powerful visual medium of cinema was an immensely popular alternative. But when broadcasting became visual, the option for its social advantages outweighed the immediate technical deficits.

The transition to television broadcasting would have occurred quite generally in the late 1930s and early 1940s, if the war had not intervened. Public television services had begun in Britain in 1936 and in the United States in 1939, but with still very expensive receivers. The full investment in transmission and reception facilities did not occur until the late 1940s and early 1950s, but the growth was thereafter very rapid. The key social tendencies which had led to the definition of broadcasting were by then even more pronounced. There was significantly higher investment in the privatized home, and the social and physical distances between these homes and the decisive political and productive centers of the society had become much greater. Broadcasting, as it had developed in radio, seemed an inevitable model: the central transmitters and the domestic sets.

Television then went through some of the same phases as radio. Essentially, again, the technology of transmission and reception developed before the content, and important parts of the content were and have remained by-products of the technology rather than independent enterprises.

As late as the introduction of color, 'colorful' programs were being devised to persuade people to buy color sets. In the earliest stages there was the familiar parasitism on existing events: a coronation, a major sporting event, theaters. A comparable parasitism on the cinema was slower to show itself, until the decline of the cinema altered the terms of trade; it is now very widespread, most evidently in the United States. But again, as in radio, the end of the first general decade brought significant independent television production. By the middle and late 1950s, as in radio in the middle and late 1930s, new kinds of programs were being made for television and there were very important advances in the productive use of the medium, including, as again at a comparable stage in radio, some kinds of original work.

Yet the complex social and technical definition of broadcasting led to inevitable difficulties, especially in the productive field. What television could do relatively cheaply was to transmit something that was in any case happening or had happened. In news, sports, and some similar areas it could provide a service of transmission at comparatively low cost. But in every kind of new work, which it had to produce, it became a very expensive medium, within the broadcasting model. It was never as expensive as film, but the cinema, as a distributive medium, could directly control its revenues. It was, on the other hand, implicit in broadcasting that given the tunable receiver all programs could be received without immediate charge. There could have been and can still be a socially financed system of production and distribution within which local and specific charges would be unnecessary; the BBC, based on the license system for domestic receivers, came nearest to this. But short of monopoly, which still exists in some state-controlled systems, the problems of investment for production, in any broadcasting system, are severe.

Thus within the broadcasting model there was this deep contradiction, of centralized transmission and privatized reception. One economic response was licensing. Another, less direct, was commercial sponsorship and then supportive advertising. But the crisis of production control and financing has been endemic in broadcasting precisely because of the social and technical model that was adopted and that has become so deeply established. The problem is masked, rather than solved, by the fact that as a transmitting technology—its function largely limited to relay and commentary on other events—some balance could be struck; a limited revenue could finance this limited service. But many of the creative possibilities of television have been frustrated precisely by this apparent solution, and this has far more than local effects on producers and on the balance of programs. When there has been such heavy investment in a particular model of social communications, there is a restraining complex of financial institutions, of cultural expectations, and of specific technical developments, which though it can be seen, superficially, as the effect of a technology is in fact a social complex of a new and central kind.

Hans Magnus Enzensberger

Constituents of a Theory of the Media

*"If you should think this is Utopian, then I would ask
you to consider why it is Utopian."*
          –Bertolt Brecht, Theory of Radio

With the development of the electronic media, the industry that shapes consciousness has become the pacemaker for the social and economic development of societies in the late industrial age. It infiltrates into all other sectors of production, takes over more and more directional and control functions, and determines the standard of the prevailing technology.

In lieu of normative definitions, here is an incomplete list of new developments which have emerged in the last twenty years: news satellites, color television, cable relay television, cassettes, videotape, videotape recorders, videophones, stereophony, laser techniques, electrostatic reproduction processes, electronic high-speed printing, composing and learning machines, microfiches with electronic access, printing by radio, time-sharing computers, data banks. All these new forms of media are constantly forming new connections both with each other and with older media like printing, radio, film, television, telephone, teletype, radar, and so on. They are clearly coming together to form a universal system.

The general contradiction between productive forces and productive relationships emerges most sharply, however, when they are most advanced. By contrast, protracted structural crises, as in coal mining, can be solved merely by getting rid of a backlog, that is to say, essentially they can be solved within the terms of their own system, and a revolutionary strategy that relied on them would be shortsighted.

Monopoly capitalism develops the consciousness-shaping industry more quickly and more extensively than other sectors of production; it must at the same time fetter it. A socialist media theory has to work at this contradiction, demonstrate that it cannot be solved within the given

Originally printed in the *New Left Review*, no. 64, November/December 1970, pp. 13–36. Reprinted by permission of the *New Left Review* and Verso.

productive relationships—rapidly increasing discrepancies and potential destructive forces. "Certain demands of a prognostic nature must be made" of any such theory (Benjamin).

A "critical" inventory of the status quo is not enough. There is the danger of underestimating the growing conflicts in the media field, of neutralizing them, of interpreting them merely in terms of trade unionism or liberalism, on the lines of traditional labor struggles or as the clash of special interests (program heads/executive producers, publishers/authors, monopolies/medium-sized businesses, public corporations/private companies, etc.). An appreciation of this kind does not go far enough and remains bogged down in tactical arguments.

So far there is no Marxist theory of the media. There is therefore no strategy one can apply in this area. Uncertainty, alternations between fear and surrender, mark the attitude of the socialist Left to the new productive forces of the media industry. The ambivalence of this attitude merely mirrors the ambivalence of the media themselves without mastering it. It could only be overcome by releasing the emancipatory potential which is inherent in the new productive forces—a potential which capitalism must sabotage just as surely as Soviet revisionism, because it would endanger the rule of both systems.

The Mobilizing Power of the Media

The open secret of the electronic media, the decisive political factor, which has been waiting, suppressed or crippled, for its moment to come, is their mobilizing power.

When I say *mobilize* I mean *mobilize.* In a country which has had direct experience of Fascism (and Stalinism) it is perhaps still necessary to explain, or to explain again, what that means—namely, to make men more mobile than they are. As free as dancers, as aware as football players, as surprising as guerrillas. Anyone who thinks of the masses only as the object of politics cannot mobilize them. He wants to push them around. A parcel is not mobile; it can only be pushed to and fro. Marches, columns, parades, immobilize people. Propaganda, which does not release self-reliance but limits it, fits into the same pattern. It leads to depoliticization.

For the first time in history, the media are making possible mass participation in a social and socialized productive process, the practical means of which are in the hands of the masses themselves. Such a use of them would bring the communications media, which up to now have not deserved the name, into their own. In its present form, equipment like television or film does not serve communication but prevents it. It allows no reciprocal action between transmitter and receiver; technically speaking, it reduces feedback to the lowest point compatible with the system.

This state of affairs, however, cannot be justified technically. On the contrary. Electronic techniques recognize no contradiction in principle between transmitter and receiver. Every transistor radio is, by the nature of its construction, at the same time a potential transmitter; it can interact with other receivers by circuit reversal. The development from a mere distribution medium to a communications medium is technically not a problem. It is consciously prevented for understandable political reasons. The technical distinction between receivers and transmitters reflects the social division of labor into producers and consumers, which in the consciousness industry becomes of particular political importance. It is based, in the last analysis, on the basic contradiction between the ruling class and the ruled class—that is to say, between monopoly capital or monopolistic bureaucracy on the one hand and the dependent masses on the other.

This structural analogy can be worked out in detail. To the programs offered by the broadcasting cartels there correspond the politics offered by a power cartel consisting of parties constituted along authoritarian lines. In both cases marginal differences in their platforms reflect a competitive relationship which on essential questions is nonexistent. Minimal independent activity on the part of the voter/viewer is desired. As in the case with parliamentary elections under the two-party system, the feedback is reduced to indices. "Training in decision making" is reduced to the response to a single, three-point switching process: Program 1; Program 2; Switch off (abstention).

> Radio must be changed from a means of distribution to a means of communication. Radio would be the most wonderful means of communication imaginable in public life, a huge linked system—that is to say, it would be such if it were capable not only of transmitting but of receiving, of allowing the listener not only to hear but to speak, and did not isolate him but brought him into contact. Unrealizable in this social system, realizable in another, these proposals, which are, after all, only the natural consequences of technical development, help towards the propagation and shaping of that *other* system.[1]

The Orwellian Fantasy

George Orwell's bogey of a monolithic consciousness industry derives from a view of the media which is undialectical and obsolete. The possibility of total control of such a system at a central point belongs not to the future but to the past. With the aid of systems theory, a discipline which is part of bourgeois science—using, that is to say, categories which are immanent in the system—it can be demonstrated that a linked series of communications or, to use the technical term, switchable network, to the degree that it exceeds a certain critical size, can no longer be centrally controlled but only dealt with statistically. This basic "leakiness" of stochastic systems admittedly allows the calculation of probabilities based on sampling and extrapolations; but blanket supervision would demand a monitor that was bigger than the system itself. The monitoring of all telephone conversations, for instance, postulates an apparatus which would need to be n times more extensive and more complicated than that of the present telephone system. A censor's office, which carried out its work extensively, would of necessity become the largest branch of industry in its society.

But supervision on the basis of approximation can only offer inadequate instruments for the self-regulation of the whole system in accordance with the concepts of those who govern it. It postulates a high degree of internal stability. If this precarious balance is upset, then crisis measures based on statistical methods of control are useless. Interference can penetrate the leaky nexus of the media, spreading and multiplying there with the utmost speed, by resonance. The regime so threatened will in such cases, insofar as it is still capable of action, use force and adopt police or military methods.

A state of emergency is therefore the only alternative to leakage in the consciousness industry; but it cannot be maintained in the long run. Societies in the late industrial age rely on the free exchange of information; the "objective pressures" to which their controllers constantly appeal are thus turned against them. Every attempt to suppress the random factors, each diminution of the average flow and each distortion of the information structure must, in the long run, lead to an embolism.

The electronic media have not only built up the information network intensively, they have also spread it extensively. The radio wars of the 1950s demonstrated that in the realm of communications, national sovereignty is condemned to wither away. The further development of satellites will deal it the coup de grâce. Quarantine regulations for information, such as were promulgated by Fascism and Stalinism, are only possible today at the cost of deliberate industrial regression.

Example. The Soviet bureaucracy, that is to say the most widespread and complicated bureaucracy in the world, has to deny itself almost entirely an elementary piece of organizational equipment, the duplicating machine, because this instrument potentially makes everyone a printer. The political risk involved, the possibility of a leakage in the information network, is accepted only at the highest levels, at exposed switchpoints in political, military, and scientific areas. It is clear that Soviet society has to pay an immense price for the suppression of its own productive resources—clumsy procedures, misinformation, *faux frais*. The phenomenon incidentally has its analogue in the capitalist West, if in a diluted form. The technically most advanced electrostatic copying machine, which operates with ordinary paper—which cannot, that is to say, be supervised and is independent of suppliers—is the property of a monopoly (Xerox), on principle it is not sold but rented. The rates themselves ensure that it does not get into the wrong hands. The equipment crops up as if by magic where economic and political power is concentrated. Political control of the equipment goes hand in hand with maximization of profits for the manufacturer. Admittedly this control, as opposed to Soviet methods, is by no means "watertight" for the reasons indicated.

The problem of censorship thus enters a new historical stage. The struggle for the freedom of the press and freedom of ideas has, up until now, been mainly an argument within the bourgeoisie itself; for the masses, freedom to express opinions was a fiction since they were, from the beginning, barred from the means of production—above all from the press—and thus were unable to join in freedom of expression from the start. Today censorship is threatened by the productive forces of the consciousness industry which is already, to some extent, gaining the upper hand over the prevailing relations of production. Long before the latter are overthrown, the contradiction between what is possible and what actually exists will become acute.

Cultural Archaism in the Left Critique

The New Left of the 1960s has reduced the development of the media to a single concept—that of manipulation. This concept was originally extremely useful for heuristic purposes and has made possible a great many individual analytical investigations, but it now threatens to degenerate into a mere slogan which conceals more than it is able to illuminate, and therefore itself requires analysis.

The current theory of manipulation on the Left is essentially defensive; its effects can lead the movement into defeatism. Subjectively speaking, behind the tendency to go on the defensive lies a sense of impotence. Objectively, it corresponds to the absolutely correct view that the decisive means of production are in enemy hands. But to react to this state of affairs with moral indignation is naive. There is in general an undertone of lamentation when people speak of manipulation which points to idealistic expectations—as if the class enemy had ever stuck to the promises of fair play it occasionally utters. The liberal superstition that in political and social questions there is such a thing as pure, unmanipulated truth seems to enjoy remarkable currency among the socialist Left. It is the unspoken basic premise of the manipulation thesis.

This thesis provides no incentive to push ahead. A socialist perspective which does not go beyond attacking existing property relationships is limited. The expropriation of Springer is a desirable goal but it would be good to know to whom the media should be handed over. The Party? To judge by all experience of that solution, it is not a possible alternative. It is perhaps no accident that the Left has not yet produced an analysis of the pattern of manipulation in countries with socialist regimes.

The manipulation thesis also serves to exculpate oneself. To cast the enemy in the role of the devil is to conceal the weakness and lack of perspective in one's own agitation. If the latter leads to self-isolation instead of mobilizing the masses, then its failure is attributed holus-bolus to the overwhelming power of the media.

The theory of repressive tolerance has also permeated discussion of the media by the Left. This concept, which was formulated by its author with the utmost care, has also, when whittled away in an undialectical manner, become a vehicle for resignation. Admittedly, when an office equipment firm can attempt to recruit sales staff with the picture of Che Guevara and the text *We would have hired him,* the temptation to withdraw is great. But fear of handling shit is a luxury a sewerman cannot necessarily afford.

The electronic media do away with cleanliness; they are by their nature "dirty." That is part of their productive power. In terms of structure, they are antisectarian—a further reason why the Left, insofar as it is not prepared to reexamine its traditions, has little idea what to do with them. The desire for a cleanly defined "line" and for the suppression of "deviations" is anachronistic and now serves only one's own need for security. It weakens one's own position by irrational purges, exclusions, and fragmentation, instead of strengthening it by rational discussion.

These resistances and fears are strengthened by a series of cultural factors which, for the most part, operate unconsciously, and which are to be explained by the social history of the participants in today's Left movement—namely their bourgeois class background. It often seems as if it were precisely because of their progressive potential that the media are felt to be an immense threatening power; because for the first time they present a basic challenge to bourgeois culture and thereby to the privileges of the bourgeois intelligentsia—a challenge far more radical than any self-doubt this social group can display. In the New Left's opposition to the media, old bourgeois fears such as the fear of "the masses" seem to be reappearing along with equally old bourgeois longings for preindustrial times dressed up in progressive clothing.

At the very beginning of the student revolt, during the Free Speech Movement at Berkeley, the computer was a favorite target for aggression. Interest in the Third World is not always free from motives based on antagonism toward civilization which has its source in conservative culture critique. During the May events in Paris, the reversion to archaic forms of production was particularly characteristic. Instead of carrying out agitation among the workers with a modern offset press, the students printed their posters on the hand presses of the École des Beaux Arts. The political slogans were hand-painted; stencils would certainly have made it possible to produce them en masse, but it would have offended the creative imagination of the authors. The ability to make proper strategic use of the most advanced media was lacking. It was not the radio headquarters that were seized by the rebels, but the Odéon Theatre, steeped in tradition.

The obverse of this fear of contact with the media is the fascination they exert on left-wing movements in the great cities. On the one hand, the comrades take refuge in outdated forms of communication and esoteric arts and crafts instead of occupying themselves with the contradiction between the present constitution of the media and their revolutionary potential; on the other hand, they cannot escape from the consciousness industry's program or from its aesthetic. This leads, subjectively, to a split between a puritanical view of political action and the area of private "leisure"; objectively, it leads to a split between politically active groups and subcultural groups.

In Western Europe the socialist movement mainly addresses itself to a public of converts through newspapers and journals which are exclusive in terms of language, content, and form. These newssheets presuppose a structure of party members and sympathizers and a situation, where the media are concerned, that roughly corresponds to the historical situation in 1900; they are obviously fixated on the *Iskra* model. Presumably the people who produce them listen to the Rolling Stones, watch occupations and strikes on television, and go to the cinema to see a Western or a Godard; only in their capacity as producers do they make an exception, and, in their analyses, the whole media sector is reduced to the slogan of "manipulation." Every foray into this territory is regarded from the start with suspicion as a step toward integration. This suspicion is not unjustified; it can, however, also mask one's own ambivalence and insecurity. Fear of being swallowed up by the system is a sign of weakness; it presupposes that capitalism could overcome any contradiction—a conviction which can easily be refused historically and is theoretically untenable.

If the socialist movement writes off the new productive forces of the consciousness industry and relegates work on the media to a subculture, then we have a vicious circle. For the Underground may be increasingly aware of the technical and aesthetic possibilities of the disc, of videotape, of the electronic camera, and so on, and is systematically exploring the terrain, but it has no political viewpoint of its own and therefore mostly falls a helpless victim to commercialism. The politically active groups then point to such cases with smug *Schaden-freude*. A process of unlearning is the result and both sides are the losers. Capitalism alone benefits from the Left's antagonism to the media, as it does from the depoliticization of the counterculture.

Democratic Manipulation

Manipulation—etymologically, "handling"—means technical treatment of a given material with a particular goal in mind. When the technical intervention is of immediate social relevance, then manipulation is a political act. In the case of the media industry, that is by definition the case.

Thus every use of the media presupposes manipulation. The most elementary processes in media production, from the choice of the medium itself to shooting, cutting, synchronization, dubbing, right up to distribution, are all operations carried out on the raw material. There is no such thing as unmanipulated writing, filming, or broadcasting. The question is therefore not whether the media are manipulated, but who manipulates them. A revolutionary plan should not require the manipulators to disappear; on the contrary, it must make everyone a manipulator.

All technical manipulations are potentially dangerous; the manipulation of the media cannot be countered, however, by old or new forms of censorship, but only by direct social control, that is to say, by the mass of the people, who will have become productive. To this end, the elimination of capitalistic property relationships is a necessary but by no means sufficient condition. There have been no historical examples up until now of the mass self-regulation learning process which is made possible by the electronic media. The Communists' fear of releasing this potential, of the mobilizing capabilities of the media, of the interaction of free producers, is one of the main reasons why even in the socialist countries, the old bourgeois culture, greatly disguised and distorted but structurally intact, continues to hold sway.

As a historical explanation, it may be pointed out that the consciousness industry in Russia at the time of the October Revolution was extraordinarily backward; their productive capacity has grown enormously since then, but the productive relationships have been artificially preserved, often by force. Then, as now, a primitively edited press, books, and theater were the key media in the Soviet Union. The development of radio, film, and television is politically arrested. Foreign stations like the BBC, the Voice of America, and the *Deutschland Welle,* therefore, not only find listeners, but are received with almost boundless faith. Archaic media like the handwritten pamphlet and poems orally transmitted play an important role.

The new media are egalitarian in structure. Anyone can take part in them by a simple switching process. The programs themselves are not material things and can be reproduced at will. In this sense the electronic media are entirely different from the older media like the book or the easel painting, the exclusive class character of which is obvious. Television programs for privileged groups are certainly technically conceivable—closed circuit television—but run counter to the structure. Potentially, the new media do away with all educational privileges and thereby with the cultural monopoly of the bourgeois intelligentsia. This is one of the reasons for the intelligentsia's resentment against the new industry. As for the "spirit" which they are endeavoring to defend against "depersonalization" and "mass culture," the sooner they abandon it the better.

Properties of the New Media

The new media are oriented toward action, not contemplation; toward the present, not tradition. Their attitude to time is completely opposed to that of bourgeois culture, which aspires to possession, that is, to extension in time, best of all, to eternity. The media produce no objects that can be hoarded and auctioned. They do away completely with "intellectual property" and liquidate the "heritage," that is to say, the class-specific handing-on of nonmaterial capital.

That does not mean to say that they have no history or that they contribute to the loss of historical consciousness. On the contrary, they make it possible for the first time to record historical material so that it can be reproduced at will. By making this material available for present-day purposes, they make it obvious to anyone using it that the writing of history is always manipulation. But the memory they hold in readiness is not the preserve of a scholarly caste. It is social. The banked information is accessible to anyone, and this accessibility is as instantaneous as its recording. It suffices to compare the model of a private library with that of a socialized data bank to recognize the structural difference between the two systems.

It is wrong to regard media equipment as mere means of consumption. It is always, in principle, also means of production and, indeed, since it is in the hands of the masses, socialized means of production. The contradiction between producers and consumers is not inherent in the electronic media; on the contrary, it has to be artificially reinforced by economic and administrative measures.

An early example of this is provided by the difference between telegraph and telephone. Whereas the former, to this day, has remained in the hands of a bureaucratic institution which can scan and file every text transmitted, the telephone is directly accessible to all users. With the aid of conference circuits, it can even make possible collective intervention in a discussion by physically remote groups.

On the other hand, those auditory and visual means of communication which rely on "wireless" are still subject to state control (legislation on wireless installations). In the face of technical developments, which long ago made local and international radio-telephony possible, and which constantly opened up new wavebands for television—in the UHF band alone, the dissemination of numerous programs in one locality is possible without interference, not to mention the possibilities offered by wired and satellite television—the prevailing laws for control of the air are anachronistic. They recall the time when the operation of a printing press was dependent on an imperial license. The socialist movements will take up the struggle for their own wavelengths and must, within the foreseeable future, build their own transmitters and relay stations.

One immediate consequence of the structural nature of the new media is that none of the regimes at present in power can release their potential. Only a free socialist society will be able to make them fully productive. A further characteristic of the most advanced media—probably the decisive one—confirms this thesis: their collective structure.

For the prospect that in the future, with the aid of the media, anyone can become a producer, would remain apolitical and limited were this productive effort to find an outlet in individual tinkering. Work on the media is possible for an individual only insofar as it remains socially and therefore aesthetically irrelevant. The collection of transparencies from the last holiday trip provides a model of this.

That is naturally what the prevailing market mechanisms have aimed at. It has long been clear from apparatus like miniature and 8 mm movie cameras, as well as the tape recorder, which are in actual fact already in the hands of the masses, that the individual, so long as he remains isolated, can become with their help at best an amateur but not a producer. Even so potent a means of production as the shortwave transmitter has been tamed in this way and reduced to a harmless and inconsequential hobby in the hands of scattered radio hams. The programs which the isolated amateur mounts are always only bad, outdated copies of what he in any case receives.

Private production for the media is no more than licensed cottage industry. Even when it is made public it remains pure compromise. To this end, the men who own the media have developed special programs which are usually called "Democratic Forum" or something of the kind. There, tucked away in the corner, "the reader (listener, viewer) has his say," which can naturally be cut short at any time. As in the case of public-opinion polling, he is only asked questions so that he may have a chance to confirm his own dependence. It is a control circuit where what is fed in has already made complete allowance for the feedback.

The concept of a license can also be used in another sense—in an economic one; the system attempts to make each participant into a concessionaire of the monopoly that develops his films or plays back his cassettes. The aim is to nip in the bud in this way the independence which video equipment, for instance, makes possible. Naturally, such tendencies go against the grain of the structure, and the new productive forces not only permit but indeed demand their reversal.

The poor, feeble, and frequently humiliating results of this licensed activity are often referred to with contempt by the professional media producers. On top of the damage suffered by the masses comes triumphant mockery because they clearly do not know how to use the media properly. The sort of thing that goes on in certain popular television shows is taken as proof that they are completely incapable of articulating on their own.

Not only does this run counter to the results of the latest psychological and pedagogical research, but it can easily be seen to be a reactionary protective formulation; the "gifted" people are quite simply defending their territories. Here we have a cultural analogue to the familiar political judgments concerning a working class which is presumed to be "stultified" and incapable of any kind of self-determination. Curiously, one may hear the view that the masses could never govern themselves out of the mouths of people who consider themselves socialists. In the best of cases, these are economists who cannot conceive of socialism as anything other than nationalization.

A Socialist Strategy

Any socialist strategy for the media must, on the contrary, strive to end the isolation of the individual participants from the social learning and production process. This is impossible unless those concerned organize themselves. This is the political core of the question of the media. It is over this point that socialist concepts part company with the neo-liberal and technocratic ones. Anyone who expects to be emancipated by technological hardware, or by a system of hardware however structured, is the victim of an obscure belief in progress. Anyone who imagines that freedom for the media will be established if only everyone is busy transmitting and receiving is the dupe of a liberalism which, decked out in contemporary colors, merely peddles the faded concepts of a preordained harmony of social interests.

In the face of such illusions, what must be firmly held on to is that the proper use of the media demands organization and makes it possible. Every production that deals with the interests of the producers postulates a collective method of production. It is itself already a form of self-organization of social needs. Tape recorders, ordinary cameras, and movie cameras are already extensively owned by wage earners. The question is why these means of production do not turn up at factories, in schools, in the offices of the bureaucracy, in short, everywhere where there is social conflict. By producing aggressive forms of publicity which were their own, the masses could secure evidence of their daily experiences and draw effective lessons from them.

Naturally, bourgeois society defends itself against such prospects with a battery of legal measures. It bases itself on the law of trespass, on commercial and official secrecy. While its secret services penetrate everywhere and plug in to the most intimate conversations, it pleads a touching concern for confidentiality, and makes a sensitive display of worrying about the question of privacy when all that is private is the interest of the exploiters. Only a collective, organized effort can tear down these paper walls.

Communication networks which are constructed for such purposes can, over and above their primary function, provide politically interesting organizational models. In the socialist movements the dialectic of discipline and spontaneity, centralism and decentralization, authoritarian leadership and anti-authoritarian disintegration has long ago reached deadlock. Networklike communications models built on the principle of reversibility of circuits might give indications of how to overcome this situation: a mass newspaper, written and distributed by its readers, a video network of politically active groups.

More radically than any good intention, more lastingly than existential flight from one's own class, the media, once they have come into their own, destroy the private production methods of bourgeois intellectuals. Only in productive work and learning processes can their individualism be broken down in such a way that it is transformed from morally based (that is to say, as individual as ever) self-sacrifice to a new kind of political self-understanding and behavior.

An all-too-widely disseminated thesis maintains that present-day capitalism lives by the exploitation of unreal needs. That is at best a half-truth. The results obtained by popular American sociologists like Vance Packard are not un-useful but limited. What they have to say about the stimulation of needs through advertising and artificial obsolescence can in any case not be adequately explained by the hypnotic pull exerted on the wage earners by mass consumption. The hypothesis of "consumer terror" corresponds to the prejudices of a middle class, which considers itself politically enlightened, against the allegedly integrated proletariat, which has become petty bourgeois and corrupt. The attractive power of mass consumption is based not on the dictates of false needs, but on the falsification and exploitation of quite real and legitimate ones without which the parasitic process of advertising would be redundant. A socialist movement ought not to denounce these needs, but take them seriously, investigate them, and make them politically productive.

That is also valid for the consciousness industry. The electronic media do not owe their irresistible power to any sleight-of-hand but to the elemental power of deep social needs which come through even in the present depraved form of these media.

Precisely because no one bothers about them, the interests of the masses have remained a relatively unknown field, at least insofar as they are historically new. They certainly extend far beyond those goals which the traditional working-class movement represented. Just as in the field of production, the industry which produces goods and the consciousness industry merge more and more, so too, subjectively, where needs are concerned, material and nonmaterial factors are closely interwoven. In the process old psychosocial themes are firmly embedded—social prestige, identification patterns—but powerful new themes emerge which are utopian in nature. From a materialistic point of view, neither the one nor the other must be supposed.

Henri Lefèbvre has proposed the concept of the *spectacle,* the exhibition, the show, to fit the present form of mass consumption. Goods and shop windows, traffic and advertisements, stores and the world of communications, news and packaging, architecture and media production come together to form a totality, a permanent theater, which dominates not only the public city centers but also private interiors. The expression "beautiful living" makes the most commonplace objects of general use into props for this universal festival, in which the fetishistic nature of the commodities triumphs completely over their use value. The swindle these festivals perpetrate is, and remains, a swindle within the present social structure. But it is the harbinger of something else. Consumption as spectacle contains the promise that want will disappear. The deceptive, brutal, and obscene features of this festival derive from the fact that there can be no question of a real fulfillment of its promise. But so long as scarcity holds sway, use-value remains a decisive category which can only be abolished by trickery. Yet trickery on such a scale is only conceivable if it is based on mass need. This need—it is a utopian one—is there. It is the desire for a new ecology, for a breaking down of environmental barriers, for an aesthetic which is not limited to the sphere of "the artistic." These desires are not—or are not primarily—internalized rules of the game as played by the capitalist system. They have physiological roots and can no longer be suppressed. Consumption as spectacle is—in parody form— the anticipation of a utopian situation.

The promises of the media demonstrate the same ambivalence. They are an answer to the mass need for nonmaterial variety and mobility—which at present finds its material realization in private car ownership and tourism—and they exploit it. Other collective wishes, which capital often recognizes more quickly and evaluates more correctly than its opponents, but naturally only so as to trap them and rob them of their explosive force, are just as powerful, just as unequivocally emancipatory: the need to take part in the social process on a local, national, and international scale; the need for new forms of interaction, for release from ignorance and tutelage; the need for self-determination. "Be everywhere!" is one of the most successful slogans of the media industry. The readers' parliament of *Bild-Zeitung* (the Springer Press mass publication) was direct democracy used against the interests of the *demos.* "Open spaces" and "free time" are concepts which corral and neutralize the urgent wishes of the masses.

There is corresponding acceptance by the media of utopian stories: for example, the story of the young Italo-American who hijacked a passenger plane to get home from California to Rome was taken up without protest even by the reactionary mass press and undoubtedly correctly understood by its readers. The identification is based on what has become a general need. Nobody can understand why such journeys should be reserved for politicians, functionaries, and businessmen. The role of the pop star could be analyzed from a similar angle; in it the authoritarian and emancipatory factors are mingled in an extraordinary way. It is perhaps not unimportant that beat music offers groups, not individuals, as identification models. In the productions of the Rolling Stones (and in the manner of their production) the utopian content is apparent. Events like the Woodstock Festival, the concerts in Hyde Park, on the Isle of Wight, and at Altamont, California, develop a mobilizing power which the political Left can only envy.

It is absolutely clear that, within the present social forms, the consciousness industry can satisfy none of the needs on which it lives and which it must fan, except in the illusory form of games. The point, however, is not to demolish its promises but to take them literally and to show that they can be met only through a cultural revolution. Socialists and socialist regimes which multiply the frustration of the masses by declaring their needs to be false, become the accomplices of the system they have undertaken to fight.

Summary

Repressive use of media	*Emancipatory use of media*
Centrally controlled program	Decentralized program
One transmitter, many receivers	Each receiver a potential transmitter
Immobilization of isolated individuals	Mobilization of the masses
Passive consumer behavior	Interaction of those involved, feedback
Depoliticization	A political learning process
Production by specialists	Collective production
Control by property owners or bureaucracy	Social control by self-organization

The Subversive Power of the New Media

As far as the objectively subversive potentialities of the electronic media are concerned, both sides in the international class struggle—except for the fatalistic adherents of the thesis of manipulation in the metropoles—

are of one mind. Frantz Fanon was the first to draw attention to the fact that the transistor receiver was one of the most important weapons in the Third World's fight for freedom. Albert Hertzog, ex-Minister of the South African Republic and the mouthpiece of the right wing of the ruling party, is of the opinion that "television will lead to the ruin of the white man in South Africa."[2] American imperialism has recognized the situation. It attempts to meet the "revolution of rising expectations" in Latin America— that is what its ideologues call it—by scattering its own transmitters all over the continent and into the remotest regions of the Amazon basin, and by distributing single-frequency transistors to the native population. The attacks of the Nixon administration on the capitalist media in the United States reveal its understanding that their reporting, however one-sided and distorted, has become a decisive factor in mobilizing people against the war in Vietnam. Whereas only twenty-five years ago the French massacres in Madagascar, with almost 100,000 dead, became known only to the readers of *Le Monde* under the heading of "Other News" and therefore remained unnoticed and without sequel in the capital city, today the media drag colonial wars into the centers of imperialism.

The direct mobilizing potentialities of the media become still more clear when they are consciously used for subversive ends. Their presence is a factor that immensely increases the demonstrative nature of any political act. The student movements in the United States, in Japan, and in Western Europe soon recognized this and, to begin with, achieved considerable momentary success with the aid of the media. These effects have worn off. Naive trust in the magical power of reproduction cannot replace organizational work; only active and coherent groups can force the media to comply with the logic of their actions. That can be demonstrated from the example of the Tupamaros in Uruguay, whose revolutionary practice has implicit in it publicity for their actions. Thus the actors became authors. The abduction of the American ambassador in Rio de Janeiro was planned with a view to its impact on the media. It was a television production. The Arab guerrillas proceed in the same way. The first to experiment with these techniques internationally were the Cubans. Fidel appreciated the revolutionary potential of the media correctly from the first. Today illegal political action demands at one and the same time maximum security and maximum publicity.

Revolutionary situations always bring with them discontinuous, spontaneous changes brought about by the masses in the existing aggregate of the media. How far the changes thus brought about take root and how permanent they are demonstrates the extent to which a cultural revolution is successful. The situation in the media is the most accurate and sensitive barometer for the rise of bureaucratic or Bonapartist anticyclones. So long as the cultural revolution has the initiative, the social imagination of the masses overcomes even technical backwardness and transforms the function of the old media so that their structures are exploded.

With our work the Revolution has achieved a colossal labor of propaganda and enlightenment. We ripped up the traditional book into single pages, magnified these a hundred times, printed them in color and stuck them up as posters in the streets. . . . Our lack of printing equipment and the necessity for speed meant that, though the best work was hand-printed, the most rewarding was standardized, lapidary and adapted to the simplest mechanical form of reproduction. Thus State Decrees were printed as rolled-up illustrated leaflets, and Army Orders as illustrated pamphlets.[3]

In the 1920s, the Russian film reached a standard that was far in advance of the available productive forces. Pudovkin's *Kinoglas* and Dziga Vertov's *Kinopravda* were no "newsreels" but political television magazine programs *avant l'écran*. The campaign against illiteracy in Cuba broke through the linear, exclusive, and isolating structure of the medium of the book. In the China of the Cultural Revolution, wall newspapers functioned like an electronic mass medium—at least in the big towns. The resistance of the Czechoslovak population to the Soviet invasion gave rise to spontaneous productivity on the part of the masses, which ignored the institutional barriers of the media. Such situations are exceptional. It is precisely their utopian nature, which reaches out beyond the existing productive forces (it follows that the productive relationships are not to be permanently overthrown), that makes them precarious, leads to reversals and defeats. They demonstrate all the more clearly what enormous political and cultural energies are hidden in the enchained masses and with what imagination they are able, at the moment of liberation, to realize all the opportunities offered by the new media.

The Media: An Empty Category of Marxist Theory

That the Marxist Left should argue theoretically and act practically from the standpoint of the most advanced productive forces in their society, that they should develop in depth all the liberating factors immanent in these forces and use them strategically, is no academic expectation but a political necessity. However, with a single great exception, that of Walter Benjamin (and in his footsteps, Brecht), Marxists have not understood the consciousness industry and have been aware only of its bourgeois-capitalist dark side and not of its socialist possibilities. An author like George Lukács is a perfect example of this theoretical and practical backwardness. Nor are the works of Horkheimer and Adorno free of a nostalgia which clings to early bourgeois media.

Their view of the cultural industry cannot be discussed here. Much more typical of Marxism between the two wars is the position of Lukács, which can be seen very clearly from an early essay in "Old Culture and New Culture."[4] "Anything that culture produces" can, according to Lukács, "have real cultural value only *if it is in itself* valuable, if the creation of each individual product is from the standpoint of its maker and a single, finite process. It must, moreover, be a process conditioned by the *human* potentialities and capabilities of the creator. The most typical example of such a process is the work of art, where the entire genesis of the work is exclusively the result of the artist's labor and each detail of the work that emerges is determined by the individual qualities of the artist. In highly developed mechanical industry, on the other hand, any connection between the product and the creator is abolished. *The human being serves the machine, he adapts to it.* Production becomes completely independent of the human potentialities and capabilities of the worker." These "forces which destroy culture" impair the work's "truth to the material," its "level," and deal the final blow to the "work as an end in itself." There is no more question of "the organic unity of the products of culture, its harmonious, joy-giving being." Capitalist culture must lack "the simple and natural harmony and beauty of the old culture—culture in the true, literal sense of the world." Fortunately things need not remain so. The "culture of proletarian society," although "in the context of such scientific research as is possible at this time" nothing more can be said about it, will certainly remedy these ills. Lukács asks himself "which are the cultural values which, in accordance with the nature of this context, *can be taken over from the old society* by the new *and further developed.*" Answer: Not the inhuman machines but "the idea of mankind as an end in itself, the basic idea of the new culture," for it is "the inheritance of the classical idealism of the nineteenth century." Quite right. "This is where the philistine concept of *art* turns up with all its deadly obtuseness—an idea to which all technical considerations are foreign and which feels that with the provocative appearance of the new technology its end has come."[5]

These nostalgic backward glances at the landscape of the last century, these reactionary ideals, are already the forerunners of socialist realism, which mercilessly galvanized and then buried those very "cultural values" which Lukács rode out to rescue. Unfortunately, in the process, the Soviet cultural revolution was thrown to the wolves; but this aesthete can in any case hardly have thought any more highly of it than did J. V. Stalin.

The inadequate understanding which Marxists have shown of the media and the questionable use they have made of them has produced a vacuum in Western industrialized countries into which a stream of non-Marxist hypothesis and practices has consequently flowed. From the Cabaret Voltaire to Andy Warhol's Factory, from the silent film comedians to the Beatles, from the first comic-strip artists to the present managers of the Underground, the apolitical have made much more radical progress in dealing with the media than any grouping of the Left. (Exception—

Münzenberg.) Innocents have put themselves in the forefront of the new productive forces on the basis of mere institutions with which communism—to its detriment—has not wished to concern itself. Today this apolitical avant-garde has found its ventriloquist and prophet in Marshall McLuhan, an author who admittedly lacks any analytical categories for the understanding of social processes, but whose confused books serve as a quarry of undigested observations for the media industry. Certainly his little finger has experienced more of the productive power of the new media than all the ideological commissions of the Communist Party and their endless resolutions and directives put together.

Incapable of any theoretical construction, McLuhan does not present his material as a concept but as the common denominator of a reactionary doctrine of salvation. He admittedly did not invent but was the first to formulate explicitly a mystique of the media which dissolves all political problems in smoke—the same smoke that gets in the eyes of his followers. It promises the salvation of man through the technology of television and indeed of television as it is practiced today. Now McLuhan's attempt to stand Marx on his head is not exactly new. He shares with his numerous predecessors the determination to suppress all problems of the economic base, their idealistic tendencies, and their belittling of the class struggle in the naive terms of a vague humanism. A new Rousseau—like all copies, only a pale version of the old—he preaches the gospel of the new primitive man who, naturally on a higher level, must return to prehistoric tribal existence in the "global village."

It is scarcely worthwhile to deal with such concepts. This charlatan's most famous saying—"the medium is the message"—perhaps deserves more attention. In spite of its provocative idiocy, it betrays more than its author knows. It reveals in the most accurate way the tautological nature of the mystique of the media. The one remarkable thing about the television set, according to him, is that it moves—a thesis which in view of the nature of American programs has, admittedly, something attractive about it.

The complementary mistake consists in the widespread illusion that media are neutral instruments by which any "messages" one pleases can be transmitted without regard for their structure or for the structure of the medium. In the East European countries the television newsreaders read fifteen-minute long conference communiqués and Central Committee resolutions which are not even suitable for printing in a newspaper, clearly under the delusion that they might fascinate a public of millions.

The sentence, "the medium is the message," transmits yet another message, however, and a much more important one. It tells us that the bourgeoisie does indeed have all possible means at its disposal to communicate something to us, but that it has nothing more to say. It is ideologically sterile. Its intention to hold on to the control of the means of production at any price, while being incapable of making the socially necessary use of them, is here expressed with complete frankness in the superstructure. It wants the media *as such* and *to no purpose.*

This wish has been shared for decades and given symbolical expression by an artistic avant-garde whose program logically admits only the alternative of negative signals and amorphous noise. Example: the already outdated "literature of silence," Warhol's films in which every thing can happen at once or nothing at all, and John Cage's forty-five-minute-long *Lecture on Nothing* (1959).

The Achievement of Benjamin

The revolution in the conditions of production in the superstructure has made the traditional aesthetic theory unusable, completely unhinging its fundamental categories and destroying its "standards." The theory of knowledge on which it was based is outmoded. In the electronic media, a radically altered relationship between subject and object emerges with which the old critical concepts cannot deal. The idea of the self-sufficient work of art collapsed long ago. The long-drawn discussion over the death of art proceeds in a circle so long as it does not examine critically the aesthetic concept on which it is based, so long as it employs criteria which no longer correspond to the state of the productive forces. When constructing an aesthetic adapted to the changed situation, one must take as a starting point the work of the only Marxist theoretician who recognized the liberating potential of the mew media. Thirty-five years ago, that is to say, at a time when the consciousness industry was relatively undeveloped, Walter Benjamin subjected this phenomenon to a penetrating dialectical-materialist analysis. His approach has not been matched by any theory since then, much less further developed.

One might generalize by saying: the technique of reproduction detaches the reproduced object from the domain of tradition. By making many reproductions it substitutes a plurality of copies for a unique existence. And in permitting the reproduction to meet the beholder or listener in his own particular situation, it reactivates the object reproduced. These two processes lead to a tremendous shattering of tradition which is the obverse of the contemporary crisis and renewal of mankind. Both processes are intimately connected with the contemporary mass movements. Their most powerful agent is the film. Its social significance, particularly in its most positive form, is inconceivable without its destructive, cathartic aspect, that is, the liquidation of the traditional value of the cultural heritage.

For the first time in world history, mechanical reproduction emancipates the work of art from its parasitical dependence on ritual. To an ever greater degree the work of art reproduced becomes the work of art designed for reproducibility. . . . But the instant the criterion of authenticity ceases to be applicable to artistic production, the total function of art is reversed. Instead of being based on ritual, it begins to be based on another practice— politics. . . . Today, by the absolute emphasis on its exhibition value, the work of art becomes a creation with entirely new functions, among which the one we are conscious of, the artistic function, later may be recognized as incidental.[6]

The trends which Benjamin recognized in his day in the film and the true import of which he grasped theoretically, have become patent today with the rapid development of the consciousness industry. What used to be called art, has now, in the strict Hegelian sense, been dialectically surpassed by and in the media. The quarrel about the end of art is otiose so long as this end is not understood dialectically. Artistic productivity reveals itself to be the extreme marginal case of a much more widespread productivity, and it is socially important only insofar as it surrenders all pretensions to autonomy and recognizes itself to be a marginal case. Wherever the professional producers make a virtue out of the necessity of their specialist skills and even derive a privileged status from them, their experience and knowledge have become useless. This means that as far as an aesthetic theory is concerned, a radical change in perspectives is needed. Instead of looking at the productions of the new media from the point of view of the older modes of production we must, on the contrary, analyze the products of the traditional "artistic" media from the standpoint of modern conditions of production.

Earlier much futile thought had been devoted to the question of whether photography is an art. The primary question—whether the very invention of photography had not transformed the entire nature of art—was not raised. Soon the film theoreticians asked the same ill-considered question with regard to the film. But the difficulties which photography caused traditional aesthetics were mere child's play as compared to those raised by the film.[7]

The panic aroused by such a shift in perspectives is understandable. The process not only changes the old burdensome craft secrets in the superstructure into white elephants, it also conceals a genuinely destructive element. It is, in a word, risky. But the only chance for the aesthetic tradition lies in its dialectical supersession. In the same way, classical physics has survived as a marginal special case within the framework of a much more comprehensive theory.

This state of affairs can be identified in individual cases in all the traditional artistic disciplines. Their present-day developments remain incomprehensible so long as one attempts to deduce them from their own prehistory. On the other hand, their usefulness or otherwise can be judged as soon as one regards them as special cases in a general aesthetic of the media. Some indications of the possible critical approaches which stem from this will be made below, taking literature as an example.

The Supersession of Written Culture

Written literature has, historically speaking, played a dominant role for only a few centuries. Even today, the predominance of the book has an episodic air. An incomparably longer time preceded it in which literature was oral. Now it is being succeeded by the age of the electronic media, which tend once more to make people speak. At its period of fullest development, the book to some extent usurped the place of the more primitive but generally more accessible methods of production of the past; on the other hand, it was a stand-in for future methods which make it possible for everyone to become a producer.

The revolutionary role of the printed book has been described often enough and it would be absurd to deny it. From the point of view of its structure as a medium, written literature, like the bourgeoisie who produced it and whom it served, was progressive. (See the *Communist Manifesto.*) On the analogy of the economic development of capitalism, which was indispensable for the development of the Industrial Revolution, the nonmaterial productive forces could not have developed without their own capital accumulation. (We also owe the accumulation of *Das Kapital* and its teachings to the medium of the book.)

Nevertheless, almost everybody speaks better than he writes. (This also applies to authors.) Writing is a highly formalized technique which, in purely physiological terms, demands a peculiarly rigid bodily posture. To this there corresponds the high degree of social specialization that it demands. Professional writers have always tended to think in caste terms. The class character of their work is unquestionable, even in the age of universal compulsory education. The whole process is extraordinarily beset with taboos. Spelling mistakes, which are completely immaterial in terms of communication, are punished by the social disqualification of the writer. The rules that govern this technique have a normative power attributed to them for which there is no rational basis. Intimidation through the written word has remained a widespread and class-specific phenomenon even in advanced industrial societies.

These alienating factors cannot be eradicated from written literature. They are reinforced by the methods by which society transmits its writing techniques. While people learn to speak very early, and mostly in psychologically favorable conditions, learning to write forms an important part of authoritarian socialization by the school ("good writing" as a kind of breaking-in). This sets its stamp forever on written communication—on its tone, its syntax, and its whole style. (This also applies to the text on this page.)

The formalization of written language permits and encourages the repression of opposition. In speech, unresolved contradictions betray themselves by pauses, hesitations, slips of the tongue, repetitions, anacoluthons, quite apart from phrasing, mimicry, gesticulation, pace, and volume. The aesthetic of written literature scorns such involuntary factors as "mistakes." It demands, explicitly or implicitly, the smoothing out of contradictions, rationalization, regularization of the spoken form irrespective of content. Even as a child, the writer is urged to hide his unsolved problems behind a protective screen of correctness.

Structurally, the printed book is a medium that operates as a monologue, isolating producer and reader. Feedback and interaction are extremely limited, demand elaborate procedures, and only in the rarest cases lead to corrections. Once an edition has been printed it cannot be corrected; at best it can be pulped. The control circuit in the case of literary criticism is extremely cumbersome and elitist. It excludes the public on principle.

None of the characteristics that distinguish written and printed literature apply to the electronic media. Microphone and camera abolish the class character of the mode of production (not of the production itself). The normative rules become unimportant. Oral interviews, arguments, demonstrations, neither demand nor allow orthography or "good writing." The television screen exposes the aesthetic smoothing out of contradictions as camouflage. Admittedly, swarms of liars appear on it, but anyone can see from a long way off that they are peddling something. As at present constituted, radio, film, and television are burdened to excess with authoritarian characteristics, the characteristics of the monologue, which they have inherited from older methods of production—and that is no accident. These outworn elements in today's media aesthetics are demanded by the social relations. They do not follow from the structure of the media. On the contrary, they go against it, for the structure demands interaction.

It is extremely improbable, however, that writing as a special technique will disappear in the foreseeable future. That goes for the book as well, the practical advantages of which for many purposes remain obvious. It is admittedly less handy and it takes up more room than other storage systems, but up to now it offers simpler methods of access than, for example, the microfilm or the tape bank. It ought to be integrated into the system as a marginal case and thereby forfeit its aura of cult and ritual.

This can be deduced from technological developments. Electronics are noticeably taking over writing: teleprinters, reading machines,

high-speed transmissions, automatic photographic and electronic composition, automatic writing devices, typesetters, electrostatic processes, ampex
libraries, cassette encyclopedias, photocopiers and magnetic copiers, speed-
printers.

The outstanding Russian media expert El Lissitsky, incidentally,
demanded an "electro-library" as far back as 1923—a request which, given
the technical conditions of the time, must have seemed ridiculous or at
least incomprehensible. This is how far this man's imagination reached
into the future:

I draw the following analogy:

Inventions in the field of verbal traffic	Inventions in the field of general traffic
Articulated language	Upright gait
Writing	The wheel
Gutenberg's printing press	Carts drawn by animal power
?	The automobile
?	The airplane

I have produced this analogy to prove that so long as the book remains a
palpable object, i.e. so long as it is not replaced by auto-vocalizing and
kino-vocalizing representations, we must look to the field of the manufacture of books for basic innovations in the near future.

There are signs at hand suggesting that this basic innovation is likely
to come from the neighborhood of the collotype.[8]

Today, writing has in many cases already become a secondary technique, a means of transcribing orally recorded speech: tape-recorded pro-
ceedings, attempts at speech-pattern recognition, and the conversion of
speech into writing.

The ineffectiveness of literary criticism when faced with so-called
documentary literature is an indication of how far the critics' thinking has
lagged behind the stage of the productive forces. It stems from the fact that
the media have eliminated one of the most fundamental categories of aesthetics up to now—fiction. The fiction/nonfiction argument has been laid
to rest just as was the nineteenth century's favorite dialectic of "art" and
"life." In his day, Benjamin demonstrated that the "apparatus" (the concept of the medium was not yet available to him) abolishes authenticity.
In the productions of the consciousness industry, the difference between
the "genuine" original and the reproduction disappears—"that aspect of
reality which is not dependent on the apparatus has now become its most
artificial aspect." The process of reproduction reacts on the object reproduced and alters it fundamentally. The efforts of this have not yet been
adequately explained epistemologically. The categorical uncertainties to

which it gives rise also affect the concept of the documentary. Strictly speaking, it has shrunk to its legal dimensions. A document is something the "forging"—that is, the reproduction—of which is punishable by imprisonment. This definition naturally has no theoretical meaning. The reason is that a reproduction, to the extent that its technical quality is good enough, cannot be distinguished in any way from the original, irrespective of whether it is a painting, a passport, or a blank note. The legal concept of the documentary record is only pragmatically useful; it serves only to protect economic interests.

The productions of the electronic media, by their nature, evade such distinctions as those between documentary and feature films. They are in every case explicitly determined by the given situation. The producer can never pretend, like the traditional novelist, "to stand above things." He is therefore partisan from the start. This fact finds formal expression in his techniques. Cutting, editing, dubbing—these are techniques for conscious manipulation without which the use of the new media is inconceivable. It is precisely in these work processes that their productive power reveals itself—and here it is completely immaterial whether one is dealing with the production of a reportage or a play. The material, whether "documentary" or "fiction," is in each case only a prototype, a half-finished article, and the more closely one examines its origins, the more blurred the difference becomes. (Develop more precisely. The reality in which a camera turns up is always faked, e.g., the moon landing.)

The Desacralization of Art

The media also do away with the old category of works of art which can only be considered as separate objects, not as independent of their material infrastructure. The media do not produce such objects. They create programs. Their production is in the nature of a process. That does not mean only (or not primarily) that there is no foreseeable end to the program—a fact which, in view of what we are at present presented with, admittedly makes a certain hostility to the media understandable. It means, above all, that the media program is open to its own consequences without structural limitations. (This is not an empirical description but a demand. A demand which admittedly is not made of the medium from without; it is a consequence of its nature, from which the much-vaunted open form can be derived—and not as a modification of it—from an old aesthetic.) The programs of the consciousness industry must subsume into themselves their own results, the reactions and the corrections which they call forth, otherwise they are already out-of-date. They are therefore to be thought of not as means of consumption but as means of their own production.

It is characteristic of artistic avant-gardes that they have, so to speak, a presentiment of the potentiality of media which still lie in the future. "It has always been one of the most important tasks of art to give rise to a demand, the time for the complete satisfaction of which has not yet come. The history of every art form has critical periods when that form strives towards effects which can only be easily achieved if the technical norm is changed, that is to say, in a new art form. The artistic extravagances and crudities which arise in this way, for instance in the so-called decadent period, really stem from art's richest historical source of power. Dadaism in the end teemed with such barbarisms. We can only now recognize the nature of its striving. Dadaism was attempting to achieve those effects which the public today seeks in film with the means of painting (or of literature)."[9] This is where the prognostic value of otherwise inessential productions, such as happenings, flux, and mixed-media shows, is to be found. There are writers who in their work show an awareness of the fact that media with the characteristics of the monologue today have only a residual use-value. Many of them admittedly draw fairly shortsighted conclusions from this glimpse of the truth. For example, they offer the user the opportunity to arrange the material provided by arbitrary permutations. Every reader as it were should write his own book. When carried to extremes, such attempts to produce interaction, even when it goes against the structure of the medium employed, are nothing more than invitations to freewheel. Mere noise permits of no articulated interactions. Short cuts, of the kind that concept art peddles, are based on the banal and false conclusion that the development of the productive forces renders all work superfluous. With the same justification, one could leave a computer to its own devices on the assumption that a random generator will organize material production by itself. Fortunately, cybernetics experts are not given to such childish games.

For the old-fashioned "artist"—let us call him the author—it follows from these reflections that he must see it as his goal to make himself redundant as a specialist in much the same way as a teacher of literacy only fulfills his task when he is no longer necessary. Like every learning process, this process too is reciprocal. The specialist will learn as much or more from the nonspecialists as the other way round. Only then can he contrive to make himself dispensable.

Meanwhile, his social usefulness can best be measured by the degree to which he is capable of using the liberating factors in the media and bringing them to fruition. The tactical contradictions in which he must become involved in the process can neither be denied nor covered up in any way. But strategically his role is clear. The author has to work as the agent of the masses. He can lose himself in them only when they themselves become authors, the authors of history.

"Pessimism of the intelligence, optimism of the will" (Antonio Gramsci).

Hans Magnus Enzensberger

NOTES

1. Bertolt Brecht, *Theory of Radio* (1932), in *Gesammelte Werke*, Band VIII, pp. 129ff., 134.

2. *Der Spiegel*, October 20, 1969.

3. El Lissitsky, "The Future of the Book," *New Left Review* 41, p. 42.

4. *Kommunismus, Zeitschrift der Kommunistischen Internationale für die Länder Südosteuropas*, 1920, pp. 1538–49.

5. Walter Benjamin, "Kleine Geschichte der Photographie," in *Das Kunstwerk im Zeitalter seiner technischen Reproduzierbarkeit* (Frankfurt, 1963), p. 69.

6. Walter Benjamin, "The Work of Art in the Age of Mechanical Reproduction," *Illuminations* (New York, 1969), pp. 221–25.

7. Ibid., p. 227.

8. Lissitsky, "The Future of the Book," p. 40.

9. Benjamin, "The Work of Art in the Age of Mechanical Reproduction," p. 237.

Brian Winston

Breakages Limited

The Past Is Prologue

The suggestion that we are *not* in the midst of monumental and increasingly frequent change in telecommunications runs so counter to our whole underlying philosophy of progress, as well as the particular rhetoric of the 'information revolution,' that it must surely be doubted by right-thinking persons. But the position taken here, rather, is that Western civilization over the past three centuries has displayed, despite enormous changes in detail, fundamental continuity. While it is impossible to predict that such continuity will be sustained over either the short or the long term, it is contended that any discontinuities, should they occur, will not *primarily* be attributable to telecommunications technologies. Other more traditionally disruptive social forces—disaffected proletarian youth, for example—rather than communications, will make greater contributions to such upheavals as might occur. This is to deny telecommunications the role of engine of change and also thereby to deny the possibility that a revolution in information technology (however that be defined) is any species of general revolution.

'Revolution' is used here in its commonly understood sense of alteration and change, rather than in its original technical sense of recurrence or turning. This is the meaning, with its modern connotation of rapid political change, intended by those who coined the phrase 'information revolution.'

> *Revolution* and *revolutionary* and *revolutionize* have of course also come to be used, outside of political contexts, to indicate fundamental changes, or fundamentally new developments, in a very wide range of activities. It can seem curious to read of 'a *revolution* in shopping habits' or of the '*revolution* in transport' and of course there are cases when this is simply the language of publicity to describe some 'dynamic' new product. But in some

From *Misunderstanding Media* (Cambridge, Mass.: Harvard University Press, 1985). Reprinted by permission of the author, Harvard University Press, and Routledge.

ways this is at least no more strange than the association of *revolution* with
VIOLENCE, since one of the crucial tendencies of the word was simply
towards important or fundamental change. Once the factory system and
the new technology of the late eighteenth century and early nineteenth
century had been called, by analogy with the French Revolution, the IN-
DUSTRIAL *Revolution,* one basis for description of new institutions and
new technologies as *revolutionary* had been laid. (capitals in original)[1]

Revolution, in whatever sense it is used, implies movement, and
in these developed usages, that means movement through time. The concept of the 'information revolution' is therefore in essence historical; and
the critique of the concept offered in this chapter is also grounded in the
past, a past limited to the particular circumstances surrounding the application, over the last two centuries, of science to the human communication process. We shall argue that there is nothing in this history to indicate
that significant major changes have not been accommodated by preexisting
social formations, and that 'revolution' is therefore quite the wrong word
to apply to the current situation. Indeed, it is possible to see this historical record as being regular enough, if the above premise of continuity is
accepted, to serve as a model for all such communication technologies,
certainly past and present and, probably, in the short-term future too.

The pattern of change in telecommunications, although historical (which is to say, diachronic), can also be expressed as a field in which
three elements—science, technology, and society—intersect. The relationship between these three elements can be elucidated by reference to another conceptual model—one taken from Saussurian linguistics.

Utterance is, for Saussure, the surface expression of a deep-seated
mental competence. In Chomskyan terms, each utterance is a performance dependent on this competence. By analogy, then, these communication technologies are also performances but of a sort of scientific
competence. Technology can be seen as standing in a structural relationship to science—as it were—utterances of a scientific language, performances of a scientific competence. In the linguistic model the link between
competence and performance is achieved by the operation of transformations which move the utterance from deep to surface level. These
movements are rule-governed and it is the rules—grammar in language—
which enable a speaker to generate comprehensible but unique utterances.
In the model proposed for technological change these transformations are
not claimed to be so regular as to be rule-governed. The notion of transformation in our model, on the contrary, allows the model to accommo-
date less predictable factors. Transformations address the operation of
factors external to the actual performance of technology, factors which
work to transform a scientifically grounded notion into a widely diffused
device.

Phase One: Scientific Competence

The development of telecommunications devices can be seen as a series of performances ('utterances') by technologists in response to the first phase of the model—*the ground of scientific competence.* The centuries-old investigations of electromagnetic phenomena and photokinesics are the two fundamental lines of scientific inquiry which make up this first phase. The possibilities of using electricity for signaling march, from the mid-eighteenth century on, virtually hand in hand with the growth of the scientific understanding of electricity. Similarly, the discovery of photography involved knowledge of the different effects light has on various substances, a scientific agenda item from at least the Middle Ages on. The propensity of certain solids to conduct sounds seems to have been known in ancient times and was certainly a well-observed phenomenon by the late eighteenth century. The photoelectric responses of selenium were known more than a century ago.

The First Transformation: Ideation

The first of these transformations is the most local, which is to say it occurs within the laboratory. To continue with the linguistic metaphor *the ideation transformation* is akin to the processes whereby a transformation at the level of competence takes place, in the human brain, so that utterance, performance, can be generated. Ideation occurs when the technologist envisages the device—gets the idea, formulates the problems involved, and hypothesizes a solution. Those mysterious mental forces—creativity, intuition, imagination, what has been called 'the will to think'—are subsumed by *ideation.*

Although the technological idea will be grounded in scientific competence, it will not necessarily relate directly to science any more than a conscious understanding of linguistic competence is needed to generate utterance. Rather, just as in language a formal understanding of the deep structure of linguistic competence is not a prerequisite of utterance, so too a lack of formal scientific competence is no bar to technological performance. But the technologist will, at some level, have absorbed the science, just as a speaker, at some level, has absorbed grammar.

The *ideation transformation* interacts with the first phase and occurs concurrently with it; in telecommunications, transformations never precede science since the formulation of technological problems has always followed agenda set by scientific inquiry—and this, contrary to received opinion which sees the primacy of science in technological

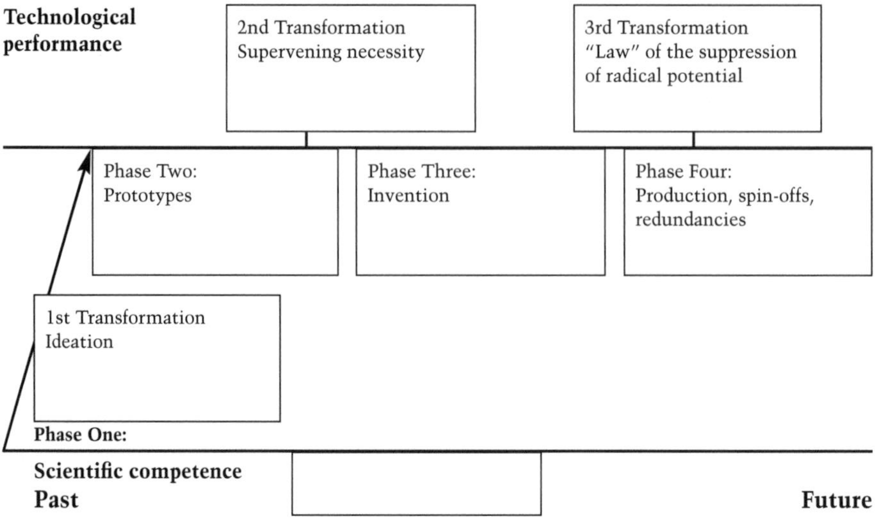

Figure 1. The model

Note
Transformations in general: The three transformations

In the model (see Fig. 1) phases, such as the phase of *scientific competence* just described, are acted upon and transformed.

(i) The first of these *transformations* moves the technology from the phase of *scientific competence* into the phase of *technological performance*. The first transformation (which will be designated *the ideation transformation*) thus moves from science to technology, its effect being to activate the technologist.

The two subsequent *transformations* alter the work of the technologist, or, to use the terminology of the model, the ongoing work of *technological performance*.

(ii) The second transformation (the transformation occasioned by *supervening social necessity*) pushes the work of the technologist from prototypes into what is popularly conceived of as 'invention.'

(iii) The third (a transformation which will be called the *'law' of the suppression of radical potential*) moves from the *invention* of devices to their diffusion.

Each transformation takes the technology further from the realm of pure science and closer to the everyday world of actual generally used devices.

We shall examine each of these three transformations in a little greater detail as they occur in the model, beginning with the *ideation transformation*.

research as a recent development, has always been true of electrical telecommunications.

Television was first triggered by a series of scientific advances. A Frenchman hypothesized the telephone in 1854, more than twenty years before Bell. A German thought of the telegraph in the last years of the eighteenth century, three decades before the first working device. Bell Laboratory workers began worrying about the transistor in the 1930s when solid-state amplifiers had already been envisaged for a decade. Some of these thinkers went on to test their ideas technically; many did not. But more often than not their work was known to those who set about building devices.

Notes
Technological performances in general
The three phases of technological performance

Before proceeding, we must now examine, in general, the last recurring element of the model—*technological performances*.

Ideation transforms the processes of science into the testing of solutions—the building of devices which is the business of *technological performance*. This will go on until the device is widely diffused and even beyond, as spin-offs and refinements are developed. Performance is triggered by a transformation (*ideation*) and each successive transformation alters the nature of what the performance produces.

(i) The second phase, after scientific competence, is designated the phase of *technological performance—prototypes*. During this phase, the technologists begin to build devices working toward fulfilling the plans which emerged from the *ideation transformation*.

(ii) The third phase, *technological performance—invention*, is, from the perspective of the technologist, exactly similar but the operation of *supervening social necessity* (the second transformation) is catalytic. So within the laboratory the work continues as it did in the prototype phase, but the second transformation—*supervening necessity*—means the devices now produced are *inventions*. This third phase, then, shall be designated *technological performance—invention*.

(iii) The operation of the next transformation, the *'law' of the suppression of radical potential*, similarly affects the last phase of technological performance. *Supervening social necessity* guarantees that the *invention* will be produced. The above 'law' operates as a constraint on that production. This final transformation occasions a tripartite phase of *technological performance—production, spin-offs, and redundant devices or redundancies*, which reflects the effects of the contradictions which are at work.

These phases of *technological performance* are all discussed below as they occur, beginning with the *prototype* phase.

Phase Two: Technological Performance—Prototypes

The Four Classes of Prototypes

The solution to a problem raised by possibilities in the advance of science has been proposed in the ideation transformation. Devices must now be built.

Prototypes, such devices, can be of four distinct classes.

(i) The prototype can be *rejected* because a supervening necessity has not yet operated and no possible use for the device is seen. Rowland's demonstration of a working telegraph in 1816 would be an example of this. The British naval authorities, understanding that the semaphore was the only machine to use in long-distance signaling, simply refused to acknowledge the superiority of the electromagnetic machine. Nearly every technology has its Rowland.

(ii) The prototype can be *accepted* because the early and incomplete operation of a supervening necessity has created a partial need which the prototype partially fills. The daguerreotype photographic process, which was widely used between 1850 and 1862, is among the clearest examples of this accepted group. It was eventually superseded by processes which used negatives, the essential mark of modern photography. Another example can be found in the development by AT&T of nongeostationary communications satellites. These were built and used a few years before the introduction, by Hughes, of the current geostationary device. The efficiency of Hollerith punch-card calculators, introduced at the turn of the century but increasingly sophisticated in the years after the First World War, can be said to have been so well accepted that the development of the electronic computer was delayed.

(iii) *Parallel* prototypes. These will occur when the device which will become the parallel prototype is already in existence solving another technological problem. Its potential use for a secondary purpose is realized only after the operation of a supervening necessity. Various devices existed in the last two decades of the nineteenth century to demonstrate the validity of electromagnetic wave theory. Distinguished physicists such as Hertz and Lodge are associated with these demonstration machines. They were in fact a species of radio but were not seen as such. Their existence is, however, of importance in tracing the work of Marconi, Popov, and others which led to radio. The cathode ray tube, before Rozing, would be another example.

(iv) Finally, in this second phase of technological performance, there can be *partial* prototypes which are machines designed to perform effectively in a given area but which do not. The telephonic apparatus developed by Reis in the 1860s and, arguably, Bell's earliest machines were of this type. Baird and Jenkins's mechanical televisions were also partial prototypes.

These then are the four prototypes of the second phase—rejected, accepted, parallel, and partial. This classification is without prejudice to the efficacy of the devices. Except for partial prototypes, which simply did not work very well, the other three classes of prototype in this phase all worked, more rather than less. The degree of their subsequent diffusion, though, depends more on the operation of the supervening necessity transformation than on their efficiency. An accepted prototype is a device which effectively fulfills the potential of the technology but, because the full power of the supervening necessity has not yet been called into play, there is still room for development. The rejected prototype might work just as well as the device eventually 'invented' but will achieve no measure of diffusion because there is no externally determined reason for its development. The parallel prototype is a similar case. The initial thrust of the technology is directed toward purposes other than those which eventually emerge. The effectiveness of this prototype in solving the problem for which it was originally designed has nothing to do with its effectiveness as a device in the second area. It is, in effect, a species of spin-off.

The Second Transformation: Supervening Necessity

The Three Broad Types of Supervening Necessities

Just as ideation worked upon the ground of scientific competence to create these different classes of prototypes, so more general supervening social necessities can now work upon these prototypes to move them out of the laboratories. In the nature of the case this second transformation, being the social, is more amorphous than the first. There is no limitation on the forces that can act as supervening necessities; they can be the objective requirements of changed circumstances or the subjective whims of perceived needs.

(i)   The least difficult group of supervening necessities to determine is occasioned by *the consequences of other technological innovation.* For instance, it was the railways that transformed telegraphic prototypes into a widely diffused device. Before railways there was no demonstrable need for such devices. Single-track rail systems, however, required, as an urgent matter of safety, instantaneous signaling. Similarly the radio came into its own with the development of the dreadnought battleship. Here, for the first time, naval battle plans called for ships to steam out of sight of each other, thus rendering the traditional signaling methods useless.

(ii) *Social forces working directly on the processes of innovation,* rather than being, as above, mediated through another technology, constitute a second, more vexed, group of supervening necessities. The rise of the modern business corporation created today's office, the architecture of the building that houses it, and the key machines, telephone and typewriter, that make it function. Such offices were the site of the first telephones and typewriters. In the middle decades of the nineteenth century the possibility of the limited liability company was established for the first time in law. The legal development of the modern corporation thus mothers telephony. The growing urban mass and its entertainment needs are the supervening necessity for the cinema. The desire of all classes to emulate the ability of the rich to have images created of themselves and for their walls focused the search for an effective photographic process.

(iii) *The commercial need for new products* and other commercial considerations would form a third group of necessities—less effective in guaranteeing diffusion and producing less significant innovation than the consequences of social change or other technological advance. Super 8 mm film, Polaroid movies, 16 rpm records, and the compact audio disk (CD) can stand for the host of devices to which commerce makes us heir.

The action of a supervening social necessity does not account for the entire development and reception of a technology. Instead, it transforms the circumstances in which the technologist labors, creating a fertile ground for effective innovation.

Phase Three: Technological Performance–Invention

The Fifth Class of Prototype

It follows from the above that there must be the possibility of a fifth class of prototype, one that is either synchronous with or subsequent to the operation of a supervening necessity. The production of such machines is the business of further technological performance and leads to what is commonly called the invention. Since the difference between the devices now produced and the previous group of prototypes is the operation of a widespread transformation (social necessity), it is likely and, indeed history reveals, common that such creations will occur in a number of places synchronously. There is, therefore, no mystery in the synchronicity of invention. This third phase will be designated *technological performance—invention.*

Of all the phases of the model this is the best known. Herein are

to be found all the heroes of communication technology's Hall of Fame—
Bell and Shockley, Marconi and Land, Rosen and Hoff, Morse, Zworykin,
Mauchly and Eckert.

The Third Transformation:
The 'Law' of the Suppression of Radical Potential

The *invention* now moves into the marketplace. Yet acceptance is never
straightforward, however 'needed' the technology. As a society we are schiz-
ophrenic about machines. On the one hand, although perhaps with an increasingly jaundiced eye, we still believe in the inevitability of progress. But
on the other hand we control every advance by conforming its context to
preexisting social patterns. The same authorities and institutions, the same
capital, the same research effort which created today's world is trying also
to create tomorrow's. A technologically induced hara-kiri on the part of
these institutions, whereby a business 'invents' a device which puts it out
of business, is obviously impossible. But what is equally true, although less
obvious, is the difficulty of inventing something to put other businesses out
of business; and the bigger the threatened business the more difficult it is.

Progress is made while going down the up escalator (or, as optimists
might argue, up the down). This jerky advance into the future can be seen
constantly repeated in telecommunications history. Its daily cavortings
can be read in the trade press. This is the measure of *the 'law' of the suppression of radical potential.* In the model this 'law' is proposed as a third
transformation, wherein general social constraints operate to limit the potential of the device radically to disrupt preexisting social formations.

Understanding the interaction of the positive effects of supervening necessity and the brake of the 'law' of the suppression of radical potential is crucial to a proper overview of how media develop. Constraints
operate firstly to preserve essential formations such as business entities
and other institutions and secondly to slow the rate of diffusion so that
the social fabric can absorb the new machine. Such functions, far from at-
rophying this century, have increased. Whatever the general perception,
there has been no speedup in the measurable rate of change. If anything,
there has been a significant diminution in the cut-throat nature of the
marketplace because the desire for stable trading circumstances, coupled
with external restrictions and monopolistic tendencies, works to contain
the crudest manifestations of the profit motive.

Two caveats must be entered as to the chosen designation of this
third and crucial transformation. Beyond the proper and necessary caution
required when postulating historical laws, *'law'* here is apostrophized to
indicate that although the phenomenon under discussion can be found

appearing in the histories of all telecommunications technologies it is not so regularly as to always manifest itself in the same form with equal force at the same point of development. It is recurrent enough to be a 'law' but not certain enough in its operation to be a law. Thus it is not a law, a universal hypothesis, in the Hempelian sense in that it does not assert that

> in every case where an event of a specified kind *C* occurs at a certain place and time, an event of a specified kind *E* will occur at a place and time which is related in a specified manner to the place and time of the occurrence of the first event.[2]

Second, *suppression* must be read in a particular way. As Lewis Carroll said, it is "rather a hard word." Here it is not meant to convey the idea of overt authoritarian prohibition or to indicate the presence of any form of conspiracy, conscious or unconscious; rather *suppression* is used in the more scientific senses given by the *OED*, namely,

> to hinder from passage or discharge; to stop or arrest the flow of; (in Botany) absence or non-development of some part or organ normally or typically present.

It is possible that even with these caveats the word is still too 'hard' to cover the sense of a technology's potential simply being dissipated by the actions of individuals and institutions. However, a 'law' of dissipation or retardation would be too soft to convey the strength of the forces at work in the process of technological innovation and diffusion.

The most obvious proof of the existence of a 'law' of *suppression*, then, is the continuation, despite the bombardments of technology, of all the institutions of our culture. To the conservative, such changes as have occurred loom very large; but any sort of informed historical vision creates a more balanced picture of their true size and scope.

The 'law' of the suppression of radical potential explains the delay of the introduction of television into the United States, which lasted at least seven years, excluding the years of war. It explains the period, from around 1880 to the eve of the First World War, during which the exercise and control of the telephone (in both the United States and the United Kingdom) was worked out while its penetration was much reduced. It accounts for the delays holding up the long-playing record for a generation and the videocassette recorder for more than decade. It underlies the halt in the growth of all but through-air propagation of audiovisual signals for more than twenty years in most countries. It is far more powerful than the concept of 'development cycles' determining, through an examination of business alone, the factors and time involved in diffusing an innovation. This 'law' works in the broadest possible way to ensure the survival, however battered, of family, home and workplace, church, president and queen, and above all, it preserves the great corporation as the primary institution of our society. The 'law' of the suppression of radical potential is the 'law' by which Bernard Shaw's mighty company 'Breakages Limited' prospers.

> Every new invention is bought up and suppressed by Breakages Limited. Every breakdown, every accident, every smash and crash, is a job for them. But for them we should have unbreakable glass, unbreakable steel, imperishable materials of all sorts.[3]

'Breakages Limited' is but a poetic version of a phenomenon obvious to anyone studying technological change—that there are endless impediments to our forward progress and not all of these can be laid at the door of the innate conservatism of a culture which is, on the contrary, dedicated to change. More is at work than, as McLuhan put it, our driving into the future with our eyes looking in the rearview mirror.

Phase Four: Technological Performance— Production, Spin-offs, Redundancies

The 'law' of the suppression of radical potential is the final transformation and it occasions this, *technological performance—production, spin-offs, and redundancies*, the last phase of the model.

(i) Of the three distinct activities covered by this phase, the least problematic is that of *production*. The acceptance of the device is to a certain extent guaranteed by the operation of the supervening necessity. Much attention has been paid by economists to the symptomatic study of diffusion at both a macro and micro level with the result that the most scholarly literature available on innovation is skewed away from the processes previously described in our model in favor of a concentration on the production and marketing phases. The problems of moving prototypes into production and marketing will therefore be peripheral to this study.

But in the course of this movement the device can be modified, extended, refined; alternative solutions can appear as rival technologies. Such developments can themselves, as in the prototype phase, be either accepted or rejected.

(ii) If such a development is accepted, diffused, it is a *spin-off*. Pac-Man, for instance, is an accepted extension of microchip technology which was certainly not developed with that specific purpose in mind. Similarly the CD record, actually a video technology, appears to be becoming

publicly accepted. These are *spin-offs*, accepted products of technological performance synchronous or subsequent to the original device's diffusion.

(iii) If, on the other hand, the technological performances of this post-
production stage are rejected, as, for instance, the Polaroid instant
movie film was or as nonlaser videodisk has been, they can be described as *redundancies* which then suffer the same fate as partial
prototypes.

Necessities and Constraints

'Revolutions' in telecommunications technologies cannot be readily accommodated in this model. But then neither can they be read easily into
the historical record nor our present situation. Using the terms established above, the current 'revolution' can only be deemed to be occurring
if the various transformations are ignored and the phases of technological
performance are misread. Such ignorance and misreading leads to the systematic misunderstanding of media involved in the concept of the 'information revolution.'

Understanding the history and current positions of telecommunications properly depends on an examination of the operation of social necessities and constraints *in vacuo*. The above brief description of the
proposed model references the history of telecommunications of the past
two centuries.

This is not the place to inscribe an account of the debate as to the
efficacy of what Popper has called 'historicism' or the propounding of 'historical prophecies.' Many, in seeking to understand the pattern of Clio's
garments, have been tempted into predicting, on the basis of that understanding, what she will wear tomorrow. The present text moves in that
company largely in response to a dominant tendency in the literature on
electronic communication devices, both popular and scholarly, which uses
an erroneous history, both implicitly and explicitly, as a predictive tool.
Unfettered by much understanding of the past beyond the anecdotal, many
currently propound insights into our future in the form of trajectories
from our past. In this literature, the historical implications of the word
'revolution' are not denied; instead a supposed technologically determined
transformational movement from 'then' to 'now' is celebrated. The purpose of this chapter is not only to explicate the 'then' by inscribing a more
balanced account of what actually occurred in the telecommunications
past but also to offer an interpretation, necessarily revisionist, of those
occurrences. The model described above arises out of a consideration of
the events of these histories.

NOTES

1. Raymond Williams, *Keywords* (New York: Oxford University Press, 1976), pp. 229 et seq.

2. C. G. Hempel, "The Function of General Laws in History," *Journal of Philosophy*, vol. 34, no. 1, January 1942, p. 35.

3. G. B. Shaw, 'Lysistrata' in *The Applecart, The Complete Plays* (London: Odhams Press, 1937), p. 1027.

Bill Nichols

The Work of Culture in the Age of Cybernetic Systems

The computer is more than an object: it is also an icon and a metaphor that suggests new ways of thinking about ourselves and our environment, new ways of constructing images of what it means to be human and to live in a humanoid world. Cybernetic systems include an entire array of machines and apparatuses that exhibit computational power. Such systems contain a dynamic, even if limited, quotient of intelligence. Telephone networks, communication satellites, radar systems, programmable laser video disks, robots, biogenetically engineered cells, rocket guidance systems, videotex networks—all exhibit a capacity to process information and execute actions. They are all "cybernetic" in that they are self-regulating mechanisms or systems within predefined limits and in relation to predefined tasks. Just as the camera has come to symbolize the entirety of the photographic and cinematic processes, the computer has come to symbolize the entire spectrum of networks, systems, and devices that exemplify cybernetic or "automated but intelligent" behavior.

This chapter traverses a field of inquiry that Walter Benjamin has crossed before, most notably in his 1936 essay, "The Work of Art in the Age of Mechanical Reproduction." My intention, in fact, is to carry Benjamin's inquiry forward and to ask how cybernetic systems, symbolized by the computer, represent a set of transformations in our conception of and relation to self and reality of a magnitude commensurate with the transformations in the conception of and relation to self and reality wrought by mechanical reproduction and symbolized by the camera. This intention necessarily encounters the dilemma of a profound ambivalence directed toward that which constitutes our imaginary Other, in this case not a mothering parent but those systems of artificial intelligence I have set out to examine here. Such ambivalence certainly permeates Benjamin's essay and is at best dialectical, and at worst, simply contradictory. Put more positively, those systems against which we test and measure the boundaries

Originally printed in *Screen*, vol. 29, no. 1, Winter 1988. Reprinted by permission of *Screen* and the author.

of our own identity require subjection to a double hermeneutic of suspicion and revelation in which we must acknowledge the negative, currently dominant, tendency toward control, and the positive, more latent potential toward collectivity.[1] It will be in terms of law that the dominance of control over collectivity can be most vividly analyzed.

In summary, what I want to do is recall a few of the salient points in Benjamin's original essay, contrast characteristics of cybernetic systems with those of mechanical reproduction, establish a central metaphor with which to understand these cybernetic systems, and then ask how this metaphor acquires the force of the real—how different institutions legitimate their practices, recalibrate their rationale, and modulate their image in light of this metaphor. In particular, I want to ask how the preoccupations of a cybernetic imagination have gained institutional legitimacy in areas such as the law. In this case, like others, a tension can be seen to exist between the liberating potential of the cybernetic imagination and the ideological tendency to preserve the existing form of social relations. I will focus on the *work* of culture—its processes, operations, and procedures—and I will assume that culture is of the essence: I include within it text and practices, art and actions that give concrete embodiment to the relation we have to existing conditions to a dominant mode of production, and the various relations of production it sustains. Language, discourse, and messages are central. Their style and rhetoric are basic. Around each "fact" and every "datum," all realities and evidence, everything "out there," a persuasive, affective tissue of discourse accrues. It is in and through this signifying tissue, arranged in discursive formations and institutional arenas, that struggle takes place and semiosis occurs.

Mechanical Reproduction and Film Culture

Benjamin argues for correspondences among three types of changes: in the economic mode of production, in the nature of art, and in categories of perception. At the base of industrial society lies the assembly line and mass production. Technological innovation allows these processes to extend into the domain of art, separating off from its traditional ritual (or "cult") value a new and distinct market (or "exhibition") value. The transformation also strips art of its 'aura' by which Benjamin means its authenticity, its attachment to the domain of tradition:

> The authenticity of a thing is the essence of all that is transmissible from its beginning, ranging from its substantive duration to its testimony to the history which it has experienced.[2]

The aura of an object compels attention. Whether a work of art or natural landscape, we confront it in one place and only one place. We discover its use value in the exercise of ritual, in that place, with that object, or in the contemplation of the object for its uniqueness. The object in possession of aura, natural or historical, inanimate or human, engages us as if it had "the power to look back in return."[3]

One thing mechanical reproduction cannot, by definition, reproduce is authenticity. This is at the heart of the change it effects in the work of art. "Mechanical reproduction emancipates the work of art from its parasitical dependence on ritual" (p. 224). The former basis in ritual yields to a new basis for art in politics, particularly, for Benjamin, the politics of the masses and mass movements, where Fascism represents an ever-present danger. The possibilities for thoroughgoing emancipation are held in check by the economic system surrounding the means of mechanical reproduction, especially in film where "illusion-promoting spectacles and dubious speculations" (p. 232) deflect us from the camera's ability to introduce us to "unconscious optics" that reveal those forms of interaction our eyes neglect.

> The act of reaching for a lighter or a spoon is familiar routine, yet we hardly know what goes on between hand and metal, not to mention how this fluctuates with our moods. Here the camera intervenes with the resources of its lowerings and liftings, its interpretations and isolations, its extensions and accelerations, its enlargements and reductions. (p. 237)

Objects without aura substitute mystique. In a remarkable, prescient passage, relegated to a footnote, Benjamin elaborates how political practice opens the way for a strange transformation of the actor when democracies encounter the crisis of Fascism. Mechanical reproduction allows the actor an unlimited public rather than the delimited one of the stage or, for the politician, Parliament. "Though their tasks may be different, the change affects equally the actor and the ruler. . . . This results in a new selection, a selection before the equipment (of mechanical reproduction) from which the star and the dictator emerge victoriously" (p. 247).

Alterations like the replacement of aura with mystique coincide with the third major change posited by Benjamin, change in categories of perception. The question of whether film or photography is an art is here secondary to the question of whether art itself has not been radically transformed in form and function. A radical change in the nature of art implies that our very ways of seeing the world have also changed: "During long periods of history, the mode of human sense perception changes with humanity's entire mode of existence" (p. 222).

Mechanical reproduction makes *copies* of visible subjects, like paintings, mountain ranges, even human beings, which until then had been thought of as unique and irreplaceable. It brings the upheavals of the Industrial Revolution to a culmination. The ubiquitous copy also serves as an externalized manifestation of the work of industrial capitalism itself. It paves the way for seeing, and recognizing, the nature and extent of the very changes mechanical reproduction itself produces.

What element of film most strongly testifies to this new form of machine-age perception? For Benjamin it is that element which best achieves what Dadaism has aspired to: "changes of place and focus which periodically assail the spectator." Film achieves these changes through montage, or editing. Montage rips things from their original place in an assigned sequence and reassembles them in ever-changing combinations that make the contemplation invited by a painting impossible. Montage multiplies the potential of collage to couple two realities on a single plane that apparently does not suit them into the juxtaposition of an infinite series of realities. As George Bataille proclaimed, "Transgression does not negate an interdiction, it transcends and completes it." In this spirit, montage transcends and completes the project of the Dadaists in their conscious determination to strip aura from the work of art and of the early French ethnographers who delighted in the strange juxtapositions of artifacts from different cultures.

Montage has a liberating potential, prying art away from ritual and toward the arena of political engagement. Montage gives back to the worker a view of the world as malleable. Benjamin writes:

> Man's need to expose himself to shock effects is his adjustment to the changes threatening him. The film corresponds to profound changes in the apperceptive apparatus—changes that are experienced on an individual scale by the man in the street in big-city traffic, on a historical scale by every present-day citizen. (p. 250)

> By close-ups of the things around us, by focusing on hidden details of familiar objects, by exploring commonplace milieus under the ingenious guidance of the camera, the film, on the one hand, extends our comprehension of the necessities which rule our lives; on the other hand, it manages to assure us of an immense and unexpected field of action. Our taverns and our metropolitan streets, our offices and furnished rooms, our railroad stations and our factories appeared to have us locked up hopelessly. Then came the film and burst this prison-world asunder by the dynamite of the tenth of a second, so that now, in the midst of its far-flung ruins and debris, we calmly and adventurously go traveling. (p. 236)

Mechanical reproduction involves the appropriation of an original, although with film even the notion of an original fades: that which is filmed has been organized in order to be filmed. This process of appropriation engenders a vocabulary: the "take" or "camera shot" used to "shoot" a scene where both stopping a take and editing are called a "cut." The violent reordering of the physical world and its meanings provides the shock effects Benjamin finds necessary if we are to come to terms with the age of mechanical reproduction. The explosive, violent potential described by Benjamin and celebrated by Brecht is what the dominant cinema must muffle, defuse, and contain. And what explosive potential can be located in the computer and its cybernetic systems for the elimination of drudgery and toil, for the promotion of collectivity and affinity, for interconnectedness, systemic networking and shared decision making, this, too, must be defused and contained by the industries of information which localize, condense, and consolidate this potential democratization of power into hierarchies of control.

"Montage—the connecting of dissimilars to shock an audience into insight—becomes for Benjamin a major principle to artistic production in a technological age."[4]

Developing new ways of seeing to the point where they become habitual is not ideological for Benjamin but transformative. They are not the habits of old ways but new; they are skills which are difficult to acquire precisely because they are in opposition to ideology. The tasks before us "at the turning points of history" cannot be met by contemplation. "They are mastered gradually by habit, under the guidance of tactile appropriation" (p. 240). The shocks needed in order to adjust to threatening changes may be co-opted by the spectacles a culture industry provides. For Benjamin the only recourse is to use those skills he himself adopted: the new habits of a sensibility trained to disassemble and reconstruct reality, of a writing style intended to relieve idlers of their convictions, of a working class trained not only to produce and reproduce the existing relations of production but to reproduce those very relations in a new, liberating form. "To see culture and its norms—beauty, truth, reality—as artificial arrangements, susceptible to detached analysis and comparison with other possible dispositions" becomes the vantage point not only of the surrealist but the revolutionary.[5]

The process of adopting new ways of seeing that consequently propose new forms of social organization becomes a paradoxical, or dialectical, process when the transformations that spawn new habits, new vision, are themselves endangered and substantially recuperated by the existing form of social organization which they contain the potential to overcome. But the process goes forward all the same. It does so less in terms of a culture of mechanical reproduction, which has reached a point similar to that of a tradition rooted in Benjamin's time, than in terms of a culture of electronic dissemination and computation.

We might then ask in what ways is our "sense of reality" being adjusted by new means of electronic computation and digital communication? Do these technological changes introduce new forms of culture into the relations of production at the same time as the "shock of the new" helps emancipate us from the acceptance of social relations and cultural forms as natural, obvious, or timeless? The distinction between an industrial capitalism, even in its "late" phase of monopoly concentration, and an information society that does not "produce" so much as "process" its basic forms of economic resource has become an increasingly familiar distinction for us. Have cybernetic systems brought about changes in our perception of the world that hold liberating potential? Is it conceivable, for example, that contemporary transformations in the economic structure of capitalism, attended by technological change, institute a less individuated, more communal form of perception similar to that which was attendant upon face-to-face ritual and aura but which is now mediated by anonymous circuitry and the simulation of direct encounter? Does montage now have its equivalent in interactive simulations and simulated interactions experienced according to predefined constraints? Does the work of art in the age of postmodernism lead, at least potentially, to apperceptions of the 'deep structure' of postindustrial society comparable to the apperceptive discoveries occasioned by mechanical reproduction in the age of industrial capitalism?

Cybernetic Systems and Electronic Culture

We can put Benjamin's arguments, summarized cursorily here, in another perspective by highlighting some of the characteristics associated with early, entrepreneurial capitalism, monopoly capitalism, and multinational or postindustrial capitalism:

Entrepreneurial capitalism	Monopoly capitalism	Multinational capitalism
steam and locomotive power	electricity and petrochemical power	microelectronics and nuclear energy
property rights	corporate rights	copyright and patents
nature as Other/conquest of nature	aliens as Other/conquest of Third World	knowledge as Other/ conquest of intelligence
nationalism	imperialism	multinationalism
working-class vanguard	consumer-group vanguard	affinity-group vanguard
Tuberculosis	Cancer	AIDS
contamination by nature	contamination by an aberrant self	deficiency of self (collapse of system that distinguishes self from environment)
isolation of self from threatening environment	isolation of aberrant tissue from self	isolation of self by artificial life support
vulnerability to invasive agents	vulnerability to self-consumption	vulnerability to systemic collapse
heightened individuation	heightened schizophrenia	heightened sense of paranoia
realism	modernism	postmodernism
film	television	computer

mechanical reproduction	instantaneous broadcast	logico-iconic simulations
reproducible instances	ubiquitous occurrences	processes of absorption and feedback
the copy	the event	the chip (and VDT display)
subtext of possession	subtext of mediation	subtext of control
image and representation	collage and juxtaposition	simulacra

Simulacra introduce the key question of how the control of information moves toward control of sensory experience, interpretation, intelligence, and knowledge. The power of the simulation moves to the heart of the cybernetic matter. It posits the simulation as an imaginary Other which serves as the measure of our own identity and, in doing so, prompts the same form of intense ambivalence that the mothering parent once did: a guarantee of identity based on what can never be made part of oneself. In early capitalism, the human was defined in relation to an animal world that evoked fascination and attraction, repulsion and resentment. The human animal was similar to but different from all other animals. In monopoly capitalism, the human was defined in relation to a machine world that evoked its own distinctive blend of ambivalence. The human machine was similar to but different from all other machines. In postindustrial capitalism, the human is defined in relation to cybernetic systems—computers, biogenetically engineered organisms, ecosystems, expert systems, robots, androids, and cyborgs—all of which evoke those forms of ambivalence reserved for the Other that is the measure of ourselves. The human cyborg is similar to but different from all other cyborgs. Through these transformations questions of difference persist. Human identity remains at stake, subject to change, vulnerable to challenge and modification as the very metaphors prompted by the imaginary Others that give it form themselves change. The metaphor that's meant (that's taken as real) becomes the simulation. The simulation displaces any antecedent reality, any aura, any referent to history. Frames collapse. What had been fixed comes unhinged. New identities, ambivalently adopted, prevail.

The very concept of a text, whether unique or one of myriad copies, for example, underpins almost all discussion of cultural forms including film, photography, and their analogue in an age of electronic communication, television (where the idea of "flow" becomes an important amendment). But in cybernetic systems, the concept of "text" itself undergoes substantial slippage. Although a textual element can still be isolated, computer-based systems are primarily interactive rather than one-way, open-ended rather than fixed. Dialogue, regulated and disseminated by digital computation, deemphasizes authorship in favor of "messages-in-circuit"[6] that take fixed but effervescent, continually variable form. The link between message and substrate is loosened: words on a printed page are irradicable; text on a video display terminal (VDT) is readily altered.

The text conveys the sense of being addressed to us. The message-in-circuit is both addressed to and addressable by us; the mode is fundamentally interactive, or dialogic. That which is most textual in nature—the fixed, read-only-memory (ROM), and software programs—no longer addresses us. Such texts are machine-addressable. They direct those operational procedures that ultimately give the impression that the computer responds personally to us, simulating the processes of conversation or of interaction with another intelligence to effect a desired outcome. Like face-to-face encounter, cybernetic systems offer (and demand) almost immediate response. This is a major part of their hazard in the workplace and their fascination outside it. The temporal flow and once-only quality of face-to-face encounter becomes embedded within a system ready to restore, alter, modify, or transform any given moment to us at any time. Cybernetic interactions can become intensely demanding, more so than we might imagine from our experience with texts, even powerfully engaging ones. Reactions must be almost instantaneous, grooved into eye and finger reflexes until they are automatic. This is the bane of the "automated workplace" and the joy of the video game. Experienced video-game players describe their play as an interactive ritual that becomes totally self-absorbing. As David, a lawyer in his mid-thirties interviewed by Sherry Turkle, puts it,

> At the risk of sounding, uh, ridiculous, if you will, it's almost a Zen type of thing. . . . When I can direct myself totally but not feel directed at all. You're totally absorbed and it's all happening there. . . . You either get through this little maze so that the creature doesn't swallow you up or you don't. And if you can focus your attention on that, and if you can really learn what you're supposed to do, then you really are in relationship with the game.[7]

The enhanced ability to test the environment, which Benjamin celebrated in film ("The camera director in the studio occupies a place identical with that of the examiner during aptitude tests," p. 246) certainly continues with cybernetic communication.[8] The computer's dialogic mode carries the art of the "what if" even further than the camera eye has done, extending beyond the "what if I could see more than the human eye can see" to "what if I can render palpable those possible transformations of existing states that the individual mind can scarcely contemplate?"

If mechanical reproduction centers on the question of reproducibility and renders authenticity and the original problematic, cybernetic simulation renders experience, and the real itself, problematic. Instead of reproducing, and altering, our relation to an original work, cybernetic communication simulates, and alters, our relation to our environment and mind. As Jean Baudrillard argues, "Instead of facilitating communication, it (information, the message-in-circuit) exhausts itself in the *staging* of communication . . . this is the gigantic simulation process with which we are familiar."[9]

Instead of a representation of social practices recoded into the conventions and signs of another language or sign-system, like the cinema, we encounter simulacra that represent a new form of social practice in their own right and represent nothing. The photographic image, as Roland Barthes proposed, suggests "having been there" of what it represents, of what is present-in-absentia. The computer simulation suggests only a "being here" and "having come from nowhere" of what it presents, drawing on those genetic-like algorithms that allow it to bring its simulation into existence, sui generis. Among other things, computer systems simulate the dialogical and other qualities of life itself. The individual becomes nothing but an ahistorical position within a chain of discourse marked exhaustively by those shifters that place him or her within speech acts ("I," "here," "now," "you," "there," "then"). In face-to-face encounter this "I" all speakers share can be inflected to represent some part of the self not caught by words. To respond to the query, "How are you?" by saying "Not *too* bad," rather than "Fine," suggests something about a particular state of mind or style of expression and opens onto the domains of feeling and empathy. What cannot be represented in language directly (the bodily, living "me" that writes or utters words) can significantly inflect speech, and dialogue, despite its enforced exclusion from any literal representation.

In cybernetic systems, though, "I" and "you" are strictly relational propositions attached to no substantive body, no living individuality. In place of human intersubjectivity we discover a systems interface, a boundary between cyborgs that selectively passes information but without introducing questions of consciousness or the unconscious, desire or will, empathy or conscience, saved in simulated forms.

Even exceptions like ELIZA, a program designed to simulate a therapeutic encounter, prove the rule. "I" and "you" function as partners in therapy only as long as the predefined boundaries are observed. As Sherry Turkle notes, if you introduce the word "mother" into your exchange, and then say, "Let's discuss paths toward nuclear disarmament," ELIZA might well offer the nonsense reply, "Why are you telling me that your mother makes paths toward nuclear disarmament?"[10] Simulations like these may bring with them the shock of recognizing the reification of a fundamental social process, but they also position us squarely within the realm of communication and exchange cleanly evacuated of the intersubjective complexities of direct encounter. Cybernetic systems give form, external expression, to processes of the mind (through messages-in-circuit) such that the very ground of social cohesion and consciousness becomes mediated through a computational apparatus. Cybernetic interaction achieves with an other (an intelligent apparatus) the simulation of social process itself.

Cybernetic dialogue may offer freedom from many of the apparent risks inherent in direct encounter; it offers the illusion of control. This use of intelligence provides a lure that seems to be much more attractive to men than women. At first there may seem to be a gain, particularly regarding the question of the look or gaze. Looking is an intensely charged act, one significantly neglected by Benjamin, but stressed in recent feminist critiques of dominant Hollywood cinema. There looking is posed as a primarily masculine act and "to-be-looked-at-ness" a feminine state, reinforced, in the cinema, by the camera's own voyeuristic gaze, editing patterns that prompt identification with masculine activism and feminine passivity, and a star system that institutionalizes these uses of the look through an iconography of the physical body.[11] This entire issue becomes circumvented in cybernetic systems that simulate dialogic interaction, or face-to-face encounter, but exclude not only the physical self or its visual representation but also the cinematic apparatus that may place the representation of sexual difference within a male-dominant hierarchy.

Correct in so far as it goes, the case for the circumvention of the sexist coding of the gaze overlooks another form of hierarchical sexual coding that revolves around the question of whether a fascination with cybernetic systems is not itself a gender-related (i.e., a primarily masculine) phenomenon (excluding from consideration an even more obvious gender coding that gives almost all video games, for example, a strong aura of aggressive militaristic activity). The questions that we pose about the sexist nature of the gaze within the cinematic text and the implications this has for the position we occupy in relation to such texts, may not be wholly excluded so much as displaced. A (predominantly masculine) fascination with the *control* of simulated interactions replaces a (predominantly masculine) fascination with the to-be-looked-at-ness of a projected image. Simulated intersubjectivity as a product of automated but intelligent systems invokes its own peculiar psychodynamic. Mechanical reproduction issues an invitation to the fetishist—a special relationship to the images of actors or politicians in place of any more direct association. The fetish *object*— the image of the other that takes the place of the other—becomes the center of attention while fetishistic viewers look on from their anonymous and voyeuristic, seeing-but-unseen sanctuary in the audience. But the output of computational systems stresses simulation, interaction, and process itself. Engagement with this *process* becomes the object of fetishization rather than representations whose own status as produced objects has been masked. Cybernetic interaction emphasizes the fetishist rather than the fetish object: instead of a taxonomy of stars we find a galaxy of computer freaks. The consequence of systems without aura, systems that replace direct encounter and realize otherwise inconceivable projections and possibilities, is a fetishism of such systems and processes of control themselves. Fascination resides in the subordination of human volition to the operating constraints of the larger system. We can talk to a system whose responsiveness grants us an awesome feeling of power. But as Paul Edwards observes, "Though individuals . . . certainly make decisions and set goals, as links in the chain of command they are allowed no choices regarding the ultimate purposes and values of the system. Their 'choices' are . . . . always the permutations and combinations of a predefined set."[12]

The desire to exercise a sense of control over a complex but predefined logical universe replaces the desire to view the image of another over which the viewer can imagine himself to have a measure of control. The explosive power of the dynamite of the tenth of a second extolled by Benjamin is contained within the channels of a psychopathology that leaves exempt from apperception, or control, the mechanisms that place ultimate control on the side of the cinematic apparatus or cybernetic system. These mechanisms—the relay of gazes among the camera, characters, and viewer, the absorption into a simulacrum with complex problems and eloquent solutions—are the ground upon which engagement occurs and are not addressable within the constraints of the system itself. It is here, at this point, that dynamite must be applied.

This is even more difficult with computers and cybernetics than with cameras and the cinema. Benjamin himself noted how strenuous a task it is in film to mask the means of production, to keep the camera and its supporting paraphernalia and crew from intruding upon the fiction. Exposure of this other scene, the one behind the camera, is a constant hazard and carries the risk of shattering the suspension of disbelief. Only those alignments between camera and spectator that preserve the illusion of a fictional world without camera, lights, directors, studio sets, and so on are acceptable. Benjamin comments, perhaps with more of a surrealist's delight in strange juxtapositions than a Marxist's, "The equipment-free aspect of reality here (in films) has become the height of artifice; the sight of immediate reality has become an orchid in the land of technology" (p. 233).

With the contemporary prison-house of language, in Frederic Jameson's apt phrase, the orchid of immediate reality, like the mechanical bird seen at the end of *Blue Velvet*, appears to have been placed permanently under glass; but for Benjamin, neither the process by which an illusionistic world is produced nor the narrative strategies associated with it receive extended consideration. For him, the reminders of the productive process were readily apparent, not least through the strenuous efforts needed to mask them. The "other scene" where fantasies and fictions actually become conceptually and mechanically produced may be repressed but is not obliterated. If not immediately visible, it lurks just out of sight in the offscreen space where the extension of a fictional world somewhere collides with the world of the camera apparatus in one dimension and the world of the viewer in another. It retains the potential to intrude at every cut or edit; it threatens to reveal itself in every lurch of implausibility or sleight of hand with which a narrative attempts to achieve the sense of an ending.

With cybernetic systems, this other scene from which complex rule-governed universes actually get produced recedes farther from sight. The governing procedures no longer address us in order to elicit a suspension of belief; they address the cybernetic system, the microprocessor of the computer, in order to absorb us into their operation. The other scene has vanished into logic circuits and memory chips, into "machine language" and interface cards. The chip replaces the copy. Just as the mechanical reproduction of copies revealed the power of industrial capitalism to reorganize and reassemble the world around us, rendering it as commodity art, the automated intelligence of chips reveals the power of postindustrial capitalism to simulate and replace the world around us, rendering not only exterior realm but also its interior ones of consciousness, intelligence, thought, and intersubjectivity as commodity experience.

The chip is pure surface, pure simulation of thought. Its material surface is its meaning without history, without depth, without aura, affect, or feeling. The copy reproduces the world, the chip simulates it. It is the difference between being able to remake the world and being able to efface it. The microelectronic chip draws us into a realm, a design for living, that fosters a fetishized relationship with the simulation as a new reality all its own based on the capacity to control, within the domain of the simulation, what had once eluded control beyond it. The orchids of immediate reality that Benjamin was wont to admire have become the paper flowers of the cybernetic simulation.

Electronic simulation instead of mechanical reproduction. Fetishistic addiction to a process of logical simulation rather than a fascination with a fetishized object of desire. Desire for the dialogic or interactive and the illusion of control versus desire for the fixed but unattainable and the illusion of possession. Narrative and realism draw us into relations of identification with the actions and qualities of characters. Emulation is possible, as well as self-enhancement. Aesthetic pleasure allows for a revision of the world from which a work of art arises. Reinforcing what is or proposing what might be, the work of art remains susceptible to a double hermeneutic of suspicion and revelation. Mechanical reproduction changes the terms decidedly, but the metonymic or indexical relationship between representational art and the social world to which it refers remains a fundamental consideration.

By contrast, cybernetic simulations offer the possibility of completely replacing any direct connection with the experiential realm beyond their bounds. Like the cinema, this project, too, has its origins in the expansion of nineteenth-century industrialism. The emblematic precursors of the cyborg—the machine as self-regulating system—were those animate, self-regulating systems that offered a source of enhancement even museums could not equal: the zoo and the botanical garden.

At the opening of the first large-scale fair or exhibition, the Great Exhibition of 1851, Queen Victoria spoke of "the greatest day in our history

[when] the whole world of nature and art was collected at the call of the queen of cities." Those permanent exhibitions—the zoo and botanical garden—introduced a new form of vicarious experience quite distinct from the aesthetic experience of original art or mechanically reproduced copies. The zoo brings back alive evidence of a world we could not otherwise know, now under apparent control. It offers experience at a remove that is fundamentally different as a result of having been uprooted from its original context. The indifferent, unthreatened, and unthreatening gaze of captive animals provides eloquent testimony to the difference between the zoo and the natural habitat to which it refers. The difference in the significance of what appears to be the same thing, the gaze, indicates that the change in context has introduced a new system of meaning, a new discourse or language.

Instead of the shocks of montage that offer a "true means of exercise" appropriate to the "profound changes in the apperceptive apparatus" under industrial capitalism, the zoo and botanical garden exhibit a predefined, self-regulating world with no reality outside its own boundaries. These worlds may then become the limit of our understanding of those worlds to which they refer but of which we seldom have direct knowledge. "Wildlife" or "the African savannah" is its simulation inside the zoo or garden or diorama. Absorption with these simulacra and the sense of control they afford may be an alternative means of exercise appropriate to the apperceptive changes required by a service and information economy.

Computer-based systems extend the possibilities inherent in the zoo and garden much further. The ideal simulation would be a perfect replica, now *controlled* by whomever controls the algorithms of simulation—a state imaginatively rendered in films like *The Stepford Wives* or *Blade Runner* and apparently already achieved in relation to certain biogenetically engineered microorganisms. Who designs and controls these greater systems and for what purpose becomes a question of central importance.

The Cybernetic Metaphor: Transformations of Self and Reality

The problems of tracking antiaircraft weapons against extremely fast targets prompted the research and development of intelligent mechanisms capable of predicting future states or positions far faster than the human brain could do.[13] The main priorities were speed, efficiency, and reliability; that is, fast-acting, error-free systems. ENIAC (Electronic Numerical Integrator and Computer), the first high-powered digital computer, was designed to address precisely this problem by performing ballistic computations at enormous speed and allowing the outcome to be translated into adjustments in the firing trajectory of antiaircraft guns.

"The men [*sic*] who assembled to solve problems of this order and who formalized their approach into the research paradigms of information theory and cognitive psychology through the Macy Foundation Conferences, represent a who's who of cybernetics: John von Neuman, Oswald Weblen, Vannevar Bush, Norbert Wiener, Warren McCulloch, Gregory Bateson and Claude Shannon, among others." Such research ushers in the central metaphors of the cybernetic imagination: not only the human as an automated but intelligent system, but also automated, intelligent systems as human, not only the simulation of reality but the reality of the simulation. These metaphors take form around the question, the still unanswered question, put by John Stroud at the Sixth Macy Conference:

> We know as much as possible about how the associated gear bringing the information to the tracker [of an anti-aircraft gun] operates and how all the gear from the tracker to the gun operates. So we have the human operator surrounded on both sides by very precisely known mechanisms and the question comes up. "What kind of machine have we placed in the middle?"[14]

This question of "the machine in the middle" and the simulation as reality dovetails with Jean Baudrillard's recent suggestion that the staging powers of simulation establish a hyperreality we only half accept but seldom refuse: "Hyperreality of communication of meaning: by dint of being more real than the real itself, reality is destroyed."[15]

Such metaphors, then, become more than a discovery of similarity, they ultimately propose an identity. Norbert Wiener's term "cyborg" (cybernetic organism) encapsulates the new identity which, instead of seeing humans reduced to automata, sees simulacra which encompass the human elevated to the organic. Consequently, the human cognitive apparatus (itself a hypothetical construct patterned after the cybernetic model of automated intelligence) is expected to negotiate the world by means of simulation.

Our cognitive apparatus treats the real as though it consisted of those properties exhibited by simulacra. The real becomes simulation. Simulacra, in turn, serve as the mythopoeic impetus for that sense of the real we posit beyond the simulation. A sobering example of what is at stake follows from the Reagonomic conceptualization of war. The Strategic Defense Initiative (SDI) represents a vast Battle of the Cyborgs video game where players compete to save the world from nuclear holocaust. Reagan's simulated warfare would turn the electromagnetic force fields of 1950s science-fiction films that shielded monsters and creatures from the arsenal of human destructive power into plowshares beyond the ozone. Star Wars would be the safe-sex version of international conflict: not one drop of our enemy's perilous bodily fluids, none of their nuclear ejaculations, will come into contact with the free world.

Reagan's simulation of war as a replacement for the reality of war does not depend entirely on SDI. We have already seen it at work in the invasion of Grenada and the raid on Libya. Each time, we have had the evocation of the reality of war: the iconography of heroic fighters, embattled leaders, brave decisions, powerful technology, and concerted effort rolled into the image of military victory, an image of quick, decisive action that defines the "American will."

These simulacra of war, though, are fought with an imaginary enemy, in the Lacanian sense, and in the commonsense meaning of an enemy posited within those permutations allowed by a predefined set of assumptions and foreign-policy options: a Grenadian or Libyan "threat" appears on the video screens of America's political leadership. Long experience with the communist menace leads to prompt and sure recognition. Ronny pulls the trigger. These simulations lack the full-blown, catastrophic consequences of real war, but this does not diminish the reality of this particular simulation nor the force with which it is mapped onto a historical "reality" it simultaneously effaces. Individuals find their lives irreversibly altered, people are wounded, many die. These indelible punctuation marks across the face of the real, however, fall into place according to a discourse empowered to make the metaphoric reality of the simulation a basic fact of existence.

A more complex example of what it means to live not only in the society of the spectacle but also in the society of the simulacrum involves the preservation/simulation of life via artificial life-support systems. In such an environment, the presence of life hinges on the presence of "vital signs." Their manifestation serves as testimony to the otherwise inaccessible presence of life itself, even though life in this state stands in relation to the "immediate reality" of life as the zoo stands in relation to nature. The important issue here is that the power of cybernetic simulations prompts a redefinition of such fundamental terms as life and reality, just as, for Benjamin, mechanical reproduction alters the very conception of art and the standards by which we know it. Casting the issue in terms of whether existence within the limits of an artificial life-support system should be considered "life" obscures the issue in the same way that asking whether film and photography are "art" does. In each case a presumption is made about a fixed, or ontologically given, nature of life or art, rather than recognizing how that very presumption has been radically overturned.

And from preserving life artificially, it is a small step to creating life by the same means. There is, for example, the case of Baby M. Surrogate mothering, as a term, already demonstrates the reality of the simulation: the actual mothering agent—the woman who bears the child—becomes a *surrogate,* thought of, not as a mother, but as an incubator or "rented uterus," as one of the trial's medical "experts" called Mary Beth Whitehead. The *real* surrogate mother, the woman who will assume the role of mother for a child not borne of her own flesh, becomes the real mother, legally and familiarly. The law upholds the priority of the simulation and the power of those who can control this system of surrogacy—measured by class and gender, for it is clearly upper-class males (Judge Harvey Sorkow and the father, William Stern) who mobilized and sanctioned this particular piece of simulation, largely, it would seem, given the alternative of adoption, to preserve a very real, albeit fantastic preoccupation with a patriarchal blood line.

Here we have the simulation of a nuclear family—a denucleated, artificial simulation made and sanctioned as real, bona fide. The trial evoked the reality of the prototypical bourgeois family: well-educated, socially responsible, emotionally stable, and economically solvent, in contrast to the lower middle-class Whitehead household. The trial judgment renders as legal verdict the same moral lesson that Cecil Hepworth's 1905 film, *Rescued by Rover*, presents as artistic theme: the propriety of the dominant class, the menace of an unprincipled, jealous, and possessive lower class, the crucial importance of narrative donors like the faithful Rover and of social agents like the patronizing Sorkow, and the central role of the husband as the patriarch able to preside over the constitution and reconstitution of his family. Now replayed as simulation, the morality play takes on a reality of its own. People suffer, wounds are inflicted. Lives are irreversibly altered, or even created. Baby M is a child conceived as a product to be sold to fill a position within the signifying discourse of patriarchy.

The role of the judge in this case was, of course, crucial to its outcome. His centrality signals the importance of the material, discursive struggle being waged within the realm of the law. Nicos Poulantzas argues that the juridical-political is the dominant or articulating region in ideological struggle today. Law establishes and upholds the conceptual frame in which subjects, "free and equal" with "rights" and "duties," engage on a playing field made level by legal recourse and due process. These fundamental concepts of *individuals* with the right to enter into and withdraw from relations and obligations to others underpin, he argues, the work of other ideologically important regions in civil society.[16]

Whether the juridical-political is truly the fulcrum of ideological contestation or not, it is clearly a central area of conflict and one in which some of the basic changes in our conception of the human/computer, reality/simulation metaphors get fought out. Reconceptualizations of copyright and patent law, brought on by computer chip design, computer software, and biogenetic engineering, give evidence of the process by which a dominant ideology seeks to preserve itself in the face of historical change.

Conceptual metaphors take on tangible embodiment through discursive practices and institutional apparatuses. Such practices give a metaphor historical weight and ideological power. Tangible embodiment has always been a conscious goal of the cybernetic imagination where abstract concepts become embedded in the logic and circuitry of a material substrate deployed to achieve specific forms of result such as a computer, an antiaircraft tracking system, or an assembly-line robot. These material objects, endowed with automated but intelligent capacities, enter our culture as, among other things, commodities. As a peculiar category of object these cyborgs require clarification of their legal status. What proprietary rights pertain to them? Can they be copyrighted, patented, protected by trade secrets acts; can they themselves as automated but intelligent entities, claim legal rights that had previously been reserved for humans or other living things on a model akin to that which has been applied to annual research?

The answers to such questions do not fall from the sky. They are the result of struggle, of a clash of forces, and of the efforts, faltering or eloquent, of those whose task it is to make and adjudicate the law. New categories of objects do not necessarily gain the protection of patent or copyright law. One reason for this is that federal law in the United States (where most of my research on this question took place) and the Constitution both enshrine the right of individuals to private ownership of the means of production while also enjoining against undue forms of monopoly control. The Constitution states, "The Congress shall have power . . . to promote the progress of science and useful arts, by securing for limited times to authors and inventors the exclusive right to their respective writings and discoveries." Hence the protection of intellectual property (copyright and trademark registration) or industrial and technological property (patents) carves out a proprietary niche within the broader principle of a "free flow" of ideas and open access to "natural" sources of wealth.

The cybernetic organism, of course, confounds the distinction between intellectual and technological property. Both a computer and a biogenetically designed cell "may be temporarily or permanently programmed to perform many different unrelated tasks."[17] The cybernetic metaphor, of course, allows us to treat the cell and the computer as sources of the same problem. As the author of one legal article observed, "A ribosome, like a computer, can carry out any sequence of assembly instructions and can assemble virtually unlimited numbers of different organic compounds, including those essential to life, as well as materials that have not yet been invented."[18] What legal debates have characterized the struggle for proprietary control of these cyborgs?

Regarding patents, only clearly original, unobvious, practical applications of the "laws of nature" are eligible for protection, a principle firmly established in the Telephone Cases of 1888 where the Supreme Court drew a sharp distinction between electricity itself as nonpatentable since it was a "force of nature" and the telephone where electricity was found, "a new, specific condition not found in nature and suited to the transmission of vocal or other sounds."

Recent cases have carried the issue further, asking whether "intelligent systems" can be protected by patent and, if so, what specific elements of such a system are eligible for protection. Generally, and perhaps ironically, the United States Supreme Court has been more prone to grant protection for the fabrication of new life forms, via recombinant DNA experiments, than for the development of computer software. In *Diamond v. Chakrabatry* (1980), the Supreme Court ruled in favor of patent protection for Chakrabatry, who had developed a new bacterial form capable of degrading petroleum compounds for projected use in oil-spill cleanups. In other, earlier cases, the Supreme Court withheld patent protection for computer software. In *Gottschalk v. Benson* (1972) and in *Parker v. Flook* (1979), the Court held that computer programs were merely algorithms, that is, simple, step-by-step mathematical procedures, and as such were closer to basic principles or concepts than to original and unobvious applications. These decisions helped prompt recourse to a legislative remedy for an untenable situation (for those with a vested interest in the marketability of computer programs); in 1980 Congress passed the Software Act, granting some of the protection the judicial branch had been reluctant to offer but still leaving many issues unsettled. A Semiconductor Chip Protection Act followed in 1984 with a new sui generis form of protection for chip masks (the templates from which chips are made). Neither copyright nor patent, this protection applies for ten years (less than copyright) and demands less originality of design than does patent law. In this case, the law itself replicates the "having come from nowhere" quality of the simulation. The *Minnesota Law Review* 70 (December 1985) is devoted to a symposium on this new form of legal protection for intellectual but also industrial property.

The Software Act began the erosion of a basic distinction between copyright and patent by suggesting that useful objects were eligible for copyright. In judicial cases such as *Diamond v. Diehr* (1981), the Court held that "when a claim containing a mathematical formula implements or applies that formula in a structure or process which, when considered as a whole, is performing a function which the patent laws were designed to protect (for example, transforming or reducing an article to a different state of things), then the claim satisfies the requirements of [the copyright law]."

This finding ran against the grain of the long-standing *White-Smith Music Publishing Co. v. Apollo Co.* decision of 1908, where the Supreme Court ruled that a player-piano roll was ineligible for the copyright protection accorded to the sheet music it duplicated. The roll was considered part of a machine rather than the expression of an idea. The distinction was formulated according to the code of the visible: a copyrightable text must be visually perceptible to the human eye and must "give to every person *seeing* it the idea created by the original."[19]

Copyright had the purpose of providing economic incentive to bring new ideas to the marketplace. Copyright does not protect ideas, processes, procedures, systems, or methods, only a specific embodiment of such things. (A book on embroidery could receive copyright but the process of embroidery itself could not.) Similarly, copyright cannot protect useful objects or inventions. If an object has an intrinsically utilitarian function, it cannot receive copyright. Useful objects can be patented, if they are original enough, or protected by trade secrets acts. For example, a fabric design could receive copyright as a specific, concrete rendition of form. It would be an "original work of authorship" fixed in the tangible medium of cloth and the "author" would have the right to display it as an ornamental or artistic object without fear of mutation. But the same fabric design, once embodied in a dress, can no longer be copyrighted since it is now primarily a utilitarian object. Neither the dress, nor any part of it, can receive copyright. Others would be free to imitate its appearance since the basic goal (according to a somewhat non-fashion-conscious law) is to produce a utilitarian object meant to provide protection from the elements and a degree of privacy for the body inside it.

What then of a video game? Is this an original work of authorship? Is it utilitarian in essence? And if it is eligible for copyright, what element or aspect of it, exactly, shall receive the copyright? The process of mechanical reproduction had assured that the copyright registration of one particular copy of a work would automatically ensure protection for all its duplicates. Even traditional games like *Monopoly*, which might produce different outcomes at each playing, were identical to one another in their physical and visible parts. But the only visible part of a video game is its video display. The display is highly ephemeral and varies in detail with each play of the game. For a game like Pac-Man, the notion of pursuit or pursuit through a maze would be too general. Like the notion of the Western or the soap opera, it is too broad for copyright eligibility. Instead the key question is whether a general idea, like pursuit, is given concrete, distinctive, *expression.* The working out of this distinction, though, lends insight into the degree of difference between mechanical reproduction and cybernetic systems perceived by the U.S. judicial system.

For video games like Pac-Man, a copyright procedure has developed that gives protection to the outward manifestation of the underlying software programs. Registration of a copyright does not involve depositing the algorithms structuring the software of the ROM (read-only memory) chip in which it is stored. Instead, registration requires the deposit of a videotape of the game in the play mode.[20]

Referring to requirements that copyright is for "original works of authorship fixed in any tangible medium," federal district courts have found that creativity directed to the end of presenting a video display constitutes recognizable authorship and "fixation" occurs in the *repetition* of specific aspects of the visual scenes from one playing of a game to the next. But fixing precisely what constitutes repetition when subtle variations are also in play is not a simple matter. For example, in *Atari v. North Amer-*

ican Phillips Consumer Electronics Corp. (1981), one district court denied infringement of Atari's Pac-Man by the defendant's K. C. Munchkin. The decision rested on a series of particular differences between the games despite overall similarities. In elaboration, the court noted that the Munchkin character, unlike Pac-Man, "initially faces the viewer rather than showing a profile." K. C. Munchkin moves in profile but when he stops, "he turns around to face the viewer with another smile." Thus the central character is made to have a personality which the central character in Pac-Man does not have. K. C. Munchkin has munchers which are "spookier" than the goblins in Pac-Man. Their legs are longer and move more dramatically, their eyes are vacant—all features absent from Pac-Man.

This opinion, however, was overturned in *Atari vs. North American Phillips* (1982). The seventh circuit court found Pac-Man's expressive distinctiveness to lie in the articulation of a particular kind of pursuit by means of "gobbler" and "ghost-figures," thereby granting broad protection to the game by likening it to a film genre or subgenre. The circuit court found the Munchkin's actions of gobbling and disappearing to be "blatantly similar," and went on to cut through to the basic source of the game's appeal and marketability.

> Video-games, unlike an artist's painting or even other audio visual works, appeal to an audience that is fairly undiscriminating insofar as their concern about more subtle differences in artistic expression. The main attraction of a game such as Pac-Man lies in the stimulation provided by the intensity of the competition. A person who is entranced by the play of the game, "would be disposed to overlook" many of the minor differences in detail and "regard their aesthetic appeal as the same."[21]

In this decision, the court stresses the process of absorption and feedback sustained by an automated but intelligent system that can simulate the reality of pursuit. The decision represents quite a remarkable set of observations. The fetishization of the image as object of desire transforms into a fetishization of a process as object of desire. This throws as much emphasis on the mental state of the participant as on the exact visual qualities of the representation ("A person who is entranced by the play of the game").

In these cases the courts have clearly recognized the need to guarantee the exclusive rights of authors and inventors (and of the corporations that employ them) to the fruits of their discoveries. Simultaneously, this recognition has served to legitimate the cybernetic metaphor and to renormalize the political-legal apparatus in relation to the question: who shall have the right to control the cybernetic system of which we are a part? On the whole, the decisions have funneled that control back to a discrete proprietor, making what is potentially disruptive once again consonant with the social formation it threatens to disrupt.

Such decisions may require recasting the legal framework itself and its legitimizing discourse. Paula Samuelson identifies the magnitude of the transformation at work quite tellingly: "It [is] necessary to reconceptualize copyright and patent in ways that would free the systems from the historical subjects to which they have been applied. It [is] necessary to rethink the legal forms, pare them down to a more essential base, and adjust their rules accordingly. It [is] necessary to reconceive the social bargain they now reflect."[22]

If efforts to gain proprietary control of computer chip masks, software, and video games have prompted little radical challenge from the left, the same cannot be said for bacteria and babies, for the issues of proprietorship that are raised by new forms of artificial life and artificial procreation is where the "social bargain" woven into our discursive formations undergoes massive transformation.

The hidden agenda of mastery and control, the masculinist bias at work in video games, in Star Wars, in the reality of the simulation (of invasions, raids, and wars), in the masculine need for autonomy and control as it corresponds to the logic of a capitalist marketplace becomes dramatically obvious when we look at the artificial reproduction of human life. The human as a metaphorical, automated, but intelligent system becomes quite literal when the human organism is itself a product of planned engineering.

Gametes, embryos, and fetuses become, like other forms of engineered intelligence that have gained legal status, babies-to-be, subject now to the rules and procedures of commodity exchange. Human life, like Baby M herself, becomes in every sense a commodity to be contracted for, subject to the proprietary control of those who rent the uterus, or the test tube, where such entities undergo gestation.

As one expert in the engineering of human prototypes put it, reproduction in the laboratory is willed, chosen, purposed, and controlled, and is, therefore, more human than coitus with all its vagaries and elements of chance.[23] Such engineering affirms the "contractor's" rights to "take positive steps to enhance the possibility that offspring will have desired characteristics, as well as the converse right to abort or terminate offspring with undesired or undesirable characteristics."[24] But what is more fundamentally at stake does not seem to be personal choice, but power and economics. These opportunities shift reproduction from family life, private space, and domestic relations to the realm of production itself by means of the medical expert, clinical space, and commodity relations. The shift allows men who previously enjoyed the privilege of paying for their sexual pleasure without the fear of consequence the added opportunity of paying for their hereditary preferences without the fear of sexual pleasure.

Such "engineered fetuses" and babies become so much like real human beings that their origin as commodities, bought and sold, may be readily obscured. They become the perfect cyborg. As with other instances in which a metaphor becomes operative and extends across the face of a culture, we have to ask who benefits and who suffers? We have to ask what is at stake and how might struggle and contestation occur? What tools are at our disposal and to what conception of the human do we adhere that can call into question the reification, the commodification, the patterns of mastery and control that the human as cyborg, the cyborg as human, the simulation of reality, and the reality of the simulation make evident?

Like the normalization of the cybernetic metaphor as scientific paradigm or the judicial legitimization of the private ownership of cybernetic systems (even when their substrate happens to be a living organism), the justification for hierarchical control of the cybernetic apparatus takes a rhetorical form because it is, in essence, an ideological argument. Dissent arises largely from those who appear destined to be controlled by the "liberating force" of new cybernetic technologies. But in no arena will the technologies themselves be determining. In each instance of ideological contestation, what we discover is that the ambivalences regarding cybernetic technology require resolution on more fundamental ground: that domain devoted to a social theory of power.

Purpose, System, Power: Transformative Potential Versus Conservative Practice

Liberation from any literal reference beyond the simulation, like liberation from a cultural tradition bound to aura and ritual, brings the actual process of constructing meaning, and social reality, into sharper focus. This liberation also undercuts the Renaissance concept of the individual. "Clear and distinct" people may be a prerequisite for an industrial economy based on the sale of labor power, but mutually dependent cyborgs may be a higher priority for a postindustrial postmodern economy. In an age of cybernetic systems, the very foundation of Western culture and the very heart of its metaphysical tradition, the individual, with his or her inherent dilemmas of free will versus determinism, autonomy versus dependence, and so on, may very well be destined to stand as a vestigial trace of concepts and traditions which are no longer pertinent.

The testing Benjamin found possible with mechanical reproduction—the ability to take things apart and reassemble them, using, in film, montage, the "dynamite of the tenth of a second"—extends yet further with cybernetic systems: what had been mere possibilities or probabilities manifest themselves in the simulation. The dynamite of nanoseconds explodes the limits of our own mental landscape. What falls open to apperception is not just the relativism of social order and how, through recombination, liberation from imposed order is possible, but also the set of systemic principles governing order itself, its dependence on messages-in-circuit, regulated at higher levels to conform to predefined constraints. We discover how, by redefining those constraints, liberation from them is possible. Cybernetic systems and the cyborg as human metaphor refute a heritage that celebrates individual free will and subjectivity.

If there is liberating potential in this, it clearly is not in seeing ourselves as cogs in a machine or elements of a vast simulation, but rather in seeing ourselves as part of a larger whole that is self-regulating and capable of long-term survival. At present this larger whole remains dominated by parts that achieve hegemony. But the very apperception of the cybernetic connection, where system governs parts, where the social collectivity of mind governs the autonomous ego of individualism, may also provide the adaptive concepts needed to decenter control and overturn hierarchy.

Conscious purpose guides the invention and legitimization of cybernetic systems. For the most part, this purpose has served the logic of capitalism, commodity exchange, control and hierarchy. Desire for short-term gain or immediate results gives priority to the criteria of predictability, reliability, and quantifiability. Ironically, the survival of the system as a whole (the sum total of system plus environment on a global scale) takes a subordinate position to more immediate concerns. We remain largely unconscious of that total system that conscious purpose obscures. Our consciousness of something indicates the presence of a problem in need of solution, and cybernetic systems theory has mainly solved the problem of capitalist systems that exploit and deplete their human and natural environment, rather than conserving both themselves and their environment.

Anthony Wilden makes a highly germane observation about the zero-sum game, *Monopoly*. The goal of the game is to win by controlling the relevant environment, the properties, and the capital they generate. But *Monopoly* and its intensification of rational, conscious purpose masks a logic in the form of being "merely a game" that is deadly when applied to the open ecosystem. Wilden writes, "We usually fail to see that Monopoly supports the ideology of competition by basing itself on a logical and ecological absurdity. It is assumed that the winning player, having consumed all the resources of all the opponents, can actually survive the end of the game. In fact this is impossible. . . . The Monopoly winner [must] die because in the context of the resources provided by the game, the winner has consumed them all, leaving no environment (no other players) to feed on."[25]

"There is the discovery," Gregory Bateson writes in one of his more apocalyptic essays, "that man is only a part of larger systems and that the part can never control the whole."[26] The cybernetic metaphor invites the testing of the purpose and logic of any given system against the goals of the larger ecosystem where the unit of survival is the adaptive organism-in-relation-to-its-environment, not the monadic individual or any other part construing itself as autonomous or "whole."[27] "Transgression does not negate an interdiction; it transcends and completes it." The transgressive and liberating potential which Bataille found in the violation of taboos and prohibitions, and which Benjamin found in the potential of mechanically reproduced works of art, persists in yet another form. The cybernetic metaphor contains the germ of an enhanced future inside a prevailing model that substitutes part for whole, simulation for real, cyborg for human, conscious purpose for the decentered goal-seeking of the totality—system plus environment. The task is not to overthrow the prevailing cybernetic model but to transgress its predefined interdictions and limits, using the dynamite of the apperceptive powers it has itself brought into being.

NOTES

1. The concept of the double hermeneutic derives from Fredric Jameson, *The Political Unconscious* (Ithaca: Cornell University Press, 1981), especially the final chapter.

2. Walter Benjamin, "The Work of Art in the Age of Mechanical Reproduction," in *Illuminations,* tr. Harry Zohn (New York: Schocken Books, 1969), p. 221. Further page references from the essay are given in the text.

3. Walter Benjamin, *Schriften,* 2 vols. (Frankfurt: Suhrkamp Verlag, 1955), I, p. 461. Translated in Fredric Jameson, *Marxism and Form* (Princeton: Princeton University Press, 1971), p. 77.

4. Terry Eagleton, *Marxism and Literary Criticism* (Berkeley: University of California Press, 1976), p. 63.

5. This quote is from James Clifford, "On Ethnographic Surrealism," *Comparative Studies in Society and History,* vol. 23, 4 (October 1981): 559–64, where he offers an excellent description of the confluences between surrealism and certain tendencies within early ethnography in 1920s France.

6. See, for example, the essays in Part III, "Form and Pathology in Relationship" by Gregory Bateson, *Steps to an Ecology of Mind* (New York: Ballantine Books, 1972), where the phrase is introduced and applied to various situations.

7. Quoted in Sherry Turkle, *The Second Self: Computers and the Human Spirit* (New York: Simon & Schuster, 1984), p. 86.

8. Steven J. Heims, *John von Neuman and Norbert Wiener: From Mathematics to the Technology of Life and Death* (Cambridge, Mass.: MIT Press, 1980), describes how research on antiaircraft guidance systems led Julian Bigelow and Norbert Wiener to develop a mathematical theory "for predicting the future as best one can on the basis of incomplete information about the past" (p. 183). For an overview of the history of cybernetic theory and cognitive psychology in the context of its military-industrial origins, see Paul N. Edwards, "Formalized Warfare," unpublished ms. (1984), History of Consciousness program, University of California, Santa Cruz.

9. Jean Baudrillard, "The Implosion of Meaning in the Media and the Implications of the Social in the Masses," in Kathleen Woodward, ed., *The Myths of Information* (Madison: Coda Press, 1980), p. 139.

10. Sherry Turkle, p. 264.

11. See Laura Mulvey, "Visual Pleasure and Narrative Cinema," *Screen,* vol. 16, 3 (Autumn 1975): 6–18.

12. Paul N. Edwards, p. 59.

13. See, for example, Paul N. Edwards, for a more detailed account of this synergism between the development of cybernetics and military needs. For a cybernetic theory of alcoholism and schizophrenia, see Gregory Bateson, and Watzlawick, Beavin, and Jackson's study of human interaction in a system framework in *Pragmatics of Human Communication.*

14. John Stroud, "Psychological Moments in Perception—Discussions," in H. Van Foersta, et al., eds., *Cybernetics: Circular Causal and Feedback Mechanisms in Biological and*

Social Systems, Transactions of the Sixth Macy Conference (New York: Josiah Macy Foundation, 1949), pp. 27–28.

15. Jean Baudrillard, p. 139.

16. See Nicos Poulantzas, *Political Power and Social Class* (London: New Left Books, 1975), pp. 211–14.

17. James J. Myrick and James A. Sprowl, "Patent Law for Programmed Computers and Programmed Life Forms," *American Bar Association Journal,* no. 68 (August 1982): 120.

18. Myrick and Sprowl, p. 121. Some other relevant articles include: "Biotechnology: Patent Law Developments in Great Britain and the United States," *Boston College International and Comparative Law Review,* no. 6 (Spring 1983): 563–90; "Can a Computer be an Author? Copyright Aspects of Artificial Intelligence," *Communication Entertainment Law Journal,* 4 (Summer 1983): 707–47; Peter Aufrichtig, "Copyright Protection for Computer Programs in Read-Only Memory Chips," *Hofstra Law Review* II (February 1983): 320–70; "Patents on Algorithms, Discoveries and Scientific Principles," *Idea* 24 (1983): 21–39; S. Hewitt, "Protection of Works Created by Use of Computers," *New Law Journal,* 133 (March 11, 1983): 235–37; E. N. Kramsky, "Video Games: Our Legal System Grapples with a Social Phenomenon," *Journal of the Patent Office Society,* 64 (June 1982): 335–51.

19. This case's relevance for computer software litigation is discussed in Peter Aufrichtig's "Copyright Protection for Computer Programs in Read-Only Memory Chips": 320–70.

20. E. N. Kramsky, p. 342.

21. 214 US PQ 33t 7th Cir, 1982, pp. 33, 42, 43.

22. Paula Samuelson, "Creating a New Kind of Intellectual Property: Applying the Lessons of the Chip Law to Computer Programs," *Minnesota Law Review* 70 (December 1985): 502.

23. Cited in Christine Overall, "'Pluck a Fetus from Its Womb': A Critique of Current Attitudes Toward the Embryo/Fetus," *University of Western Ontario Law Review,* vol. 24, 1 (1986): 6–7.

24. Overall, p. 7.

25. Anthony Wilden, "Changing Frames of Order: Cybernetics and the Machina Mundi," in Kathleen Woodward, ed., *The Myths of Information,* p. 240.

26. Gregory Bateson, "Conscious Purpose and Nature," in *Steps to an Ecology of Mind,* p. 437.

27. Gregory Bateson, "Style, Grace and Information in Primitive Art," in *Steps to an Ecology of Mind,* p. 145.

Producing Technoculture

"At Oracle, even if you were a good programmer, you were usually labeled as a babe who can code."
–Katrina Garnett, CEO, Crosswords Software

"You're living a Web lifestyle when you just take it for granted that any purchase you make, any new thing you'd want to plan, like a trip, you turn to the Web as part of that process. . . . I've been bold enough to say that in the next decade, the majority of Americans will be living the Web lifestyle. It'll just be there."
–Bill Gates, Microsoft, after redirecting a $300 billion company with 20,000 employees to compete in Internet space

"Telecom–even in the era of fiber optics–has always lacked the vision people. On the superhighway, the aesthetic capital of Hollywood and the programming expertise of its emerging clones will be no less bankable–and leveraged–than in the past. . . . The infrastructural players on the highway (the Malones and Cases, the TCIs and AOLs) are already in place, as are the softwarehousers of canned experience (the film and television industries, the Igers and Katzenbergs, the ABCs and Disneys), without which such infrastructure is just so much costly real-estate. The genius of the televised summit was that it allowed controlling interests, typically relegated invisibly offscreen to the board rooms of industry, to act as the highway's live and on-screen historiographers and aestheticians–and to do so long before America was either fully wired or on-line."
–John Caldwell, Televisuality

Arthur Kroker
Michael A. Weinstein

The Theory of the Virtual Class

Wired Shut

Wired *intends to profit from the Internet. And so do a lot of others.* "People are going to have to realize that the Net is another medium, and it has to be sponsored commercially and it has to play by the rules of the marketplace," *says John Battelle,* Wired's *28-year-old managing editor.* "You're still going to have sponsorship, advertising, the rules of the game, because it's just necessary to make commerce work." "I think that a lot of what some of the original Net god-utopians were thinking," *continued Battelle,* "is that there was just going to be this sort of huge anarchist, utopian, bliss medium, where there are no rules and everything is just sort of open. That's a great thought, but it's not going to work. And when the Time Warners get on the Net in a hard fashion it's going to be the people who first create the commerce and the environment, like* Wired, *that will be the market leaders."*
—Andrew Leonard, "Hot-Wired," The Bay Guardian

The twentieth century ends with the growth of cyberauthoritarianism, a stridently protechnotopia movement, particularly in the mass media, typified by an obsession to the point of hysteria with emergent technologies, and with a consistent and very deliberate attempt to shut down, silence, and exclude any perspectives critical of technotopia. Not a wired culture, but a virtual culture that is wired shut: compulsively fixated on digital technology as a source of salvation from the reality of a lonely culture and radical social disconnection from everyday life, and determined to exclude

From *Data Trash: The Theory of the Virtual Class* (New York: St. Martin's Press, 1994). Reprinted by permission of the authors, St. Martin's Press, and New World Perspectives.

117

from public debate any perspective that is not a cheerleader for the coming-to-be of the fully realized technological society. The virtual class is populated by would-be astronauts who never got the chance to go to the moon, and they do not easily accept criticism of this new *Apollo* project for the body telematic.

This is unfortunate since it is less a matter of being pro- or anti-technology, but of developing a critical perspective on the ethics of virtuality. When technology mutates into virtuality, the direction of political debate becomes clarified. If we cannot escape the hard-wiring of (our) bodies into wireless culture, then how can we inscribe primary ethical concerns onto the will to virtuality? How can we turn the virtual horizon in the direction of substantive human values: aesthetic creativity, social solidarity, democratic discourse, and economic justice? To link the relentless drive to cyberspace with ethical concerns is, of course, to give the lie to technological liberalism. To insist, that is, that the coming-to-be of the will to virtuality, and with it the emergence of our doubled fate as either body dumps or hyper-texted bodies, virtualizers or data trash, does not relax the traditional human injunction to give primacy to the ethical ends of the technological purposes we choose (or the will to virtuality that chooses us).

Privileging the question of ethics via virtuality lays bare the impulse to nihilism that is central to the virtual class. For it, the drive to planetary mastery represented by the will to virtuality relegates the ethical suasion to the electronic trashbin. Claiming with monumental hubris to be already beyond good and evil, it assumes perfect equivalency between the will to virtuality and the will to the (virtual) good. If the good is equivalent to the disintegration of experience into cybernetic interactivity or to the disappearance of memory and solitary reflection into massive Sunstations of archived information, then the virtual class is the leading exponent of the era of telematic ethics. Far from having abandoned ethical concerns, the virtual class has patched a coherent, dynamic, and comprehensive system of ethics onto the hard-line processors of the will to virtuality. Against economic justice, the virtual class practices a mixture of predatory capitalism and gung-ho technocratic rationalizations for laying waste to social concerns for employment, with insistent demands for "restructuring economies," "public policies of labor adjustment," and "deficit cutting," all aimed at maximal profitability. Against democratic discourse, the virtual class institutes anew the authoritarian mind, projecting its class interests onto cyberspace from which vantage point it crushes any and all dissent to the prevailing orthodoxies of technotopia. For the virtual class, politics is about absolute control over intellectual property by means of warlike strategies of communication, control, and command. Against social solidarity, the virtual class promotes a grisly form of raw social materialism, whereby social experience is reduced to its prosthetic after-effects: the body becomes a passive archive to be processed, entertained, and stockpiled by the seduction-apertures of the virtual reality complex. And finally, against aesthetic creativity, the virtual class promotes the value of pattern-maintenance (of its own choosing), whereby human intelligence is reduced to a circulating medium of cybernetic exchange floating in the interfaces of the cultural animation machines. Key to the success of the virtual class is its promotion of a radically diminished vision of human experience and of a disintegrated conception of the human good: for virtualizers, the good is ultimately that which disappears human subjectivity, substituting the war-machine of cyberspace for the data trash of experience. Beyond this, the virtual class can achieve dominance today because its reduced vision of human experience consists of a digital superhighway, a fatal scene of circulation and gridlock, which corresponds to how the late-twentieth-century mind likes to see itself. *Reverse nihilism:* not the nihilistic will as projected outward onto an external object, but the nihilistic will turned inward, decomposing subjectivity, reducing the self to an object of conscience and body vivisectioning. What does it mean when the body is virtualized without a sustaining ethical vision? Can anyone be strong enough for this? What results is rage against the body: a hatred of existence that can only be satisfied by an abandonment of flesh and subjectivity and, with it, a flight into virtuality. Virtuality without ethics is a primal scene of social suicide: a site of mass cryogenics where bodies are quick-frozen for future resequencing by the archived data networks. The virtual class can be this dynamic because it is already the aftershock of the living dead: body vivisectionists and early (mind) abandoners surfing the Net on a road trip to the virtual Inferno.

"Adapt or You're Toast"

The virtual class has driven to global power along the digital superhighway. Representing perfectly the expansionary interests of the recombinant commodity-form, the virtual class has seized the imagination of contemporary culture by conceiving a techno-utopian high-speed cybernetic grid for traveling across the electronic frontier. In this mythology of the new technological frontier, contemporary society is either equipped for fast travel down the main arterial lanes of the information highway, or it simply ceases to exist as a functioning member of technotopia. As the CEOs and the specialist consultants of the virtual class triumphantly proclaim: "Adapt or you're toast."

We now live in the age of dead information, dead (electronic) space, and dead (cybernetic) rhetoric. *Dead information?* That's our co-optation as servomechanisms of the cybernetic grid (the digital superhighway) that swallows bodies, and even whole societies, into the dynamic momentum of its telematic logic. Always working on the basis of the illusion of enhanced interactivity, the digital superhighway is really about the full immersion of the flesh into its virtual double. As *dead (electronic) space,* the digital superhighway is a big real estate venture in cybernetic form, where competing claims to intellectual property rights in an array of multimedia technologies of communication are at stake. No longer capitalism under the double sign of consumer and production models, the digital superhighway represents the disappearance of capitalism into colonized virtual space. And *dead (cybernetic) rhetoric?* That's the internet's subordination to the predatory business interests of a virtual class, which might pay virtual lip service to the growth of electronic communities on a global basis, but which is devoted in actuality to shutting down the anarchy of the Net in favor of virtualized (commercial) exchange. Like a mirror image, the digital superhighway always means its opposite: not an open telematic autoroute for fast circulation across the electronic galaxy, but an immensely seductive harvesting machine for delivering bodies, culture, and labor to virtualization. The information highway is paved with (our) flesh. So consequently, *the glory of the virtual class:* cultural accommodation to technotopia is its goal, political consolidation (around the aims of the virtual class) its method, multimedia nervous systems its relay, and (our) disappearance into pure virtualities its ecstatic destiny.

That there is an inherent political contradiction between the attempt by the virtual class to liquidate the sprawling web of the internet in favor of the smooth telematic vision of the digital superhighway is apparent. The information highway is the antithesis of the Net, in much the same way as the virtual class must destroy the *public dimension* of the internet for its own survival. The informational technology of the internet as a new *force* of virtual production provides the social conditions necessary for instituting fundamentally new *relations* of electronic creation. Spontaneously and certainly against the long-range interests of the virtual class, the internet has been swamped by demands for meaning. Newly screen-radiated scholars dream up visions of a Virtual University, the population of Amsterdam goes on-line as Digital City, environmentalists become web weavers as they form a global Green cybernetic informational grid, and a new generation of fiction writers develops forms of telematic writing that mirror the crystalline structures and multiphasal connections of hypertext.

But, of course, for the virtual class, content slows the speed of virtualized exchange, and meaning becomes the antagonistic contradiction of data. Accordingly, demands for meaning must be immediately denied as just another road-kill along the virtual highway. As such, the virtual class exercises its intense obsessive-compulsive drive to subordinate society to the telematic mythology of the digital superhighway. The democratic possibilities of the internet, with its immanent appeal to new forms of global communication, might have been the seduction-strategy appropriate for the construction of the digital superhighway, but now that the cybernetic grid is firmly in control, the virtual class must move to liquidate the internet. It is an old scenario, repeated this time in virtual form. Marx understood this first: every technology releases opposing possibilities toward emancipation and domination. Like its early bourgeois predecessors at the birth of capitalism, the virtual class christens the birth of technotopia by suppressing the potentially emancipatory relations of production released by the internet in favor of the traditionally predatory force of production signified by the digital superhighway. Data is the antivirus of meaning—telematic information refuses to be slowed down by the dragweight of content. And the virtual class seeks to exterminate the *social* possibilities of the internet. These are the first lessons of the theory of the virtual class.

Information Highway/Media-Net: Virtual Pastoral Power

The "information highway" has become the key route into virtuality. The "information highway" is another term for what we call the "media-net." It's a question of whether we're cruising on a highway or being caught up in a Net, always already available for (further) processing. The "highway" is definitely an answer to "Star Wars": the communications complex takes over from the "military-industrial complex." Unlike "Star Wars," however, the "highway" has already (de-)materialized in the world behind the monitors: cyberspace. For crash theory there is an irony: the highway is a trompe l'oeil of possessive individualism covering the individual possessed by the Net, sucked into the imploded, impossible world behind the screen—related to the dubious world of ordinary perception through cyberspace.

Information Highways Versus Media-Net

The prophet-hypesters of the information highway, from President Bill Clinton, United States, to President Bill Gates, Microsoft, proclaim a revolution to a higher level of bourgeois consciousness. The highway is the utopia of the possessive individual: the possessive individual now resides in technotopia.

This is how the higher level of bourgeois consciousness comes to be in grades of perfection. First, we enter an information highway which promises the "individual" access to "information" from the universal archive instantly and about anything. The capacity of the Net to hold information is virtually infinite and, with the inevitable advances in microprocessors, its capacities to gather, combine, and relay information will be equal to any demand for access. Are you curious about anything? The answer is right at your fingertips. More seriously, do you need to know something? A touch of a button will get you what you need and eventually your brain waves alone (telekinesis fantasy) will do it. Here is the world as information completely at the beck and call of the possessive individual (the individual, that is, who is *possessed* by information). Here, everyone is a god who, if they are not omniscient all at once, can at least entertain whatever information that they wish to have at any time they wish to have it. Information is not the kind of thing that has to be shared. If everyone all at once wanted to know who won the Stanley Cup in 1968 they could have the information simultaneously: cyberspace as the site of Unamuno's panarchy, where each one is king.

At the next grade of perfection, the highway not only provides access to that which is already given, but allows the "individual" to "interact" with other "individuals," to create a society in cyberspace. The freedom to access information will be matched by the freedom to access individuals anywhere and at any time, since eventually everyone will be wired. The hybridization of television, telephone, and computer will produce every possible refinement of mediated presence, allowing interactors an unprecedented range of options for finely adjusting the distance of their relations. Through the use of profiles, data banks, and bulletin boards people will be able to connect with exactly those who will give them the most satisfaction, with whom they share interests, opinions, projects, and sexual preferences, and for whom they have need. Just as "individuals" will be able to access the realm of "information" (anything from their financial and insurance records to any movie ever made), they will also be able to access the domain of "human" communicators to find the ones who are best suited to them. As Bill Gates of Microsoft puts it: "The opportunity for people to reach out and share is amazing."[1]

The information highway as technotopia is the place where "individuals" command information for whatever purpose they entertain and find others with whom to combine to pursue those purposes. As Gates puts it, it is "empowering stuff." Technotopia is the seduction by which the flesh is drawn into the Net. What seduces is the fantasy of "empowerment," the center of the contemporary possessive individualist complex. By having whatever information one wants instantly and without effort, and by being linked to appropriate associates one saves an immense amount of time and energy, and is more likely to make better decisions for oneself. Who can complain about having more information, especially if it can be accessed easily and appropriately by a system of selectors that gives you what you ask for and nothing else, or even better, that knows you so well that it gives you what you really want (need?) (is good for you?), but did not even realize that you wanted?

The information highway means the death of the (human) agent and the triumph of the expert program, the wisdom that the greatest specialist would give you. Expert programs to diagnose you. Medical tests performed at home while you are hooked up to a computer that are interpreted by an expert program. In order to serve you, the "highway" will demand information from you. The selector systems will have to get to know you, scan you, monitor you, give you periodic tests. The expert program will be the new center for pastoral power. This is, of course, still enacted under capitalism. You will have to pay for information with money and there will be plenty of restrictions on its accessibility. Leave that as a contradiction of the virtual class between the capitalist organization of the highway and its technotopian vision: a contradiction within possessive individualism. More important, you will pay for information with information; indeed, you will be information.

The highway becomes the Net. What appears as "empowerment" is a trompe l'oeil, a seduction, an entrapment in a Baudrillardian loop in which the Net elicits information from the "user" and gives it back in what the selectors say is an appropriate form for that user. The great agent of possibility becomes the master tool of normalization, now a micronormalization with high specificity . . . perhaps uniqueness! Each "individual" has a unique disciplinary solution to hold them fast to the Net, where they are dumped for image processing and image reception." The information highway is the way by which bodies are drawn into cyberspace through the seduction of empowerment.

Bourgeois masculinity has always been prepubescent: the thoughts of little boys thinking about what they would do if they controlled the world, but now the world is cyberspace. The dream of being the god of cyberspace—public ideology as the fantasy of prepubescent males: a regression from sex to an autistic power drive.

Against the Virtual Class

The virtual class holds on to its worldview with cynicism or with vicious naivete. It is a compound of late-nineteenth-century Darwinian capitalism (retroindustrial Darwinism) and tech-hype. After what has happened so far in the twentieth century and is still going on in the way of technological carnage, it is amusing to realize that there are still technofetishists filled with enthusiasm about how technology is going to fulfill their prepubescent dream, which they assume unthinkingly that everyone inevitably shares with them. Why? Is it so clear that technology cannot serve anything else than the last man as the prepubescent boy who would like nothing else but to play video games forever?

The retrochild. The virtual class is in its utopian visionary phase, filled with cyberworlds to conquer. What will it be in its consolidation phase when we are fully entrapped in the Net and it starts tightening around us? Normalization will come here too. Radically empowering computer land is the utopia of a rising class identifying its peculiar occupational psychosis with (a weird) "humanity." When we are immersed in the Net the fiction of the "possessive individual" will be discarded from the virtual class's ideology in favor of some sort of defense of cyberslavery, in which the virtual class affirms its own slavery, along with that of all the rest, to the Net. This will be the culminating moment of the ascetic priests (Nietzsche). One can only think of Jonestown. The virtual class ushers itself and everyone else into the Net to serve it as image/information resources and as image/information receptors. Wired into the command functions at work and wired into the sensibility functions when off work: the body as a function of cyberspace.

Panic Information Highway

Organizations are in a panic stampede to get on the "information highway," to be players in cyberspace. Everyone wants in on the exploitation of the new frontier and even more they don't want to be killed in the real world, which will be managed ever-increasingly from cyberspace; not to mention the efficiencies of the Net. For the moment the advantages of the Net are not that obvious once you get on, but that is only a temporary situation. The Net is filling up fast with everything imaginable and it's indefinitely expandable.

There is another kind of panic in process about the "information highway." This one from the concerned liberals who are afraid of the power of those who will determine the configuration of the highway. In his report on Bill Gates, John Seabrook provides an enlightening glimpse of Gates's character along with cautionary warnings. We are concerned with the latter, with a specimen of the liberal ideology which counts as the major ideological resistance to cyber tech-hype.

Seabrook frames his warnings within a bit of short-range futurology. There is a new kind of computer on the way that will change our lives in incalculable ways: "The new machine will be a communications device that connects people to the information highway. It will penetrate far beyond the fifteen per cent of American households that now own a computer, and it will control, or absorb, other communications machines now in people's home—the phone, the fax, the television. It will sit in the living room, not in the study."[2] The cyber command-machine: the entrance to the highway: the lip of the Net.

Seabrook notes that Bill Gates's current ambition is to have Microsoft be the source of "the standard operating-system software for the information-highway machine, just as it now supplies the standard operating-system software, called Windows, for the personal computer."[3] The standard operating-system will be the program that makes possible specific uses of the Net, all across the Net. Seabrook believes that by supplying the standard operating-system software for the "information-highway machine" Gates would gain great power: "If Gates does succeed in providing the operating system for the new machine, he will have tremendous influence over the way people communicate with one another: he, more than anyone else, will determine what it is like to use the information highway."[4]

Seabrook shows a misunderstanding here of the "influence" of the virtual class. What is the "influence" of a standard operating system? Would there be major differences among possible alternative competing operating systems for the information highway machine that would alter significantly "the way people communicate with each other"? Or, as with the phone system, is the object simply to facilitate entry into the Net? If the latter is so, no power in any conventional sense accrues to the organizational leader who wins the competition to supply the system. Gates understands this. He wrote to Seabrook that "the digital revolution is all about facilitation-creating tools to make things easy."[5] This is the gospel of the last man, not of the "technology-oriented dictator" that one of his competitors is afraid that Gates might become. There is greater power of a wholly different kind than the conventional power to order people around, in ushering people into the Net, in being the agent of technological dependency. This is the power of silent seduction, of giving accessibility to cyberspace. Bill Gates is not Zeus, casting thunderbolts, but Charon taking us across the electronic Styx into virtuality. Seabrook, the technohumanist liberal on a diversionary mission, is concerned with what goes into cyberspace. He accepts the technohype and is afraid of a technofascism that he refuses to acknowledge has already been instituted. Gates only cares that we all get into cyberspace: the seducer as great facilitator.

Gates, indeed, has no interest in the conventional politics of the communication revolution. As much as Seabrook tried to get him to acknowledge the question of power, Gates would resist. He made his position plain in commenting that the highway would have some "secondary effects that people will worry about." That is not his problem, however: "We are involved in creating a new media but it is not up to us to be the censors or referees of this media—it is up to public policy to make those decisions."[6]

"Public policy" is what goes on to get the flesh to adjust to the Net. The greater project is beyond policy, transcendent to it—that is the project of wiring bodies to the Net. That everyone will be wired to the information highway machine is an historical inevitability that puts politics in its place as a local cleanup activity around the Net. This is technotopianism in its purest and most cynical form. Compare it to that other computer entrepreneur, the retrofascist Ross Perot, who uses the wealth he has gained from the information industry to finance his appeal to a nationalistic policy. The technotopian has no such leanings, but with vicious naivete depends on liberal-fascist allies in government to protect the Net. Gates had identified himself with Technology, the greater power, the one that will finally be decisive. Through the silent seduction of the operating system.

The Virtual Class and Capitalism

The computer industry is in an intensive phase of "creative destruction," the term coined by Schumpeter and used by the Neo-Darwinian macho apologists for capitalism to refer to the economic killing fields produced by rapid technological change. The Net is being brought into actuality through the offices of ruthless capitalist competition, in which vast empires fall and rise within a single decade (Big Blue/Microsoft). Under the disciplinary liberal night watchman's protection of "private" property rights, capitalist freebooters destroy one another as they race to be the ones who actualize the Net, just like the railroads of the nineteenth century racing across the continent. This means that the virtual class retains a strict capitalist determination and that its representative social type must be a capitalist, someone who is installing the highway to win a financial competition, if nothing else. If one is not so minded in today's computer industry they will be eaten alive. You will only be able to get personal kicks and pursue your (ressentiment-laden) idealistic views of computer democracy in this industry if you sell. So you hype your ideas and your ideals become hype—that is the twisted psychology of the virtual class: not hyped ideology, but something of, by, and for the Net: ideological hype.

There are pure capitalists in the cyber industry and there are capitalists who are also visionary computer specialists. The latter, in a spirit of vicious naivete, generate the ideological hype, a messianic element, that the former take up cynically. It's the old story of the good cop and the bad cop. How come the good cop tolerates the bad cop? So much for the computer democracy of cyberpossessive individuals. The economic base of the virtual class is the entire communications industry—everywhere it reaches. As a whole, this industry processes ideological hype for capitalist ends. It is most significantly constituted by cynicism, not viciously naive vision. Yet, though a small group in numerical proportion to the whole virtual class, the visionaries are essential to cybercapitalism because they provide the ideological mediation to seduce the flesh into the Net. In this sense the cynical capitalists and the well-provided techies are merely drones, clearing the way for the Pied Piper's parade.

A frontier mentality rules the drive into cyberspace. It is one of the supreme ironies that a primitive form of capitalism, a retrocapitalism, is actualizing virtuality. The visionary cybercapitalist is a hybrid monster of social Darwinism and technopopulist individualism. It is just such an imminently reversible figure that can provide the switching mechanism back and forth between cyberspace and the collapsing space of (crashed) perception.

The most complete representative of the virtual class is the visionary capitalist who is constituted by all of its contradictions and who, therefore, secretes its ideological hype. The rest of the class tends to split the contradictions: the visionless-cynical-business capitalists and the perhaps visionary, perhaps skill-oriented, perhaps indifferent technointelligentsia of cognitive scientists, engineers, computer scientists, video game developers, and all the other communication specialists, ranged in hierarchies, but all dependent for their economic support on the drive to virtualization. Whatever contradictions there are within the virtual class—that is, the contradictions stemming from the confrontation of bourgeois and proletarian—the class as a whole supports the drive into cyberspace through the wired world. This is the way it works in post late-capitalism, where the communication complex is repeating the pattern of class collaborationism that marked the old military-industrial complex. The drive into the Net is one of those great capitalist technoprojects that depends upon a concert of interests to sustain it, as it sucks social energy into itself. The phenomenon of a collaborationist complex harboring a retro-Darwinian competition is something new, but is stabilized, in the final analysis, by a broad consensus among the capitalist components of the virtual class that the liberal-fascist state structure is deserving of support. Indeed, in the United States in the 1990s the state is the greatest producer of the ideological hype of the "information highway." The virtual class has its administration in the White House. The concerted drive into cyberspace proceeds, all in the name of economic development and a utopian imaginary of possessive individualists.

The Hypertexted Body, or Nietzsche Gets a Modem

But why be nostalgic? The old body type was always okay, but the wired body with its microflesh, multimedia channeled ports, cybernetic fingers, and bubbling neurobrain finely interfaced to the "standard operating system" of the internet is infinitely better. Not really the wired body of sci-fi with its mutant designer look, or body flesh with its ghostly reminders of nineteenth-century philosophy, but the hypertexted body as both: a wired nervous system embedded in living (dedicated) flesh.

The hypertexted body with its dedicated flesh? That is our telematic future, and it's not necessarily so bleak. Technology has always been our sheltering environment: not second-order nature, but primal nature for the twenty-first-century body. In the end, the virtual class is very old-fashioned. It clings to an antiquated historical form—capitalism—and, on its behalf, wants to shut down the creative possibilities of the internet. Dedicated flesh rebels against the virtual class. It does not want to be interfaced to the Net through modems and external software black boxes, but *actually wants to be an internet.* The virtual class wants to appropriate emergent technologies for purposes of authoritarian political control over cyberspace. It wants to drag technotopia back to the age of the primitive politics of predatory capitalism. But dedicated (geek) flesh wants something very different. Unlike the (typically European) rejection of technotopia in favor of a newly emergent nostalgia movement under the sign of "Back to Vinyl" in digital sound or "Back to Pencils" in literature, dedicated flesh wants to deeply instantiate the age of technotopia. Operating by means of the aesthetic strategy of overidentification with the feared and desired object, the hypertexted body insists that ours is already the era of postcapitalism, and even post-technology. Taking the will to virtuality seriously, it demands its telematic rights to be a functioning interfaced body: to be a multimedia thinker, to patch BUS ports on its cyberflesh as it navigates the gravity well of the internet, to create aesthetic visions equal to the pure virtualities found everywhere on the now superseded digital superhighway, and to become data to such a point of violent implosion that the body finally breaks free of the confining myth of "wired culture" and goes wireless.

The wireless body? That is the floating body, drifting around in the debris of technotopia: encrypted flesh in a sea of data. The perfect evolutionary successor to twentieth-century flesh, the wireless body fuses the speed of virtualized exchange into its cellular structure. DNA-coated data is inserted directly through spinal taps into dedicated flesh for better navigation through the treacherous shoals of the electronic galaxy. Not a body without memory or feelings, but the opposite. The wireless body is the battleground of the major political and ethical conflicts of late-twentieth- and early-twenty-first-century experience.

Perhaps the wireless body will be just a blank data dump, a floating petri dish where all the brilliant residues of technotopia are mixed together in newly recombinant forms. In this case, the wireless body would be an indefinitely reprogrammable chip: microsoft flesh where the "standard operating system" of the new electronic age comes off the top of the TV set, flips inside the body organic, and is soft-wired to a waiting vat of remaindered flesh.

But the wireless body could be, and already is, something very different. Not the body as an organic grid for passively sampling all the drifting bytes of recombinant culture, but the wireless body as a highly charged theoretical and political site: a moving field of aesthetic contestation for remapping the galactic empire of technotopia. Data flesh can speak so confidently of the possibility of multimedia democracy, of sex without secretions, and of integrated (cyber-) relationships because it has already burst through to the other side of technotopia: to that point of brilliant dissolution where the Net comes alive, and begins to speak the language of wireless bodies in a wireless world.

There are already many wireless bodies on the internet. Many data travelers on the virtual road have managed under the weight of the predatory capitalism of the virtual class and the even weightier humanist prejudices against geek flesh, to make of the internet a charmed site for fusing the particle waves of all the passing data into a new body type: hypertexted bodies circulating as "web weavers" in electronic space.

Refusing to be remaindered as flesh dumped by the virtual class, the hypertexted body bends virtuality to its own purposes. Here, the will to virtuality ceases to be one-dimensional, becoming a doubled process, grisly yet creative, spatial yet memoried, in full violent play as the hypertexted body. Always schizoid yet fully integrated, the hypertexted body swallows its modem, cuts its wired connections to the information highway, and becomes its own system-operating software, combining and remutating the surrounding data storm into new virtualities. And why not? Human flesh no longer exists, except as an incept of the wireless world. Refuse, then, nostalgia for the surpassed past of remaindered flesh, and hypertext your way to the (World Wide) Webbed body: the body that actually dances on its own data organs, sees with multimedia graphical interface screens, makes new best telefriends on the MOO, writes electronic poetry on the disappearing edges of video, sound, and text integrators, and insists on going beyond the tedious world of binary divisions to the new cybermathematics of FITS. The hypertexted body, then, is the precursor of a new world of multimedia politics, fractalized economics, incept personalities, and (cybernetically) interfaced relationships. After all, why should the virtual class monopolize digital reality? It only wants to suppress the creative possibilities of virtualization, privileging instead the tendencies of technotopia toward new and more vicious forms of cyberauthoritarianism. The virtual class only wants to subordinate digital reality to the will to capitalism. The hypertexted body responds to the challenge of virtualization by making itself a monstrous double: pure virtuality/pure flesh. Consequently, our telematic future: the wireless body on the Net as a sequenced chip microprogrammed by the virtual class for purposes of (its) maximal profitability, or the wireless body as the leading edge of critical subjectivity in the twenty-first century. If the virtual class is the posthistorical successor to the early bourgeoisie of primitive capitalism, then the hypertexted body is the internet equivalent of the Paris Commune: anarchistic, utopian, and in full revolt against the suppression of the general (tele-)human possibilities of the Net in favor of the specific (monetary) interests of the virtual class. Always already the past to the future of the hypertexted body, the virtual class is the particular interest that must be overcome by the hypertexted body of data trash if the Net is to be gatewayed by soft ethics.

Soft ethics? Nietzsche's got a modem, and he is already rewriting the last pages of *The Will to Power as The Will to Virtuality*. As the patron saint of the hypertexted body, Nietzsche is data trash to the smooth, unbroken surface of the virtual class.

Soft Ideology

So then, some road maps for following the digital route taken by the virtual class across the landscape of the body recombinant.

Map 1: The Digital Superhighway as Ruling Metaphor

The high-speed digital superhighway is the ruling metaphor of the virtual class. As the class that specializes in virtualized exchange, the information superhighway allows the virtual class to speak in the language of encrypted data, circulate through all the capillaries of digital, fiber-optic electric space, and float at hyperspeed to the point where data melt down into pure virtualities. The information superhighway is the playground of the virtual class. While defining the virtual class, it is also the privileged monopoly of global data communication.

As the language of the virtual class, the information superhighway is where the virtual class lives, dreams, works, and conspires. Not accessible to all, the information superhighway with its accelerated transfers of data, voice, and video is open only to those possessing the privileged corporate codes. And not evident to everyone, the information superhighway is also a site of global power because it remains an invisible, placeless, floating electronic space to the unvirtualized classes, to those, that is, who have been abandoned by the flight of the virtual class to the telematic future. Here, virtual power is about invisibility: the endocolonization of the unwired world of time, history, and human flesh by the electronic body.

A space-binding technological medium of communication, the information superhighway invests the electronic body of the virtual class with a new language, fit for twenty-first-century simu-flesh and fibrillated nerve tissue. Neither the late-twentieth-century language of cyberspace (with its romantic invocation of pure electronic space as the site of a "consensual hallucination") nor the traditional laboratory language of recombinant genetics, the information superhighway speaks the digital language of the world's first postflesh body. Postflesh? That is the electronic body of the virtual class: accessed by serial arrays of BUS ports, animated by its 3-D graphic interfaces, coded in its Web by its designated URLs (uniform resource locators), energized by the telematic dream of instantly disposable cybersex machines, and reduced in its bodily movements to a twitching finger (on the cyberdial). The electronic body is equipped with a surfer's consciousness, and is obsessed with its own disappearance into the inertial gridlock of high speed. A pure virtuality, the electronic body is always in flight (from itself): it constitutes a sampler spectrum of the media force fields which it navigates with the assistance of communication satellites parked in deep-space orbital trajectories. Certainly not a cyberbody, a "pure virtuality" is where the electronic body is reborn as a living, (telematically) breathing simu-flesh: a specimen of evolutionary implosion where technology merges with biology, the result being the post-flesh body of the virtual class. Not a passively engineered product of recombinant genetics, the electronic body as a pure virtuality has its neural synapses coded with an instinctual drive to cut, clone, and retranscribe the genetic strips of new media culture. Multimedia by nature, space-binding by instinct, and driven by an obsession compulsion toward its own disappearance down the information superhighway, the electronic body of the virtual class is the first mutant-body type to appear on the long-range scanners of the awaiting twenty-first century.

Map 2: The Information "Superhighway" Does Not Exist

Or maybe it is just the opposite? If the information superhighway can be the ruling sign of the virtual class it is because it has no existence other than that of an old modernist metaphor concealing the disappearance of technology into virtuality, information into data, and the highway into space-binding electronic circuitry. In this case, the concept of the information superhighway simultaneously performs a revealing and concealing function with respect to the virtual class. It reveals the deep association of this class with high-speed virtualized exchange, but it conceals the drive to global power on the part of the virtual class in favor of a comforting, romantic myth of outlaw travel across the electronic frontier.

Take apart the dense ideogram of the information superhighway to see what is inside and all the political tactics of the virtual class suddenly spill out: its promotional rhetoric, its policing methods, its doubled strategy of an ideology of facilitation and an actuality of virtualization, its ruling illusions of immersion and interactivity, and its missionary commitment to technotopia. The opposition to the virtual class also emerges: a growing political critique based on hypernostalgia ("Back to Vinyl"), reinforced by an alternative aesthetic refusal of the virtual class based on overidentification with the electronic body ("Data Trash").

Map 3: Seduce and Virtualize

Functioning as the political ideology of the virtual class, the information highway delivers up the body to virtualization. While its promotional rhetoric is cloaked in a seductive ideology of facilitation, in actuality the ruling metaphor of the information superhighway is a policing mechanism by which human flesh is gripped in the cyberjaws of virtualization. The ideology of facilitation? That is the promotional culture of the virtual class which speaks eloquently about how the expansion of the high-speed data network will facilitate every aspect of contemporary society: heightened interactivity, increased high-tech employment in a "globally competitive market," and a massive acceleration of access to knowledge. Not a democratic discourse but a deeply authoritarian one, the ideology of facilitation is always presented in the crisis-context of technological necessitarianism. As the CEOs of leading computer companies and their specialist consultants like to say: *We have no choice but to adapt or perish given the technological inevitability associated with the coming to be of technotopia.* Or, as the virtual elite summarizes the situation: *We will be jettisoned into the history dump file if we don't submit to the imperatives of digital technology.*

Map 4: The Information Elite

Monarchs of the electronic kingdom, the information elite rules the digital superhighway. Having no country except digital-land, no history but for their passing electronic traces, and no future other than the conquest of cyberculture, the information elite is a global fraternity (mostly male) of data hounds flying the virtual airways. Fueled by missionary enthusiasm for the emergent technologies of technotopia, it is at the empty center of virtual power.

But like all high priests before them, from the ancient Egyptian ecclesiastics and the Christian cardinals to the Soviet commissars, the information elite are practitioners of a dead power. A precondition for operating at the center of any power is the sacred knowledge that power is dead, that its signs are always cynical, and that the price for revealing this secret is expulsion or even death. The information elite lives under the double sign of cynicism and an eternal law of silence. If it should reveal the cynicism within or betray the secret of dead power to the uninitiated, its offending member would be executed immediately (or in the twentieth-century version, dumped from the virtual class in a classic buyout). Information is a dead sign, and the information elite is the priestly keeper of the eternal flame of the nothingness within.

Map 5: Soft Ideology

> "[Nickelodeon's] expansion into preschool territory was part of a larger, marketing strategy for the company. . . . 'We recognize that if we start getting kids to watch us at this age, we have them for life'. . . . 'That's exactly the reason why we're doing it.' In its fifteen years, Nickelodeon has conquered the marketplace for children between 6 and 11 years old."
> —New York Times, March 21, 1994

Soft television is the new horizon of the electronic body. An integrated multimedia world where the networks of cyberspace and television suddenly merge into a common telematic language. Cablesoft, Videoway, Smart TV: these are the futuristic (CompuTV) collector points for accessing, harvesting, and distributing the remainders of the virtual body.

Soft television expresses perfectly the ruling ideology of the virtual class. When the networked world of the information superhighway is finally linked to television, then the will to virtuality will be free to produce fully functioning networked bodies: cybershoppers, cyberbankers, and cybersex. Soft television is an electronic televisual space populated by body dumps where human flesh goes to be virtualized. Itself a product of the will to virtuality, soft ideology is necessarily virtual: a series of ruling illusions about the efficacy and inevitability of the virtualization of human experience. Here, the future of the hyperhuman body is translated into the language of public policy for immediate circulation through the international networks of political power. Consequently, the soft ideology of the virtual class is based on three key illusions.

The Illusion of Interactivity. Consider Microsoft's newest corporate venture, Cablesoft, which is actively promoted under the sign of enhanced interactivity. Cablesoft is a multimedia world linking the programming language of computers with television screens to produce fully integrated media. Cyber-Interactivity is, however, the opposite of *social* relationships. The human presence is reduced to a twitching finger, spastic body, and an oversaturated informational pump that surfs the channels and makes choices within strictly programmed limits. What is really "interfaced" by Cablesoft is the soft matter of the brain. It is a standard operating system for melting previously externalized technologies of communication into the human nervous system. And what is the Cablesoft brain? It is multiplatform, multimedia, and multi-disciplinary: a hypermind that has its neurosynapses fired by directly accessed signals drawn from passing data storms on the big bandwidth. The hypermind creates teleconsciousness in its wake. Imagine *Star Trek*'s image of the Borg stepping out of the television screen and patching into the Cablesoft mind. Not the interesting ("You will be assimilated") Borg of the early episodes, but the smarmy Borg of the latter episode. The "good Borg" has a veneer of individual consciousness, but an inner reality of suburban consciousness that just wants to do good for the human race. Cablesoft, then, as that point where the individual mind embedded in spinal nerve tissues disappears, and is replaced by *our* circulation as phasal moments in a new medium of cybernetic intelligence. Under the entertainment cover of the ideology of facilitation, Cablesoft promises to mind-meld (our) brains into a circulating process of cyberintelligence: a total human mind scan for the body electronic.

The Illusion of (Cyber)Knowledge.   Soft television is also sold under the sign of the "knowledge society." Technohype has it that wired culture delivers us to a vastly expanded range of human awareness. What is not said, however, is that for the virtual class, true knowledge is cold data, and the very best data of all is the willing readout of the human sensorium into the info-net. That is why there is such an immense social pressure today for everyone to get on the Net. Unlike the 1950s, with its promotion of technology under the sign of "good industrial design" for consumer society, the 1990s is typified by the glorification of virtual technology under the banner of "good body design" for the cyberculture of tomorrow. In virtual culture, knowledge is literally vacuumed from all the orifices of the body, society, and economy, downloaded into data storage banks, and then sampled and resampled across the liquid media-net, and all this in perfect synch with the expansionary momentum of the recombinant commodity-form. When knowledge is reduced to information, then consciousness is stripped of its lived connection to history, judgment, and experience. What results is the illusion of an expanded knowledge society, and the reality of virtual knowledge. Knowledge, that is, as a tightly controlled medium of cybernetic exchange where thought has a disease, and that disease is called information.

The Illusion of Expanded Choice.   Soft television has a veneer of expanded (consumer) choice, but an inner reality of growing desensitization and infantilization. A multichanneled world driven by the need for information by all the drifting cyberminds projects itself perfectly by the promise of five hundred-channel television. A channel for every firing synapse, a data stream for every retromood. If there can be such intense demand for quantum leaps of televisual information ports for the hungry cablesoft brain it is because the cybermind has already patched to a new emotional territory. Not expansive minds for expanded (soft television) choice, but a fantastic infantilization of the televisual audience, with its fever-pitch connections between (emotional) primitivism and (multimedia) hypertech. Why the charismatic appeal today of scandal television and talk show formats privileging the deterioration of the public mind? It is because virtual culture has already evolved into a new, more insidious phase of nihilism: that movement where self-hatred and self-abuse is so sharp that we willingly deliver ourselves up as the butt of the television joke. The cultural condition that makes this possible is that, like the training programs for CIA assassins with their repeated exposure of agents to brutal scenes of torture, soft television functions on the basis of desensitization. Floating corpses, live executions, rape television: all delivered under the sign of media fascination, and all with the intent of desensitizing the soft mass of the cyberaudience to the point of its humiliated complicity in the evil of the times.

Map 6: The Red Guard Meets Generation X

The editors of *AXCESS* magazine, published in San Diego, recently wrote about themselves as the "young entrepreneurs": the leading-edge members of Generation X. At about the same time, a CBC television program, entitled "Red Capitalism," interviewed former members of the Red Guard who have now become full-fledged participants of the rising Chinese entrepreneurial class. So what happens when the old ideological competition between capitalism and socialism disappears, and Generation X meets the Red Guard on the world stage, they look in the mirror of shared economic interests, and discover to their pleasant surprise that they are exactly the same (virtual) class? Perhaps this fusion of unlikely partners in a global virtual class of young entrepreneurs who are finally liberated from cold war ideology was best expressed by a high-ranking official at the Boeing Company when asked about the linkage of human rights issues with the extension of "most favored nation" status to China. He argued that there should be no relationship between politics and trade: "We are living in the age of global competition." Without a twinge of nostalgia for the disappeared rhetoric of "jobs for Americans," the official from Boeing is joined in this chorus for unimpeded free trade by multinational corporate leaders (think of the American multinational directors in China who castigated the U.S. Secretary of State for criticizing the Chinese record on human rights) and government officials (the Canadian Minister for External Affairs has recently announced a new public policy in relations with Latin and South America whereby trade is cut loose from human rights issues). A fundamental political objective of the virtual class is decoupling the linkage between free trade (virtualized exchange) and human rights. That is why the technotopians of Generation X and the ex-cadres of the Red Guard are hyperlinked by the same ideology. With the death of communism, the world has undergone a big political flip. In the glory days of the cold war, business would have justified its expansionary interests in the name of fighting the Red Menace. Today the virtual class valorizes its recombinant interests in the name of emancipating business from the shackles of (cold war) political rhetoric. Like meaning before it, human rights issues slow down the rate of circulation of virtualized exchange, and, consequently, they must be eliminated from the political history file.

The 1990s, therefore, are typified by the rapid decline of the hard ideologies of capitalism and communism, and by the ascendancy of the soft ideology of the virtual class. Soft ideology? That's the will to virtuality as the common language of the new managerial elites of the postcapitalist, postcommunist, and also post-technological society.

Itself a product of the will to virtuality, soft ideology is necessarily virtual: a series of ruling illusions about the efficacy and inevitability of the virtualization of experience. Here, the future of the hyperhuman body is translated into the language of public policy for immediate circulation through the international networks of political power. When the Red Guard meets the (technotopian) members of Generation X on the common ground of missionary enthusiasm for pan-capitalism, they insert themselves into the political economy of virtual reality as its leading elites. As the young entrepreneurs of Generation X, the virtual class finally has a name. Under the sign of the Red Guard gone technotopian, it also has an historical destiny—creating a new global "culture revolution" on behalf of unimpeded virtualized exchange. Finally, in the fusion of the young entrepreneurs of Generation X and the Red Guard, it has a grisly political method: sacrificing human rights on the altar of virtual (economic) expediency. We're living in the new morning of a big (ideological) sign-switch. The cold war of hard ideology may finally be over, but the new cold war of soft ideology, the one that pits the virtual class against all barriers to its global sovereignty, is just beginning.

NOTES

1. John Seabrook, "E-mail from Bill," *The New Yorker,* LXIX, 45 (January 10, 1994): 54.
2. Ibid., 49.
3. Ibid., 49–50.
4. Ibid.
5. Ibid., 52.
6. Ibid., 54.

Vivian Sobchack

The Scene of the Screen: Envisioning Cinematic and Electronic "Presence"

It is obvious that cinematic and electronic technologies of representation have had enormous impact upon our means of signification during the past century. Less obvious, however, is the similar impact these technologies have had upon the historically particular significance or "sense" we have and make of those temporal and spatial coordinates that radically inform and orient our social, individual, and bodily existences. At this point in time in the United States, whether or not we go to the movies, watch television or music videos, own a video tape recorder/player, allow our children to play video and computer games, or write our academic papers on personal computers, we are all part of a moving-image culture and we live cinematic and electronic lives. Indeed, it is not an exaggeration to claim that none of us can escape daily encounters—both direct and indirect—with the *objective* phenomena of motion picture, televisual, and computer technologies and the networks of communication and texts they produce. Nor is it an extravagance to suggest that, in the most profound, socially pervasive, and yet personal way, these objective encounters transform us as *subjects.* That is, although relatively novel as "materialities" of human communication, cinematic and electronic media have not only historically *symbolized* but also historically *constituted* a radical alteration of the forms of our culture's previous temporal and spatial consciousness and of our bodily sense of existential "presence" to the world, to ourselves, and to others.

This different sense of *subjective* and *material* "presence" both signified and supported by cinematic and electronic media emerges within and co-constitutes *objective* and *material* practices of representation and social existence. Thus, while cooperative in creating the moving-image

From *Materialities of Communications,* edited by Hans Ulrich Gumbrecht and K. Ludwig Pfeiffer; translated by William Whobrey. Reprinted with the permission of the publishers, Stanford University Press, and the author. © 1994 by the Board of Trustees of the Leland Stanford Junior University.

culture or "life-world" we now inhabit, cinematic and electronic technologies are each quite different from each other in their concrete "materiality" and particular existential significance. Each offers our lived-bodies radically different ways of "being-in-the world." Each implicates us in different structures of material investment, and—because each has a particular affinity with different cultural functions, forms, and contents—each stimulates us through differing modes of representation to different aesthetic responses and ethical responsibilities. In sum, just as the photograph did in the last century, so in this one, cinematic and electronic screens differently demand and shape our "presence" to the world and our representation in it. Each differently and objectively alters our subjectivity while each invites our complicity in formulating space, time, and bodily investment as significant personal and social experience.

These preliminary remarks are grounded in the belief that, during the last century, historical changes in our contemporary "sense" of temporality, spatiality, and existential and embodied presence cannot be considered less than a consequence of correspondent changes in our technologies of representation. However, they also must be considered something more, for as Martin Heidegger reminds us, "The essence of technology is nothing technological."[1] That is, technology never comes to its particular material specificity and function in a neutral context for neutral effect. Rather, it is always historically informed not only by its materiality but also by its political, economic, and social context, and thus always both co-constitutes and expresses cultural values. Correlatively, technology is never merely "used," never merely instrumental. It is always also "incorporated" and "lived" by the human beings who engage it within a structure of meanings and metaphors in which subject-object relations are cooperative, co-constitutive, dynamic, and reversible. It is no accident, for example, that in our now dominantly electronic (and only secondarily cinematic) culture, many human beings describe and understand their minds and bodies in terms of computer systems and programs (even as they still describe and understand their lives as movies). Nor is it trivial that computers are often described and understood in terms of human minds and/or bodies (for example, as intelligent, or as susceptible to viral infection)—and that these new "life forms" have become the cybernetic heroes of our most popular moving-image fictions (for example, *Robocop* or *Terminator II*).[2] In this sense, a qualitatively new technologic can begin to alter our perceptual orientation in and toward the world, ourselves, and others. And as it becomes culturally pervasive, it can come to profoundly inform and affect the sociologic, psychologic, and even the bio-logic by which we daily live our lives.

This power to alter our perceptions is doubly true of technologies of representation. A technological artifact like the automobile (whose technological function is not representation but transportation) has profoundly changed the temporal and spatial shape and meaning of our life-world and our own bodily and symbolic sense of ourselves.[3] However, representational technologies of photography, the motion picture, video, and computer in-form us twice over: first, like the automobile, through the specific material conditions by which they latently engage our senses at the bodily level of what might be called our *microperception,* and then again through their explicit representational function by which they engage our senses textually at the hermeneutic level of what might be called our *macroperception.*[4] Most theorists and critics of the cinematic and electronic have been drawn to macroperceptual analysis, to descriptions and interpretations of the hermeneutic-cultural contexts that inform and shape both the materiality of the technologies and their textual representations.[5] Nonetheless, "all such contexts find their fulfillment *only* within the range of microperceptual possibility."[6] We cannot reflect upon and analyze either technologies or texts without having, at some point, engaged them *immediately*—that is, through our perceptive sensorium, through the materiality (or *immanent mediation*) of our own bodies. Thus, as philosopher of technology Don Ihde puts it, while "there is no microperception (sensory-bodily) without its location within a field of macroperception," there could be "no macroperception without its microperceptual foci."[7] It is important to note, however, that since perception is constituted and organized as a bodily and sensory gestalt that is always already meaningful, a microperceptual focus is not the same as a physiological or anatomical focus. The perceiving and sensing body is always also a *lived-body*—immersed in and making social meaning as well as physical sense.

The aim of this essay, then, is to figure certain microperceptual aspects of our engagement with the technologies of cinematic and electronic representation and to suggest some ways in which our microperceptual experience of their respective material conditions informs and transforms our temporal and spatial sense of ourselves and our cultural contexts of meaning. Insofar as both the cinematic and the electronic have each been *objectively constituted* as a new and discrete technologic, each also has been *subjectively incorporated,* enabling a new perceptual mode of existential and embodied "presence." In sum, as they have mediated our engagement with the world, with others, and with ourselves, cinematic and electronic technologies have transformed us so that we presently see, sense, and make sense of ourselves as quite other than we were before them.

It should be evident at this point that the co-constitutive, reversible, and dynamic relations between objective material technologies and embodied human subjects invite a phenomenological investigation. Existential phenomenology, to use Ihde's characterization, is a "philosophical style that emphasizes a certain interpretation of human *experience* and that, in particular, concerns *perception* and *bodily activity.*"[8] Often misunderstood as purely "subjective" analysis, existential phenomenology is instead concerned with describing, thematizing, and interpreting the structures of lived spatiality, temporality, and meaning that are co-constituted dynamically as embodied human subjects perceptually engage an objective material world. It is focused, therefore, on the *relations between* the subjective and objective aspects of material, social, and personal existence and sees these relations as constitutive of the meaning and value of the phenomena under investigation.[9]

Existential phenomenology, then, attempts to describe, thematize, and interpret the *experiential* and *perceptual field* in which human beings play out a particular and meaningful structure of spatial, temporal, and bodily existence. Unlike the foundational, Husserlian transcendental phenomenology from which it emerged, existential phenomenology rejects the goal of arriving at universal and "essential" description, and "settles" for a historicized and "qualified" description as the only kind of description that is existentially possible or, indeed, desirable. It is precisely *because* rather than *in spite of* its qualifications that such a description is existentially meaningful—meaningful, that is, to human beings who are themselves particular, finite, and partial, and thus always in culture and history, always open to the world and further elaboration. Specifically, Maurice Merleau-Ponty's existential phenomenology departs from the transcendental phenomenology most associated with Edmund Husserl in that it stresses the *embodied* nature of human consciousness and views bodily existence as the original and originating *material premise* of sense and signification. We sit in a movie theater, before a television set, or in front of a computer terminal not only as *conscious* beings but also as *carnal* beings. Our vision is not abstracted from our bodies or from our other modes of perceptual access to the world. Nor does what we see merely touch the surface of our eyes. Seeing images mediated and made visible by technological vision enables us not only to see technological images but also to see technologically. As Ihde emphasizes, "the concreteness of [technological] 'hardware' in the broadest sense connects with the equal concreteness of our bodily existence," and, in this regard, "the term 'existential' in context refers to perceptual and bodily experience, to a kind of 'phenomenological materiality.'"[10]

This correspondent and objective materiality of both human subjects and worldly objects not only suggests some commensurability and possibilities of exchange between them, but also suggests that any phenomenological analysis of the existential relation between human subjects and technologies of representation must be semiological and historical even at the microperceptual level. Description must attend both to the particular materiality and modalities through which meanings are signified and to the cultural and historical situations in which materiality and meaning come to cohere in the praxis of everyday life. Like human vision, the materiality and modalities of cinematic and electronic technologies of representation are not abstractions. They are concrete and situated and institutionalized. They inform and share in the spatiotemporal structures of a wide range of interrelated cultural phenomena. Thus, in its attention to the broadly defined "material conditions" and "relations" of production (specifically, the conditions for and production of existential meaning), existential phenomenology is not incompatible with certain aspects of Marxist analysis.

In this context, we might turn to Fredric Jameson's useful discussion of three crucial and expansive historical "moments" marked by "a technological revolution within capital itself" and the particular and dominant "cultural logic" that correspondently emerges in each of them.[11] Historically situating these three "moments" in the 1840s, 1890s, and 1940s, Jameson correlates the three major technological changes that revolutionized the structure of capital—by changing market capitalism to monopoly capitalism and this to multinational capitalism—with the emergence and domination of three new "cultural logics": those axiological norms and forms of representation identified respectively as realism, modernism, and postmodernism. Extrapolating from Jameson, we can also locate within this conceptual and historical framework three correspondent technologies, forms, and institutions of visual (and aural) representation: respectively, the photographic, the cinematic, and the electronic. Each, we might argue, has been critically complicit not only in a specific "*technological* revolution within capital*," but also in a specific and radical *perceptual* revolution within the culture and the subject. That is, each has been co-constitutive of the very temporal and spatial structure of the "cultural logics" Jameson identifies as realism, modernism, and postmodernism. Writing about the nature of cultural transformation, phenomenological historian Stephen Kern suggests that some major cultural changes can be seen as "*directly* inspired by new technology," while others occur relatively independently of technology, and yet still others emerge from the new technological "metaphors and analogies" that *indirectly* alter the structures of perceptual life and thought.[12] Implicated in and informing each historically specific "technological revolution in capital" and transformation of "cultural logic," the technologically discrete nature and phenomenological impact of new "materialities" of representation co-constitute a complex cultural gestalt. In this regard, the technological "nature" of the photographic, the cinematic, and the electronic is graspable always and only in a qualified manner—that is, less as an "essence" than as a "theme."

Although I wish to emphasize the technologies of cinematic and electronic representation, those two "materialities" that constitute our current *moving*-image culture, something must first be said of that culture's grounding in the context and phenomenology of the *photographic*. The photographic is privileged in the "moment" of market capitalism—located by Jameson in the 1840s, and cooperatively informed and driven by the technological innovations of steam-powered mechanization that allowed for industrial expansion and the cultural logic of "realism." Not only did industrial expansion give rise to other forms of expansion, but expansion itself was historically unique in its unprecedented *visibility*. As Jean-Louis Comolli points out:

> The second half of the nineteenth century lives in a sort of frenzy of the visible. . . . [This is] the effect of the social multiplication of images. . . . [It is] the effect also, however, of something of a geographical extension of the field of the visible and the representable: by journies, explorations, colonisations, the whole world becomes visible at the same time that it becomes appropriatable.[13]

Thus, while the cultural logic of "realism" has been seen as primarily represented by literature (most specifically, the bourgeois novel), it is, perhaps, even more intimately bound to the mechanically achieved, empirical, and representational "evidence" of the world constituted by photography.

Until very recently, the photographic has been popularly and phenomenologically perceived as existing in a state of testimonial verisimilitude—its film emulsions analogically marked with (and objectively "capturing") material traces of the world's concrete and "real" existence.[14] Photography produced images of the world with a perfection previously rivaled only by the human eye. Thus, as Comolli suggests, with the advent of photography, the human eye loses its "immemorial privilege" and is devalued in relation to "the mechanical eye of the photographic machine," which "now sees *in its place.*"[15] This replacement of human with mechanical vision had its compensations, however—among them, the material control, containment, and actual possession of time and experience.[16] Abstracting visual experience from a temporal flow, the photographic chemically and metaphorically "fixes" its ostensible subject as an *object* for vision, and concretely reproduces it in a *material* form that can be possessed, circulated, and saved, in a form that can over time accrue an increasing rate of interest, become more *valuable* in a variety of ways. Thus, identifying the photograph as a fetish object, Comolli links it with gold, and aptly calls it "the money of the 'real'"—of "life"—the photograph's materiality assuring the possibility of its "convenient circulation and appropriation."[17]

In his phenomenological description of human vision Merleau-Ponty tells us, "To see is *to have at a distance.*"[18] This subjective activity of *visual* possession is objectified and literalized by the materiality of photography, which makes possible its *visible* possession. What you see is what you get. Indeed, this structure of objectification and empirical possession is doubled, even tripled. Not only does the photograph materially "capture" traces of the "real world," not only can the photograph itself be possessed concretely, but the photograph's culturally defined semiotic status as a mechanical reproduction (rather than a linguistic representation) also allows an unprecedentedly literal and material, and perhaps uniquely complacent form—and ethics—of self-possession. Family albums serve as "memory banks" that authenticate self, other, and experience as empirically "real" by virtue of the photograph's material existence as an object and possession with special power.[19]

In regard to the materiality of the photograph's authenticating power, it is instructive to recall one of a number of particularly relevant ironies in *Blade Runner* (Ridley Scott, 1982), a science-fiction film focusing on the ambiguous ontological status of a group of genetically manufactured "replicants." At a certain moment, Rachel, the film's putative heroine and the latest replicant prototype, disavows the revelation of her own manufactured status by pointing to a series of keepsake photographs that give "proof" to her mother's existence, to her own existence as a little girl, to her subjective memory. Upon being told that both her memory and their material extroversion "belong to someone else," she is both distraught and ontologically re-signed as someone with no "real" life, no "real" history—although she still remembers what she remembers and the photographs still sit on her piano. Indeed, the photographs are suddenly foregrounded (for the human spectator as well as the narrative's replicant) as utterly suspect. That is, when interrogated, the photographs simultaneously both reveal and lose that great material and circulatory value they commonly hold for *all* of us as the "money of the 'real.'"

The structures of objectification and material possession that constitute the photographic as both a "real" trace of personal experience and a concrete extroversion of experience that can "belong to someone else" give specific form to its temporal existence. In capturing aspects of "life itself" in a "real" object that can be possessed, copied, circulated, and saved as the "currency" of experience, the appropriable materiality and static form of photography accomplish a palpable intervention in what was popularly perceived in the mid-nineteenth century to be time's linear, orderly, and teleological flow from past to present to future. The photograph freezes and preserves the homogeneous and irreversible *momentum* of this temporal stream into the abstracted, atomized, and secured space of a *moment*. But at a cost. A moment cannot be inhabited. It cannot entertain in the abstraction of its visible space, its single and static *point* of view, the presence of a lived-body—and so it does not really invite the spectator *into* the scene (although it may invite contemplation *of* the scene). In its conquest of time, the photographic constructs a space to hold and to look at, a "thin" insubstantial space that keeps the lived-body out even as it may imaginatively catalyze—in the parallel but temporalized space of memory or desire—an animated drama.

The radical difference between the transcendental, posited moment of the photograph and the existential momentum of the cinema, between the scene to be contemplated and the scene to be lived, is foregrounded in the remarkable short film *La jetée* (Chris Marker, 1962).[20] A study of desire, memory, and time, *La jetée* is presented completely through the use of still photographs—except for one extraordinarily brief but utterly compelling sequence in which the woman who is the object of the hero's desire, lying in bed and looking toward the camera, blinks her eyes. The space between the camera's (and the spectator's) gaze becomes suddenly habitable, informed with the real possibility of bodily movement and engagement, informed with a lived temporality rather than an eternal timelessness. What, in the film, has previously been a mounting accumulation of nostalgic moments achieves substantial and present presence in its sudden accession to momentum and the consequent possibility of effective action.

As did André Bazin (1967), we might think of photography, then, as primarily a form of mummification (although, unlike Bazin, I shall argue that cinema is not).[21] While it testifies to and preserves a sense of the world and experience's real "presence," it does not preserve their present. The photographic—unlike the cinematic and the electronic—functions neither as a coming-into-being (a presence always presently constituting itself) nor as being-in-itself (an absolute presence). Rather, it functions to fix a being-that-has been (a presence in the present that is always past). Paradoxically, as it objectifies and preserves in its acts of possession, the photographic has something to do with loss, with pastness, and with death, its meanings and value intimately bound within the structure and investments of nostalgia.

Although dependent upon the photographic, the cinematic has something more to do with life, with the accumulation—not the loss—of experience. Cinematic technology *animates* the photographic and reconstitutes its visibility and verisimilitude in a difference not of degree but of kind. The *moving picture* is a visible representation not of activity finished or past, but of activity coming-into-being—and its materiality comes to be in the 1890s, the second of Jameson's transformative moments of "technological revolution within capital itself." During this moment, the combustion engine and electric power literally reenergized market capitalism into the highly controlled yet expansive structure of monopoly capitalism. Correlatively, the new culture logic of "modernism" emerged, restructuring and eventually dominating the logic of realism to more adequately represent the new perceptual experience of an age marked by the strange autonomy and energetic fluidity of, among other mechanical phenomena, the motion picture. The motion picture, while photographically verisimilar, fragments, reorders, and synthesizes time and space as animation in a completely new "cinematic" mode that finds no necessity in the objective teleologic of realism. Thus, although modernism has found its most remarked expression in the painting and photography of the futurists (who attempted to represent motion and speed in a static form) and the Cubists (who privileged multiple perspectives and simultaneity), and in the novels of James Joyce, we can see in the cinema modernism's fullest representation.[22]

Philosopher Arthur Danto tells us, "with the movies, we do not just see *that* they move, we see them *moving:* and this is because the pictures themselves move."[23] While still objectifying the subjectivity of the visual into the visible, the cinematic qualitatively transforms the photographic through a materiality that not only claims the world and others as objects for vision but also signifies its own bodily agency, intentionality, and subjectivity. Neither abstract nor static, the cinematic brings the *existential activity* of vision into visibility in what is phenomenologically experienced as an *intentional stream* of moving images—its continuous and autonomous visual production and meaningful organization of these images testifying to the objective world and, further, to an anonymous, mobile, embodied, and ethically invested *subject* of worldly space. This subject (however physically anonymous) is able to inscribe visual and bodily changes of situation, to dream, hallucinate, imagine, and re-member its habitation and experience of the world. And, as is the case with human beings, this subject's potential mobility and experience are both open-ended and bound by the existential finitude and bodily limits of its particular vision and historical coherence (that is, its narrative).

Here, again, *La jetée* is exemplary. Despite the fact that the film is made up of what strikes us as a series of discrete and still photographs rather than the "live" and animated action of human actors, even as it foregrounds the transcendental and atemporal nonbecoming of the photograph, *La jetée* nonetheless phenomenologically *projects* as a temporal flow and an existential becoming. That is, *as a whole,* the film organizes, synthesizes, and enunciates the discrete photographic images into animated and intentional coherence and, indeed, makes this temporal synthesis and animation its explicit narrative theme. What *La jetée* allegorizes in its explicit narrative, however, is the transformation of the moment to momentum that constitutes the ontology of the cinematic, and the latent background of every film.

While the technology of the cinematic is grounded, in part, in the technology of the photographic, we need to remember that "the essence of technology is nothing technological." The fact that the technology of the cinematic *necessarily* depends upon the discrete and still photograph moving intermittently (rather than continuously) through the shutters of both camera and projector does not *sufficiently* account for the materiality of the cinematic as we experience it. Unlike the photograph, a film is semiotically engaged in experience not merely as a mechanical objectification—or material *reproduction*—that is, not merely as an object for vision. Rather, the moving picture, however mechanical and photographic its origin, is semiotically experienced as also subjective and intentional, as *presenting representation* of the objective world. Thus perceived as the subject of its own vision as well as an object for our vision, a moving picture is not precisely a *thing* that (like a photograph) can be easily controlled, contained, or materially possessed. Up until very recently in what has now become a dominantly electronic culture, the spectator could share in and thereby, to a degree, interpretively alter a film's presentation and representation of embodied and enworlded experience, but could not control or contain its autonomous and ephemeral flow and rhythm, or materially possess its animated experience. Now, of course, with the advent of videotape and VCRs, the spectator can alter the film's temporality and easily possess, at least, its inanimate "body." However, the ability to control the autonomy and flow of the cinematic experience through "fast forwarding," "replaying," and "freezing"[24] and the ability to possess the film's body and animate it at will at home are functions of the materiality and technological ontology of the electronic—a materiality that increasingly dominates, appropriates, and transforms the cinematic.

In its preelectronic state and original materiality, however, the cinematic mechanically projects and makes visible for the very first time not just the objective world, but the very structure and process of subjective, embodied vision—hitherto only directly available to human beings as that invisible and private structure we each experience as "my own." That is, the materiality of the cinematic gives us concrete and empirical insight and makes objectively visible the reversible, dialectical, and social nature of our own subjective vision. Speaking of human vision, Merleau-Ponty tells us: "As soon as we see other seers . . . henceforth, through other eyes we are for ourselves fully visible. . . . For the first time, the seeing that I am is for me really visible; for the first time I appear to myself completely turned inside out under my own eyes."[25] The cinematic uniquely allows this philosophical turning, this objective insight into the subjective structure of vision, into oneself as both viewing subject and visible object, and, remarkably, into others as the same.

Again, the paradoxical status of the "more human than human" replicants in *Blade Runner* is instructive. Speaking to the biotechnologist who genetically produced and quite literally manufactured his eyes, replicant Roy Baty says with an ironic concreteness that resonates through the viewing audience even if its implications are not fully understood: "If you could only see what I've seen with your eyes." The perceptive and expressive materiality of the cinematic through which we engage this ironic articulation of the "impossible" desire for intersubjectivity is the very materiality through which this desire is visibly and objectively fulfilled.[26] Thus, rather than merely replacing human vision with mechanical vision, the cinematic mechanically functions to bring to visibility the reversible structure of human vision (the system visual/visible)—a lived-system that necessarily entails not only an enworlded object but always also an embodied and perceiving object.

Indeed, through its motor and organizational agency (achieved by the spatial immediacy of the mobile camera and the reflective and temporalizing editorial re-membering of that primary spatial experience), the cinematic inscribes and provokes a sense of existential "presence" that is as synthetically centered as it is also mobile, split, and decentering. The cinematic subject (both film and spectator) is perceived as at once introverted and extroverted, as existing in the world as both subject and object. Thus, the cinematic does not evoke the same sense of self-possession as that generated by the photographic. The cinematic subject is sensed as never completely self-possessed, for it is always partially and visibly given over to the vision of others at the same time that it visually appropriates only part of what it sees and, indeed, also cannot entirely see itself. Further, the very mobility of its vision structures the cinematic subject as always in the act of displacing itself in time, space, and the world—and thus, despite its existence as embodied and centered, always eluding its own (as well as our) containment.

The cinematic's visible inscription of the dual, reversible, and animated structure of embodied and mobile vision radically transforms the temporal and spatial structure of the photographic. Consonant with what Jameson calls the "high-modernist thematics of time and temporality," the cinematic thickens the photographic with "the elegaic mysteries of *durée* and of memory."[27] While its visible structure of "unfolding" does not challenge the dominant realist perception of objective time as an irreversibly directed stream (even flashbacks are contained by the film's vision in a forwardly directed momentum of experience), the cinematic makes time visibly *heterogeneous.* That is, we visibly perceive time as differently structured in its subjective and objective modes, and we understand that these two structures *simultaneously* exist in a demonstrable state of *discontinuity* as they are, nonetheless, actively and constantly *synthesized* in a specific lived-body experience (i.e., a personal, concrete, and spatialized history and a particularly temporalized narrative).

Cinema's animated presentation of representation constitutes its "presence" as always presently engaged in the experiential process of signifying and coming-into-being. Thus the significant value of the "streaming forward" that informs the cinematic with its specific form of temporality (and differentiates it from the atemporality of the photographic) is intimately bound to a structure not of possession, loss, pastness, and nostalgia, but of accumulation, ephemerality, and anticipation—to a "presence" in the present informed by its connection to a collective past and to a future. Visually (and aurally) presenting the subjective temporality of memory, desire, and mood through flashbacks, flash forwards, freeze framing, pixilation, reverse motion, slow motion, and fast motion, and the editorial expansion and contraction of experience, the cinema's visible (and audible) activity of *retension* and *protension* constructs a subjective temporality different from the irreversible direction and momentum of objective time, yet simultaneous with it. In so thickening the present, this temporal simultaneity also extends cinematic presence spatially—not only embracing a multiplicity of situations in such visual/visible cinematic articulations as double exposure, superimposition, montage, parallel editing, but also primally, expanding the space in every image between that Here where the enabling and embodied cinematic eye is situated and that There where its gaze locates itself in its object.

The cinema's existence as simultaneously presentational and

representational, viewing subject and visible object, present presence in-formed by both past and future, continuous becoming that synthesizes temporal heterogeneity as the conscious coherence of embodied experi-ence, transforms the thin abstracted space of the photographic into a thickened and concrete *world.* We might remember here the animated blinking of a woman's eyes in *La jetée* and how this visible motion trans-forms the photographic into the cinematic, the flat surface of a picture into the lived space of a lover's bedroom. In its capacity for movement, the cin-ema's embodied agency (the camera) thus constitutes visual/visible space as always also motor and tactile space—a space that is deep and textural, that can be materially inhabited, that provides not merely a ground for the visual/visible, but also its particular *situation.* Indeed, although it is a favored term among film theorists, there is no such abstraction as *point of view* in the cinema. Rather, there are concrete *situations of viewing—* specific and mobile engagements of embodied, enworlded, and situated subjects/objects whose visual/visible activity prospects and articulates a shifting field of vision from a world whose horizons always exceed it. The space of the cinematic, in-formed by cinematic time, is also experienced as heterogeneous—both discontiguous and contiguous, lived from within and without. Cinematic presence is multiply located—simultaneously displacing itself in the There of past and future situations yet orienting these displacements from the Here where the body at present is. That is, as the multiplicity and discontinuity of time are synthesized and centered and cohere as the *experience* of a specific lived-body, so are multiple and discontiguous spaces synopsized and located in the spatial *synthesis* of a particular *material* body. Articulated as separate shots and scenes, dis-contiguous and discontinuous times are synthetically gathered together in a coherence that is the cinematic lived-body: the camera its perceptive organ, the projector its expressive organ, the screen its discrete and mate-rial center. In sum, the cinematic exists as a visible performance of the perceptive and expressive structure of lived-body experience.

Not so the electronic, whose materiality and various forms and contents engage its spectators and "users" in a phenomenological structure of sensual and psychological experience that seems to belong to *no-body.* Born in the United States and with the nuclear age, the electronic emerges in the 1940s as the third "technological revolution within capital itself," and, according to Jameson, involved the unprecedented and "prodigious expansion of capital into hitherto uncommodified areas," including " a new and historically original penetration and colonization of Nature and the Unconscious."[28] Since that time, electronic technology has "saturated all forms of experience and become an inescapable environment, a 'tech-nosphere.'"[29] This expansive and totalizing incorporation of Nature by industrialized culture, and the specular production and commodification of the Unconscious (globally transmitted as visible and marketable "de-sire"), restructure capitalism as multinational. Correlatively, a new cultural logic identified as "postmodernism" begins to dominate modernism, and to alter our sense of existential presence.

A function of technological pervasion and dispersion, this new electronic sense of presence is intimately bound up in a centerless, network-like structure of instant stimulation and desire, rather than in a nostalgia for the past or anticipation of a future. Television, videocassettes, video tape recorder/players, video games, and personal computers all form an encompassing electronic representational system whose various forms "interface" to constitute an alternative and absolute world that uniquely incorporates the spectator/user in a spatially decentered, weakly temporalized, and quasi-disembodied state. Digital electronic technology atomizes and abstractly schematizes the analogic quality of the photographic and cinematic into discrete pixels and bits of information that are then transmitted *serially*, each bit discontinuous, discontiguous, and absolute—each bit being-in-itself even as it is part of a system.[30]

Once again we can turn to *Blade Runner* to provide illustration of how the electronic is neither photographic nor cinematic. Tracking Leon, one of the rebellious replicants, the human protagonist Deckard finds his empty rooms and discovers a photograph that seems, itself, to reveal nothing but an empty room. Using a science-fictional device, Deckard directs its electronic eye to zoom in, close up, isolate, and enlarge to impossible detail various portions of the photograph. On the one hand, it might seem that Deckard is functioning like a photographer working in his darkroom to make, through optical discovery, past experience significantly visible. (Indeed, this sequence of the film recalls the photographic blowups of an ambiguously "revealed" murder in Michelangelo Antonioni's 1966 classic, *Blow-Up*.) On the other hand, Deckard can be and has also been likened to a film director, using the electronic eye to probe photographic space intentionally and to animate a discovered narrative. Deckard's electronic eye, however, is neither photographic nor cinematic. While it constitutes a series of moving images from the static singularity of Leon's photograph and reveals to Deckard the stuff of which narrative can be made, it does so serially and in static, discrete "bits." The moving images do not move themselves and reveal no animated and intentional vision to us or to Deckard. Transmitted to what looks like a television screen, the moving images no longer quite retain the concrete and material "thingness" of the photograph, but they also do not achieve the subjective animation of the intentional and prospective vision objectively projected by the cinema. They exist less as Leon's experience than as Deckard's information.

Indeed, the electronic is phenomenologically experienced not as a discrete, intentional, and bodily centered projection in space, but rather as simultaneous, dispersed, and insubstantial transmission across a network.[31] Thus, the "presence" of electronic representation is at one remove from previous representational connections between signification and referentiality. Electronic presence asserts neither an objective possession of the world and self (as does the photographic) nor a centered and subjective spatiotempral engagement with the world and others accumulated and projected as conscious and embodied experience (as does the cinematic). Digital and schematic, abstracted both from *reproducing* the empirical objectivity of Nature that informs the photographic and from *presenting* a representation of individual subjectivity and the Unconscious that informs the cinematic, the electronic constructs a metaworld where ethical investment and value are located in *representation-in-itself.* That is, the electronic semiotically constitutes a system of *simulation*—a system that constitutes "copies" lacking an "original" origin. And, when there is no longer a phenomenologically perceived connection between signification and an "original" or "real," when, as Guy Debord tells us, "everything that was lived directly has moved away into a representation," referentiality becomes *intertextuality.*[32]

Living in a schematized and intertextual metaworld far removed from reference to a real world liberates the spectator/user from what might be termed the latter's moral and physical gravity. The materiality of the electronic digitizes *durée* and situation so that narrative, history, and a centered (and central) investment in the human lived-body become atomized and dispersed across a system that constitutes temporality not as the flow of conscious experience, but as a transmission of random information. The primary value of electronic temporality is the bit or *instant*—which (thanks to television and videotape) can be selected, combined, and instantly replayed and rerun to such a degree that the previously irreversible direction and stream of objective time seems overcome in the creation of a recursive temporal network. On the one hand, the temporal cohesion of history and narrative gives way to the temporal discretion of chronicle and episode, to music videos, to the kinds of narratives that find both causality and intentional agency incomprehensible and comic. On the other hand, temporality is dispersed and finds resolution as part of a recursive, if chaotic, structure of coincidence. Indeed, objective time in postmodern electronic culture is perceived as phenomenologically discontinuous as was subjective time in modernist cinematic culture. Temporality is constituted paradoxically as a *homogeneous* experience of *discontinuity* in which the temporal distinctions between objective and subjective experience (marked by the cinematic) disappear and time seems to turn back in on itself recursively in a structure of equivalence and reversibility. The temporal move is from *Remembrance of Things Past,* a modernist remembering of experience, to the recursive postmodernism of a *Back to the Future.*

Again "science-fiction" film is illuminating.[33] While the *Back to the Future* films are certainly apposite, Alex Cox's postmodern, parodic, and deadpan *Repo Man* (1984) more clearly manifests the phenomenologically experienced homogeneity of postmodern discontinuity. The film is constructed as both a picaresque, episodic, loose, and irresolute tale about an affectless young man involved with car repossessors, aliens from outer space, Los Angeles punks, governmental agents, and others, and a tightly bound system of coincidences. Individual scenes are connected not through narrative causality but through the connection of literally material signifiers. A dangling dashboard ornament, for example, provides the acausal and material motivation between two of the film's otherwise disparate episodes. However, the film also re-solves its acausal structure through a narrative recursivity that links all the characters and events together in what one character calls both the "cosmic unconsciousness" and a "lattice of coincidence." Emplotment in *Repo Man* becomes diffused across a vast relational network. It is no accident that the car culture of Los Angeles figures in *Repo Man* to separate and segment experience into discrete and chaotic bits (as if it were metaphysically lived only through the window of an automobile)—while the "lattice of coincidence," the "network" of the Los Angeles freeway system, reconnects experience at another and less human order of magnitude.

The postmodern and electronic "instant," in its break from the temporal structures of retension and protension, constitutes a form of absolute presence (one abstracted from the continuity that gives meaning to the system past/present/future) and changes the nature of the space it occupies. Without the temporal emphases of historical consciousness and personal history, space becomes abstract, ungrounded, and flat—a site for play and display rather than an invested situation in which action "counts" rather than computes. Such a superficial space can no longer hold the spectator/user's interest, but has to stimulate it constantly in the same way a video game does. Its flatness—a function of its lack of temporal thickness and bodily investment—has to attract spectator interest at the surface. Thus, electronic space constructs objective and superficial equivalents to depth, texture, and invested bodily movement. Saturation of color and hyperbolic attention to detail replace depth and texture at the surface of the image, while constant action and "busyness" replace the gravity that grounds and orients the movement of the lived-body with a purely spectacular, kinetically exciting, often dizzying sense of bodily freedom (and freedom from the body). In an important sense, electronic space disembodies.

What I am suggesting is that, ungrounded and uninvested as it is, electronic presence has neither a point of view nor a visual situation, such as we experience, respectively, with the photograph and the cinema. Rather, electronic presence randomly disperses its being *across* a network, its kinetic gestures describing and lighting on the surface of the screen rather than inscribing it with bodily dimension (a function of centered and intentional projection). Images on television screens and computer terminals seem neither projected nor deep. Phenomenologically, they seem, rather, somehow just there as they confront us.

The two-dimensional, binary superficiality of electronic space at once disorients and liberates the activity of consciousness from the gravitational pull and orientation of its hitherto embodied and grounded existence. All surface, electronic space cannot be inhabited. It denies or prosthetically transforms the spectator's physical human body so that subjectivity and affect free-float or free-fall or free-flow across a horizontal/vertical grid. Subjectivity is at once decentered and completely extroverted—again erasing the modernist (and cinematic) dialectic between inside and outside and its synthesis of discontinuous time and discontiguous space as conscious and embodied experience. As Jameson explains:

> The liberation . . . from the older *anomie* of the centered subject may also mean, not merely a liberation from anxiety, but a liberation from every other kind of feeling as well, since there is no longer a self present to do the feeling. This is not to say that the cultural products of the postmodern era are utterly devoid of feeling, but rather that such feelings—which it might be better and more accurate to call "intensities"—are now free-floating and impersonal, and tend to be dominated by a peculiar kind of euphoria.[34]

Brought to visibility by the electronic, this kind of euphoric "presence" is not only peculiar. At the risk of sounding reactionary, I would like to suggest that it is also dangerous. Its lack of specific interest and grounded investment in the human body and enworlded action, its saturation with the present instant, could well cost us all a future.

Phenomenological analysis does not end with the "thick" description and thematization (or qualified reduction) of the phenomenon under investigation. It aims also for an interpretation of the phenomenon that discloses, however partially, the lived meaning, significance, and non-neutral value it has for those who engage it. In terms of contemporary moving-image culture, the material differences between cinematic and electronic representation emerge as significant differences in their meaning and value. Cinema is an objective phenomenon that comes—and becomes—before us in a structure that implicates both a sensible body and a sensual and sense-making subject. In its visual address and movement, it allows us to see what seems a visual impossibility: that we are at once intentional subjects and material objects in the world, the seer and the seen. It affirms both embodied being and the world. It also shows us that, sharing materiality and the world, we are intersubjective beings.

Now, however, it is the electronic and not the cinematic that dominates the form of our cultural representations. And, unlike cinematic representation, electronic representation by its very structure phenomenologically denies the human body its fleshly presence and the world its dimension. However significant and positive its values in some regards, the electronic trivializes the human body. Indeed, at this historical moment in our particular society and culture, the lived-body is in crisis. Its struggle to assert its gravity, its differential existence and situation, its vulnerability and mortality, its vital and social investment in a concrete life-world inhabited by others is now marked in hysterical and hyperbolic responses to the disembodying effects of electronic representation. On the one hand, contemporary moving images show us the human body relentlessly and fatally interrogated, "riddled with holes" and "blown away," unable to maintain its material integrity or gravity. If the Terminator doesn't finish it off, then electronic smart bombs will. On the other hand, the current popular obsession with physical fitness manifests the wish to transform the human body into something else—a lean, mean, and immortal "machine," a cyborg that can physically interface with the electronic network and maintain material presence in the current digitized life-world of the subject. (It is no accident that body builder Arnold Schwarzenegger played the cyborg Terminator.)

Within the context of this material and technological crisis of the flesh, one can only hope that the hysteria and hyperbole surrounding it is strategic—and that through it the lived-body has, in fact, managed to reclaim our attention to forcefully argue for its existence against its simulation. For there are other subjects of electronic culture out there who prefer the simulated body and a virtual world. Indeed, they actually believe the body (contemptuously called "meat" or "wetware") is best lived only as an image or as information, and that the only hope for negotiating one's presence in our electronic life-world is to exist on a screen or to digitize and "download" one's consciousness into the neural sets of a solely electronic existence. Such an insubstantial electronic presence can ignore AIDS, homelessness, hunger, torture, and all the other ills the flesh is heir to outside the image and the datascape. Devaluing the physically lived body and the concrete materiality of the world, electronic presence suggests that we are all in danger of becoming merely ghosts in the machine.

NOTES

1. Martin Heidegger, "The Question Concerning Technology," tr. William Lovitt, in *Martin Heidegger: Basic Writings*, ed. David Farrell Krell (New York: Harper & Row, 1977), p. 317.

2. *Robocop* (1987) was directed by Paul Verhoeven; *Terminator II: Judgment Day* (1991) by James Cameron.

3. Reference here is not only to the way in which automotive transportation has changed our lived sense of distance and space, the rhythms of our temporality, and the hard currency that creates and expresses our cultural values relative to such things as class and style, but also to the way in which it has changed the very sense we have of our bodies. The vernacular expression of regret at "being without wheels" is profound, and ontologically speaks to our very real incorporation of the automobile as well as its incorporation of us.

4. These terms are derived from Don Ihde, *Technology and the Lifeworld: From Garden to Earth* (Bloomington: Indiana University Press, 1990), p. 29. Ihde distinguishes two senses of perception: "What is usually taken as sensory perception (what is immediate and focused bodily in actual seeing, hearing, etc.), I shall call microperception. But there is also what might be called a cultural, or hermeneutic, perception, which I shall call macroperception. Both belong equally to the lifeworld. And both dimensions of perception are closely linked and intertwined."

5. Two types of theory that are, to some degree, microperceptual analysis are, first, psychoanalytic accounts of the processes of cinematic identification in which cinematic technology is deconstructed to reveal its inherent "illusionism" and its retrogressive duplication of infantile and/or dream states and, second, Neo-Marxist accounts of both photography's and cinema's optical dependence upon a system of "perspective" based on an ideology of the individual subject and its appropriation of the "natural" world. One could argue, however, as I do here, that these two types of theory are not microperceptual *enough*. Although they both focus on the "technological" construction of subjectivity, they do so abstractly. That is, neither deals with the technologically constructed temporality and spatiality that *ground* subjectivity in a sensible and sense-making *body*.

6. Ihde, p. 29. (Emphasis mine.)

7. Ihde, ibid.

8. Ihde, p. 21.

9. For the history, philosophy, and method of phenomenology, see Herbert Spiegelberg's *The Phenomenological Movement: A Historical Introduction,* 2nd ed., 2 vols. (The Hague: Martinus Nijoff, 1965); David Carr, "Maurice Merleau-Ponty: Incarnate Consciousness," in *Existential Philosophers: Kierkegaard to Merleau-Ponty,* ed. George Alfred Schrader Jr. (New York: McGraw Hill, 1967), pp. 369–429; and Don Ihde, *Experimental Phenomenology: An Introduction* (New York: Paragon Books, 1979).

10. Ihde, *Technology and the Lifeworld,* p. 26.

11. Fredric Jameson, "Postmodernism, or The Cultural Logic of Late Capitalism," *New Left Review* 146 (July–August, 1984): 77.

12. Stephen Kern, *The Culture of Time and Space: 1880–1918* (Cambridge, Mass.: Harvard University Press, 1983), pp. 6–7.

13. Jean-Louis Comolli, "Machines of the Visible," in *The Cinematic Apparatus,* ed. Teresa deLauretis and Stephen Heath (New York, 180), pp. 122–23.

14. The very recent erosion of "faith" in the photographic as "evidence" of the real in the popular consciousness has been a result of the development of the *seamless electronic manipulation* of even the tiniest "bits" of the photographic image. While airbrushing and other forms of image manipulation have been around for a long while, they have left a discernable "trace" on the image; such is not the case with digital computer alterations of the photographic image. For an overview, see "Ask It No Questions: The Camera Can Lie," *New York Times,* August 12, 1990, sec. 2, pp. 1, 29.

15. Comolli, p. 123.

16. Most media theorists point out that photographic (and later cinematic) optics are structured according to a norm of perception based upon Renaissance perspective, which represented the visible as originating in and organized by an individual, centered subject. This form of representation is *naturalized* by photography and the cinema. Comolli says: "The mechanical eye, the photographic lens . . . functions . . . as a guarantor of the identity of the visible with the normality of vision . . . with the norm of visual perception" (pp. 123–24).

17. Comolli, p. 142.

18. Maurice Merleau-Ponty, "Eye and Mind," tr. Carleton Dallery, in *The Primacy of Perception,* ed. James Edie (Evanston, Ill.: Northwestern University Press, 1964), p. 166.

19. It must be noted that the term "memory bank" is analogically derived in this context from electronic (not photographic) culture. It nonetheless serves us as a way of reading backward that recognizes a literal as well as metaphorical *economy* of representation and suggests that attempts to understand the photographic in its "originality" are pervasively informed by our contemporary electronic consciousness.

20. For readers unfamiliar with the film, *La jetée* is a narrative about time, memory, and desire articulated in a recursive structure. A survivor of World War III has a recurrent memory of a woman's face and a scene at Orly airport where, as a child, he has seen a man killed. Because of this vivid memory, his postapocalyptic culture—underground, with minimal power and without hope—attempts experiments to send him back into his vivid past

so that he can, perhaps, eventually time-travel to the future. This achieved, aware he has no future in his own present, the protagonist, with the assistance of those in the future, ultimately returns to his past and the woman he loves. But his return to the scene of his original childhood memory at Orly reveals, first, that he (as an adult) has been pursued by people from his own present and, second, that his original memory was, in fact, the vision of his own adult death.

21. André Bazin, "The Ontology of the Photographic Image," in *What is Cinema?*, tr. Hugh Gray, vol. 1 (Berkeley: University of California Press, 1967), pp. 9–16.

22. James Joyce, in 1909, was "instrumental in introducing the first motion picture theater in Dublin." (See Kern, pp. 76–77.)

23. Arthur M. Danto, "Moving Pictures," *Quarterly Review of Film Studies* 4 (Winter 1979): 17.

24. In the traditional cinema, an image can be "frozen" only by replicating it many times so that it can continue moving through the projector to appear frozen on the screen.

25. Maurice Merleau-Ponty, *The Visible and the Invisible*, tr. Alphonso Lingus (Evanston, Ill.: Northwestern University Press, 1968), pp. 143–44.

26. For a complete and lengthy argument supporting this assertion, see my *The Address of the Eye: A Phenomenology of Film Experience* (Princeton, N.J.: Princeton University Press, 1992).

27. Jameson, p. 64.

28. Ibid., p. 78.

29. Brooks Landon, "Future So Bright They Gotta Wear Shades." (Unpublished manuscript.)

30. It is important to point out that although all moving images follow each other serially, each cinematic image (or frame) is projected analogically rather than digitally. That is, the image is projected *as a whole*. Electronic images, however, are transmitted digitally, each bit of what appears as a single image sent and received as a discrete piece of information.

31. "Network" was a term that came into common parlance as it described the electronic transmission of television images. Now, we speak of our social relations as "networking." In spatial terms, however, a "network" suggests the most flimsy, the least substantial, of grounds. A "network" is constituted more as a lattice between nodal points than as grounded and physical presence.

32. Guy Debord, *Society of the Spectacle* (Detroit, Mich.: Red and Black, 1983), n.p.

33. It is no accident that all of the films used illustratively here can be identified with the generic conventions and thematics of science fiction. Of all genres, science fiction has been most concerned with poetically mapping the new spatiality, temporality, and subjectivities informed and/or constituted by new technologies. As well, science-fiction cinema, in its particular materiality, has made these new poetic maps concretely visible. For elaboration of this mapping, see chap. 4, "Postfuturism" of my *Screening Space: The American Science Fiction Film* (Ungar, 1987).

34. Jameson, p. 64.

Allucquère Rosanne Stone

Sex, Death, and Machinery, or How I Fell in Love with My Prosthesis

It started this afternoon when I looked down at my boots. I was emerging from a stall in the women's room in my department. The university was closed for the holidays. The room was quite silent except for the distant rush of the air conditioning, imparting to the cramped institutional space the mechanical qualities of a submarine. I was idly adjusting my clothing, thinking of nothing in particular, when I happened to look down, and there they were: My boots. Two completely unremarkable boots. They were right where they belonged, on the ends of my legs. Presumably my feet were inside.

I felt a sudden thrill of terror.

Maybe, I suppose, the boots could have reminded me of some long-buried trauma, of the sort that Freudians believe leads to shoe fetishism. But my sudden fear was caused by something quite different. What was driving me was not the extraordinariness of the sight of my own boots, but the ordinariness of them. They were common as grass. In fact, I realized that I hadn't even thought about putting them on. They were *just there.* If you wanted to "get real ugly about it"—as they say in Austin—you might call it a moment of radical existential *Dasein,* in the same way you might say déjà vu again. I had become transparent to myself. Or rather, the *I* that I customarily express and that reflexively defines me through my chosen personal style had become part of the wallpaper.

This is hardly a serious problem for me. But I tend to see myself as an entity that has chosen to make its life career out of playing with identity. It sometimes seems as though everything in my past has been a kind of extended excuse for experiments with subject position and interaction. After all, what material is better to experiment with than one's self? Academically speaking, it's not exactly breaking new ground to say that any

From *The War of Desire and Technology at the Close of the Mechanical Age* (Cambridge, Mass.: MIT Press, 1996). Reprinted by permission of the author and MIT Press.

subject position is a mask. That's well and good, but still most people take some primary subject position for granted. When pressed, they may give lip service to the idea that perhaps even their current "root" persona is also a mask, but nobody really believes it. For all intents and purposes, your "root" persona is *you*. Take that one away, and there's nobody home.

Perhaps someone with training in drama already perceives this, but it was a revelation to me. In the social sciences, symbolic interactionists believe that the root persona is always a momentary expression of ongoing negotiations among a horde of subidentities, but this process is invisible both to the onlooker and to the persona within whom the negotiations are taking place. For me this has never been particularly true. My current *I* has been as palpably a mask to me as any of my other *I*'s have been. Perceiving that which is generally invisible as really a kind of capital has been more than a passing asset (as it were); it has been a continual education, a source of endless challenge, not to mention fear, and certainly not least, an ongoing celebration of the sacred nature of the universe of passing forms. It was for these reasons, then, that I found looking down rather complacently at my boots and not really seeing them to be so terrifying. Like an athlete who has begun to flub a long-polished series of moves, I began to wonder if I was losing my edge.

Going through life with this outlook has been a terrific asset in my chosen work, and the current rise in the number of people who engage in social interactions without ever meeting in the customary sense of the term—that is, engaging in social intercourse by means of communication technologies—has given me increasing opportunities to watch others try on their own alternative personae. And although most still see those personae as just that—alternatives to a customary "root" identity—there are some out at the margins who have always lived comfortably with the idea of floating identities, and inward from the margins there are a few who are beginning, just a bit, to question. What it is they are questioning is a good part of what this essay is about.

A bit of background may be appropriate here.

I have bad history: I am a person who fell in love with her own prostheses. Not once, but twice. Then I fell in love with somebody *else's* prosthesis.

The first time love struck was in 1950. I was hunkered down in the dark late at night, on my bed with the big iron bedstead on the second floor, listening absently to the crickets singing and helping a friend scratch around on the surface of a galena crystal that was part of a primitive radio. We were looking for one of the hot spots, places where the crystal had active sites that worked like diodes and could detect radio waves. There was nothing but silence for a long, long time, and then suddenly the earphones burst into life, and a whole new universe was raging in our heads—the ranting voice of Jean Shepherd, boiling into the atmosphere from the massive transmitter of WOR-AM, 50 kilowatts strong and only a few miles away. At that distance we could have heard the signal in our tooth fillings if we'd had any, but the transmitter might as well have been in Rangoon, for all the fragrant breath of exotic worlds it suggested. I was hooked. Hooked on technology. I could take a couple of coils of wire and a hunk of galena and send a whole part of myself out into the ether. An extension of my will, of my instrumentality . . . that's a prosthesis, all right.

The second time happened in 1955, while I was peering over the edge of a 24 × 24 recording console. As I stood on tiptoe, my nose just clearing the top of the console, from my age and vantage point the massive thing looked as wide as a football field. Knobs and switches from hell, all the way to the horizon . . . there was something about that vast forest of controls that suggested the same breath of exotic worlds that the simple coil of wire and the rickety crystal did. I was hooked again. Hooked on even bigger technology, on another extension of my instrumentality. I could create whole oceans of sound, universes of sound, could at last begin on my life's path of learning how to make people laugh, cry, and throw up in darkrooms. And I hadn't even heard it turned *on*.[1]

But the third time . . .

The third time was when Hawking came to town.

Stephen Hawking, the world-famous physicist, was giving a lecture at UC Santa Cruz. The auditorium was jammed, and the overflow crowd was being accommodated outside on the lawn. The lawn looked like a medieval fair, with people sitting on blankets and towels, others standing or milling around, all ears cocked toward the loudspeakers that were broadcasting Hawking's address across the landscape.

If you haven't seen Stephen Hawking give a talk, let me give you a quick background. Hawking has amyotrophic lateral sclerosis, which makes it virtually impossible for him to move anything more than his fingers or to speak. A friendly computer engineer put together a nice little system for him, a program that displays a menu of words, a storage buffer, and a Votrax allophone generator—that is, an artificial speech device. He selects words and phrases, the word processor stores them until he forms a paragraph, and the Votrax says it. Or he calls up a prepared file, and the Votrax says that.

So I and a zillion other people are on the lawn, listening to Hawking's speech, when I get the idea that I don't want to be outside with the PA system—what I really want to do is sneak into the auditorium, so I can actually hear Hawking give the talk.

In practice this maneuver proves not too hard. The lecture is under way, security is light—after all, it's a *physicist*, dammit, not the UC Board of Regents, for which they would have had armed guards with two-way radios—so it doesn't take long for me to worm my way into the first row.

And there is Hawking. Sitting, as he always does, in his wheelchair, utterly motionless, except for his fingers on the joystick of the laptop;

and on the floor to one side of him is the PA system microphone, nuzzling into the Votrax's tiny loudspeaker.

And a thing happens in my head. Exactly where, I say to myself, *is* Hawking? Am I any closer to him now than I was outside? Who is it doing the talking up there on stage? In an important sense, Hawking doesn't stop being Hawking at the edge of his visible body. There is the obvious physical Hawking, vividly outlined by the way our social conditioning teaches us to see a person as a person. But a serious part of Hawking extends into the box in his lap. In mirror image, a serious part of that silicon and plastic assemblage in his lap extends into him as well . . . not to mention the individual ways, displaced in time and space, in which discourses of medical technology and their physical accretions already permeate him and us. No box, no discourse; in the absence of the prosthetic, Hawking's intellect becomes a tree falling in the forest with nobody around to hear it. On the other hand, with the box his voice is auditory and simultaneously electric, in a radically different way from that of a person *speaking* into a microphone. Where *does* he stop? Where are his edges? The issues his person and his communication prostheses raise are boundary debates, borderland/*frontera* questions. Here at the close of the mechanical age, they are the things that occupy a lot of my attention.[2]

Flashback: I was Idly Looking

I was idly looking out my window, taking a break from some nasty piece of academic writing, when up the dusty, rutted hill that constitutes my driveway and bastion against the world there abruptly rode, on a nasty little Suzuki Virago, a brusque, sharp-tongued person of questionable sexuality. Doffing her helmet, she revealed herself, both verbally and physically, as Valkyrie, a postoperative m/f transgender with dark hair and piercing black eyes who evinced a pronounced affinity for black leather. She announced that there were things we had to do and places we had to go, and before I could mutter "science fiction" we were off on her bike.[3]

Valkyrie proceeded to introduce me to a small community of women in the San Francisco Bay area. Women's collectives were not new to me; I had recently studied a group of women who ran a business, housed themselves under one roof, and lived their lives according to the principles of a canonically undefined but quite powerful idea known as lesbian separatism.[4] But the group to which my new friend now introduced me did not at all fit the model I had painstakingly learned to recognize. This collective ran a business, and the business was hetero phone sex . . . not something of which my other research community, immured in radical lesbian orthodoxy, would have approved.

I was instantly entranced, and also oddly repelled. After all, I had broken bread with one of the most episcopal of women's collectives for five years, and any deviation from group norms would have been punishable in fairly horrid ways. To imagine that hetero sex could be enjoyable, not to mention profitable, was playing into the hands of the gentiles, and even to spend time with a group that supported itself in such a manner (and even joked about it) could have had mortal consequences.

For reasons best described as kismet, the phone sex workers and I became good friends. We found each other endlessly fascinating. They were intrigued by my odd history and by what I'd managed to make out of it. In turn, I was intrigued by the way they negotiated the minefields of ethics and personal integrity while maintaining a lifestyle that my other research community considered unthinkable.

After a while, we sorted out two main threads of our mutual attraction. From my point of view, the more I observed phone sex the more I realized I was observing very practical applications of data compression. Usually sex involves as many of the senses as possible. Taste, touch, smell, sight, hearing—and, for all I know, short-range psychic interactions—all work together to heighten the erotic sense. Consciously or unconsciously phone sex workers translate all the modalities of experience into audible form. In doing so they have reinvented the art of radio drama, complete down to its sound effects, including the fact that some sounds were best represented by *other* improbable sounds that they resembled only in certain iconic ways. On the radio, for example, the soundmen (they were always literally men) represented fire by crumpling cellophane, because to the audience it sounded *more like* fire than holding a microphone to a real fire did.

The sex workers did similar stuff. I made a little mental model out of this: The sex workers took an extremely complex, highly detailed set of behaviors, translated them into a single sense modality, then further boiled them down to a series of highly compressed tokens. They then squirted those tokens down a voice-grade phone line. At the other end of the line the recipient of all this effort added boiling water, so to speak, and reconstituted the tokens into a fully detailed set of images and interactions in multiple sensory modes.

Further, what was being sent back and forth over the wires wasn't just information, it was *bodies*. The majority of people assume that erotics implies bodies; a body is a part of the idea of erotic interaction and its concomitants, and the erotic sensibilities are mobilized and organized around the idea of a physical body which is the seat of the whole thing. The sex workers' descriptions were invariably and quite directly about physical bodies and what they were doing or what was being done to them.

Later I came to be troubled by this focus on bodies because of its relation to a remark of Elaine Scarry's. In a discussion of human experience in her book *The Body In Pain*, she says,

> Pain and imagining are the "framing events" within whose boundaries all other perceptual, somatic, and emotional events occur; thus, between the two extremes can be mapped the whole terrain of the human psyche. (p. 165)

By that time I had stopped thinking of the collective as a group of sex workers and had begun to think of them in rather traditional anthropological terms as *my* sex workers. I had also moved on to a more complex mode of fieldwork known as participant observation, and I was getting an education I hadn't expected. Their experience of the world, their ethical sense, the ways they interpreted concepts like work and play were becoming part of my own experience. I began to think about how I could describe them in ways that would make sense to a casual reader. As I did so, Scarry's remark returned to intrigue me because of its peculiar relationship to the social groups I was studying. It seemed to me that the sex workers' experiential world was organized in a way that was almost at right angles to Scarry's description of the continuum of pain and imagining. The world of the sex workers and their clients, I observed, was not organized along a continuum of pain and imagination but rather within an experiential field in which *pleasure* and imagination were the important attractors.

Patently it is not difficult in these times to show how phone sex interactions take place within a field of power by means of which desire comes to have a particular shape and character. In the early days of phone sex that view would have been irrefutable, but things are changing rather fast in the phone sex business; more traditional hetero and hetero-modeled interactions may still get their kick from very old patterns of asymmetrical power, but there seems little doubt that the newer forums for phone sex (as well as other forms of technologically mediated human interaction) have made asymmetrical power relationships part of a much larger and more diverse erotic and experiential tool kit.

This diversity has obvious and interesting implications for critical studies, but it does not in any way imply that a hypothetical "new erotics," if that's what I'm describing, has escaped from the bottomless gravity well of the same power structures within which we find ourselves fixed in position, regardless of what our favorite position is. It does seem to mean, though, that a good many of the people I observe are aware of the effects of those structures, even though as of this writing I see little effort to alter or transcend them. There does appear to be a central and critical reason for this lack of effort, particularly in regard to erotics, and that is that none of the people I observe who *do* erotics—even those who play with different structures of power—have yet begun to speculate on how erotics really works.

There are other areas of inquiry which are organized around what might be called an epistemological Calvinism. A recent but fairly broad area of inquiry in the social sciences into the nature and character of human-computer interaction is known as the study of computer-supported cooperative work (CSCW). Part of the informing philosophy of this discipline is the idea that all human activity can be usefully interpreted as a kind of work, and that work is the quintessential defining human capacity. This, too, I think, misses some of the most important qualities of human-computer interaction just as it does when applied to broader elements of human experience. By this I mean that a significant part of the time that humans spend in developing interactional skills is devoted not to work but to what by common understanding would be called play. Definitions of what counts as play are many and varied, generally revolving around the idea of purposive activities that do not appear to be directly goal-oriented. "Goal orientation" is, of course, a problematic phrase. There is a fine body of research addressed to the topic of play versus work activities, but it doesn't appear to have had a deep effect on CSCW and its allied disciplines. From the standpoint of cultural criticism, the issue is not one of definitions of work or play, but of how the meanings of those terms are produced and maintained. Both work and play have culture-specific meanings and purposes, and I am conducting a quite culture-specific discussion when I talk about the primacy of play in human-computer interaction (HCl, or for our purposes just "interaction") as I do here.[5]

In order to clarify this point, let me mention that there are many definitions of interaction and many opinions about what interaction is for. As I write, large industry consortiums are finalizing their standards for what they call interactive multimedia platforms. These devices usually consist of a computer, color monitor, mouse, CD-ROM drive, sound car, and pair of speakers. This electronic instantiation of a particular definition freezes the conceptual framework of interaction in a form most suitable for commercial development—the user moves the cursor to the appropriate place and clicks the mouse, which causes something to happen—or what the interactivist Michael Naimark would call, more pejoratively, poke-and-see technology. This definition of interaction has been in the wind for so long now that few researchers say much about it. It is possible to play within the constraints of such a system, but the potential for *interaction* is limited, because the machine can only respond to an on-off situation: that is, to the click of the mouse. Computer games offer a few more input modes, usually in the form of a joystick, which has two or three degrees of freedom. However, from the standpoint of kind and gentle instruction, what the game companies do with this greater potential is not very inspiring. Technologically speaking, Sega's *Sewer Shark* (1993), for example, was an amazing exercise in game design for its time, but it reinforced the feeling that interaction in a commercial frame is still a medium like television, in which the most advanced product of the technological genius of an entire species conveys Geraldo Rivera to millions of homes in breathtaking color.

I don't want to make this a paradise-lost story, but the truth is that the definitions of interactivity used by the early researchers at MIT possessed a certain poignancy that seems to have become lost in the commercial translation. One of the best definitions was set forth by Andy Lippman, who described interaction as mutual and simultaneous activity on the part of both participants, usually working toward some goal—but, he added, not necessarily. Note that from the beginning of interaction research the idea of a common goal was already in question, and in that fact inheres interaction's vast ludic dimension.[6]

There are five corollaries to Lippman's definition. One is *mutual interruptibility*, which means that each participant must be able to interrupt the other, mutually and simultaneously. Interaction, therefore, implies conversation, a complex back-and-forth exchange, the goal of which may change as the conversation unfolds.

The second is *graceful degradation*, which means that answerable questions must be handled in a way that doesn't halt the conversation: "I'll come back to that in a minute," for example.

The third is *limited look-ahead*, which means that because both parties can be interrupted there is a limit to how much of the shape of the conversation can be anticipated by either party.

The fourth is *no-default*, which means that the conversation must not have a preplanned path; it must develop fully in the interaction.

The fifth, which applies more directly to immersive environments (in which the human participant is surrounded by the simulation of a world), is that the participants should have *the impression of an infinite database*. This principle means that an immersive interactional world should give the illusion of not being much more limiting in the choices it offers than an actual world would be. In a nonimmersive context, the machine should give the impression of having about as much knowledge of the world as you do, but not necessarily more. This limitation is intended to deal with the Spock phenomenon, in which more information is sometimes offered than is conversationally appropriate.

Thus interactivity implies two conscious agencies in conversation, playfully and spontaneously developing a mutual discourse, taking cues and suggestions from each other as they proceed.

In order to better draw this out let me briefly review the origins and uses of computers. Afterward I will return to the subject of play from a slightly different perspective.

The first devices that are usually called computers were built as part of a series of projects mandated by the military during World War II. For many years, computers were large and extremely costly. They were also cranky and prone to continual breakdown, which had to do with the primitive nature of their components. They required continual maintenance by highly skilled technicians. The factors of cost, unreliability, and the need for skilled and continual attention, not to mention the undeniable aura of power that surrounded the new machines like some heady smell, combined to keep computers available only to large corporations and government organizations. These entities came already equipped with their own ideas of efficiency, with the concepts of time and motion study then in vogue in industry (of which my colleagues have written at length), and of course with the cultural abstraction known as the work ethic perpetually running in the background. Even within the organizations themselves, access to the new machines was restricted to a technological elite which, though by no means monolithic in its view of technological achievement, had not had enough time to develop much of a sense, not to mention a sensibility, of the scope and potential of the new devices.

These factors combined to keep attention focused on the uses of computers as rather gross instrumentalities of human will—that is, as number crunches and databases. Computers could extend human abilities, physically and conceptually. That is, computers were tools, like crowbars and screwdrivers, except that they primarily extended the mind rather than the muscles. Even Vannevar Bush's astonishingly prophetic "As We May Think" (1949) treated computers as a kind of superswitch. In this frame of understanding, computers were prosthetic in the specific sense of the Greek term *prosthenos*—extension. Computers assisted or augmented human intelligence and capabilities in much the same way that a machine or even another human being would; that is, as separate, discrete agencies or tools that occupied physical or conceptual spaces separate from those of the human.

It seems significant that the epistemic evolution that appeared to be gradually but inexorably making its way across Western cultures also manifested itself in a number of unexpected and quite unpredictable ways in cultural milieus far removed from the context of the Enlightenment and after. A pertinent though perhaps startling (and perhaps offensive) example is the aesthetics and philosophy of bullfighting. Prior to the schismatic work of the torero Juan Belmonte in the 1940s and 1950s, the physical area in which bullfighting took place was divided into spaces of signification called "territories of the bull" and "territories of the torero." When designing his choreography for the bullring, Belmonte raised the heretical argument that since the human possessed the only agency in the arena, territory of the bull was a polite but fictional concept; all territories were territories of the torero. The choreographic movements Belmonte developed as a result of this argument transformed the character of bullfighting. The abstraction I call attention to here is the breakdown of boundaries between two systems of agency and how that transformation affects the play of power within a field of social action. In dance, Martha Graham articulated a similar revision of shared spaces of action, but somewhat closer to the center of what might be called traditional Western culture. Graham's relocating the center of agency to a hypothesized center of the body redefined the quality of contact that was possible between two agents. Susan

Foster's theoretical and practical work on dance discusses these points in considerable detail.

All this changed in the 1960s, but the change was largely invisible both physically and conceptually. Deleuze and Guattari and Manuel De Landa and the eerie concept of the machine phylum would not arrive on the scene for some thirty years. In 1962, the young hackers at Project MAC, deep in the bowels of MIT, made hardly a ripple in corporate arenas with their invention of a peculiarly engrossing computational diversion that they called *Space War*.[7] This first computer game was still firmly identified with the military, even down to its name and playing style, but in that moment something quite new and (dare I say it) completely different had happened to the idea of computation. Still, it would not be until the 1970s that two kids in a garage in Mountain View, California, rather than a corporate giant like Sperry Rand or IBM or a government entity like the Bureau of Vital Statistics, would knock the props out from under the idea of computation-as-tool for all time.

Let me return to the discussion of work versus play once again, from the standpoint of computation and instrumentality. Viewing computers as calculatory devices that assist or mediate human work seems to be part of a Kuhnian paradigm that consists of two main elements. The first is a primary *human work ethic*; the second is a particularized view of *computers as tools*. The emergence of the work ethic has been the subject of innumerable essays, but the view of computers as tools has been so totally pervasive among those with the power to determine meaning in such forums as school policy and corporate ethics that only recently has the idea begun to be seriously challenged. The paradigm of computers as tools burst into existence, more or less, out of the Allied victory in World War II (although the Nazis were working on their own computers). A paradigm of computers as something other than number crunchers does not have a similar launching platform, but the signs of such an imminent upheaval are perspicuous. Let me provide an example.

One of the most perceptive scholars currently studying the emergent computer societies is the anthropologist Barbara Joans. She describes the community of cyberspace workers as composed of two groups that she calls Creative Outlaw Visionaries and Law and Order Practitioners. One group has the visions; the other group knows how to build stuff and get it sold. One group fools around with terminology and designs fantastic stuff; the other group gets things done and keeps the wheels turning. They talk to each other, if they talk to each other, across a vast conceptual gulf. These groups are invisible to each other, I think, because one is operating out of the older paradigm of computers as tools and the other out of the newer paradigm of computers as something else. Instead of carrying on an established work ethic, the beliefs and practices of the cultures I observe incorporate a *play* ethic—not to displace the corporate agendas that produce their paychecks, but to complexify them. This play ethic is manifest in many of the communities and situations I study. It is visible in the northern California Forth community, a group of radical programmers who have adopted for their own an unusual and controversial programming language; in the CommuniTree community, an early text-based virtual discussion group that adopted such mottos as "If you meet the electronic avatar on the road, laserblast Hir"; and in the Atari Research Lab, where a group of hackers created an artificial person who became real enough to become pro tem lab director. The people who play at these technosocial games do not do so out of any specific transformative agenda, but they have seized upon advantages afforded by differences of skill, education, and income to make space for play in the very belly of the monster that is the communication industry.

This production and insertion of a play ethic like a mutation into the corporate genome is a specifically situated activity, one that is only possible for workers of a certain type and at a certain job level. In specific, it is only possible to the communities who are perhaps best described as hackers—mostly young (although the demographic changes as the first- and second-generation hackers age), mostly educated (although the field is rife with exceptions, perhaps indicating the incapability of U.S. public schools to deal with talented individuals), mostly white (and exceptions are quite rare in the United States), and mostly male (although a truly egregious exception is part of this study). They create and use a broad variety of technological prosthetics to manifest a different view of the purpose of communication technology, and their continual and casual association with the cutting edge of that technology has molded them and their machines—separately and jointly—in novel and promising ways. In particular, because they are thoroughly accustomed to engaging in nontrivial social interactions through the use of their computers—social interactions in which they change and are changed, in which commitments are made, kept, and broken, in which they may engage in intellectual discussions, arguments, and even sex—they view computers not only as tools by also as *arenas for social experience.*

The result is a multiple view of the state of the art in communication technology. When addressing the question of what's new about networking, it's possible to give at least two answers. Let's stick with two for now.

Answer 1: Nothing The tools of networking are essentially the same as they have been since the telephone, which was the first electronic network prosthesis. Computers are engines of calculation, and their output is used for quantitative analysis. Inside the little box is information. I recently had a discussion with a colleague in which he maintained that there was nothing new about virtual reality. "When you sit and read a book," he said, "you create characters and action in your head. That's the same things as VR, without all the electronics." Missing the point, of course, but understandably.

Answer 2: Everything Computers are arenas for social experience and dramatic interaction, a type of media more like public theater, and their output is used for qualitative interaction, dialogue, and conversation. Inside the little box are *other people.*

In order for this second answer to be true, we have to rethink some assumptions about presence. Presence is currently a word that means many different things to many different people. One meaning is the sense that we are direct witnesses to something or that we ourselves are being directly apprehended. This is what we might call the straightforward meaning, the one used by many sober virtual reality researchers. Another meaning is related to agency, to the proximity of intentionality. The changes that the concept of presence is currently undergoing are embedded in much larger shifts in cultural beliefs and practices. These include repeated transgressions of the traditional concept of the body's physical envelope and of the locus of human agency. This phenomenon shows itself in such variegated forms as the appearance and growth of the modern primitive movement, and the astonishing fascination of a portion of the population with prosthetic implants. Simultaneously new companies spring up to develop and manufacture wearable and eventually implantable computers. The film *Tetsuo, the Ironman* appears, with its disturbingly florid intermingling of biology and technology. William Gibson's cyberspace and Neal Stephenson's Metaverse are both science-fiction inflections of inhabitable virtual worlds. A slow process of belief and acceptance, perhaps most clearly instantiated in the process of cultural acclimatization to the telephone, accompanied by the issues of warranting and authentication raised by the interjection into human social life of a technological object that acts as a channel or representative for absent human agencies.

In studying issues of presence, warranting, and agency, the work of theorists of dramatic interaction vis-à-vis computation, of which Brenda Laurel is an outstanding example, is invaluable. Many of the interesting debates involved in my research would not have been possible without the arguments Laurel presents in *Computers as Theatre* and elsewhere.

My first organized piece of research in the field of virtual systems involved studying a group of phone sex workers in the early 1980s. In this study I was doing two things. On one hand, I was beginning to develop some of the ideas I set forth here and, on the other, also discovering in microcosm the fascinating interplays among communication technology, the human body, and the uses of pleasure. If I were to frame some of the questions that occurred to me during that time, they might be these: How are bodies represented through technology? How is desire constructed through representation? What is the relationship of the body to self-awareness? What is the role of play in an emergent paradigm of human-computer interaction? And overall: What is happening to sociality and desire at the close of the mechanical age?

If I'm going to give in to the temptation to periodize—which I do again and again, though frequently with tongue in cheek—then I might as well take the period that follows the mechanical age and call it the virtual age. By the virtual age I don't mean the hype of virtual reality technology, which is certainly interesting enough in its own ways. Rather, I refer to the gradual change that has come over the relationship between sense of self and the body, and the relationship between individual and group, during a particular span of time. I characterize this relationship as virtual because the accustomed grounding of social interaction in the physical facticity of human bodies is changing. Partly this change seems good, and partly it seems bad. There are palpable advantages to the virtual mode in relation to the ways that the structure of cities and expectations of travel have changed with the advent of the telephone, the rise of large corporations, the invention and marketing of inexpensive tract housing, the development of the shopping mall, the commercial development and exploitation of electronic mass media, the development of the personal computer, the greening of large-scale information networks (which can be co-opted for social interaction), and the increasing miniaturization of electronic components (eventually perhaps to be extended to mechanical devices, that is, Drexler and others). There are equally palpable disadvantages to each of these deep changes in our lives. I don't want this perhaps too-familiar list to be read as either extolling or condemnation. They are the manifestations, as well as causative agents, of the social changes, ruptures, and reorganizations that they accompany. In the course of this essay I sometimes organize the manifestation of these developments as a progressus, an ensemble of events that had a beginning and that leads in a particular direction. In doing so, I nod in the direction of Deleuze and Guattari, Paul Virilio, and Manuel De Landa.[8] But I am large; I contain multitudes. At other times the story is not at all meant to be teleological, because I don't foresee the telos toward which it tends. I may make some suggestions in that regard, but they are suggestions only and do not arise from any prophetic vision. I try to leave the prophetic side of things to my academic betters in the same line of work.

In the process of articulating the gradual unfolding of the cultural and technological foundations for virtual systems, I call on the work of scholars in a number of disciplines. One factor that bears importantly on the emergence of virtual systems is a change in the character of public space and the development and articulation of particular kinds of private space. I discuss this change in the context of portions of the social world of Elizabethan England with the help of the useful and important work of Francis Barker. In her study *The Tremulous Private Body*, Barker discusses from the point of view of textuality the creation of new social spaces; of particular relevance to our concerns here is a new and progressively ramified division of social space from a predominantly public space to a congeries of spaces increasingly privatized.[9] Barker uses the physicality of this new privatized space as a link to the metaphoricality of symbolic and psychological private space that is both elicited by and is mutually supportive of its physical concomitant. In this regard the development of separate interior spaces within small dwellings—changes in philosophies of architecture and in methods of carpentry—is crucial.

The relationship of these changes to the changing concepts of interior and exterior space that enable and support the character of virtual systems is complex. In regard to the emergence of the concepts of the interiorized cultural and epistemic individual, which we are by now used to calling the sovereign subject and to seeing as perhaps the most egregious product of the Enlightenment, this too bears a complex relationship to the changes in social and architectural space within which it is embedded. In his study *Segmented Worlds and Self,* Yi-Fu Tuan calls attention to these changes in the context of studies of architecture and subjectivity. Over time, Tuan shows, we can trace the emergence of an increasing social and epistemic privatization that leads to the idea of the individual, for better or worse, as we understand it today. The development of a palpable awareness of self can be followed through the changes by means of which it is produced, beginning in the Middle Ages when information first begins to accumulate—the increasing number of family and self-portraits; the increasing popularity of mirrors; the development of autobiographical elements in literature; the evolution of seating from benches to chairs; the concept of the child as a stage in development; the ramification of multiple rooms in small dwellings; the elaboration of a theater of interiority in drama and the arts; and, most recently, psychoanalysis.

The development of a sense of individuality seems to be accompanied by a corresponding withdrawal of portions of a person's attention and energy from the public arena and their nourishment and concentration within the new arena of social action called the self. In the discourses with which we are perhaps most familiar, the self appears to be a constant, unchanging, the stable product of a moment in Western history. This seems a rather episcopal view of something that is not only better described as a process but that is also palpably in continual flux. Yet our institutions continue to be based on a fixed notion of what a self is—a local notion, a culturally delimited notion that inhabits the larger cultural infections of the mass media. It seems clear enough that the self continues to change, in fact has changed, beyond the snapshots we have of it that were taken within the last hundred years or so. The trends toward interiority and perhaps more important toward textuality that Barker reported still continue with increasing speed.

Further, they are abetted by concomitant developments in communication technology. Just as textual technologies—cheap paper, the typewriter, printing—accompanied new discourse networks and social formations, so electronic communication technologies—radio, television, computer networks—accompany the discourse networks and social formations now coming into being. These technologies, discourse networks, and social formations continue the trend toward increasing awareness of a sense of self; toward increasing physical isolation of individuals in Western and Western-influenced societies; and toward displacement of shared physical space, both public and private, by textuality and prosthetic communication—in brief, the constellation of events that define the close of the mechanical age and the unfolding or revealing of what, for lack of a better term, we might call the virtual age.

About Method

In regard to the term *virtual age,* I want it clear that when I talk about *ages, closes,* and *dawns,* it is not without being aware of what these words mean. I am grappling with the forms of historicization, and seeking—if frequently not finding—different ways to tell these stories. Pasted to one corner of my monitor screen I have a card that says,

NO CAUSES
NO EFFECTS
MUTUAL EMERGENCE

which is also an extreme position. Death and furniture, as Malcolm Ashmore said: If somebody whacks me in the head, I could rightly attribute my headache to their intervention. Larger phenomena are, of course, tricky. I don't think I can show with any assurance what "caused" the Atari Lab, but I can tell a few of the stories that surround its coming into being, each one of which is situated in a web of stories of its own. If I could walk the walk as well as talk the talk, there would be no "ages" or "dawns" in this essay, and eventually, given time, I hope to produce a different account in which the events I discuss here are more deeply situated in their context . . . and vice versa.

My chosen method of representation for this attempt—a kind of adventure narrative interspersed with forays into theory—developed out of earlier work in which I mentioned that my hypothesized ideal method would be a cross between Sharon Traweek's *Beamtimes and Lifetimes* and Leo Tolstoy's *War and Peace.*[10] This piece/*peace* is a sally in that direction. It is thoroughly experimental and subject to recall for factory modification at any time. I feel that it is only through the process of trying out various forms of representation, some experimental and some not, that I can properly grapple with the formidable challenge of finding viable pathways into academic discourse in the time of cultural studies. ("In the time of . . . " There I've done it again.)

Rather than presenting a succession of chapters explicating a common theme, I have tried to organize the work as a set of provocations whose central ideas remain more or less unstated—hovering, as I would like to imagine them, in the background. In this effort, my idealized stance as a novelist is the motivating concept. That is my *preferred, ideal* method; however, in the interests of avoiding some possibly unfortunate debates I have cheated and provided a theoretical section as well, and more explanation than I would have liked. I am still trying to move toward a methodology that Donna Haraway recently called cat's cradle. In other work I have mentioned that I prefer to thread these discourses and hold them in productive tension rather than allowing them to collapse into a univocal account, and cat's cradle describes this move perfectly. Haraway has added to my experimental statement the missing piece of community, of passing the accounts from hand to hand, perhaps turning them in different ways and threading them in new configurations, being ever mindful that we tell our stories within webs of power that distort them; and of course the important thing about a cat's cradle is that you can never let it collapse.

On Content

Although other accounts of cyberspace communities and the people who construct them are now appearing, it is possible that readers of this essay may not yet have encountered them; therefore, I include a few here. In any account of the advent of ludic interactive technology the MIT Media Lab occupies a central role as nurturer of almost all of the first generation of "reality hackers," and its founder, Nicholas Negroponte, continues to be seen as an individual with both tremendous foresight and stupendous abilities to attract capital and power.

When the first generation of young technokids left MIT in 1987 for the physical world (eschewing for now the slippery term *real*), many of them moved directly to a brand-new research lab financed by the Atari Corporation in California. The Atari Lab was headed by Alan Kay, who might have been compared to Negroponte in his ability to understand and navigate structures of power. It included among its staff the largest percentage of women in any laboratory up to that time and for a long time afterward, and this fact appears to have been due to Negroponte's influence both at the MIT Media Lab and on Alan Kay. The high attrition rates among women staff members that plagued most research labs did not affect the Media Lab in its early days. It appears that Negroponte's encouragement, his even-handedness, and possibly his personal charm helped keep a cadre of bright young women in the lab long enough for Kay to hire them. Negroponte himself moved in a web of events that enabled and constrained his choices, including his secure and prestigious directorship of *Le Grand Experiment,* the modestly named World Center for Research in Computation, just opened in Paris. The fortunes of the Atari Lab, unlike the Media Lab or the World Center, were tied to the continued success of a single company, and the glory days of Atari passed their peak shortly after its lab started work. But the days of success for both the Xerox Palo Alto Research Center and the World Center ended within a few months of the sudden fall of Atari, thus ending one of the most interesting and perhaps most promising periods in prosthetic communication research.

A "golden window" of financial support, theoretical encouragement, free imagination, and peer camaraderie was open at Atari for perhaps two years, perhaps no longer than six months, depending upon which events seem important. But in that brief period the young researchers performed astonishing feats. The thrust of their work was toward issues of presence not in terms of an hypothecated "human-machine" interface, but in *situated* technologies that addressed such issues as gender and ethnicity. The impact of this work was largely lost on Atari, because of a hidden misunderstanding among Kay, the researchers, and Atari management about the purpose of the lab. This miscommunication didn't become visible until later, and consequently the young researchers' work remained to bear fruit at other research organizations at later times. When Warner sold Atari to the notorious Tramiel brothers, known in Silicon Valley for their bloodthirsty approach to entrepreneurism, the lab in its original form was doomed.[11] Its research group, composed of brilliant young men and an unusually high proportion of brilliant young women, suddenly found themselves on the street. As they scattered, they founded the first generation of companies directly associated with the development of what would come to be called virtual reality technology. The Atari Lab remains both emblematic of and the best example of a singular moment in the emergence of a constellation of ideas concerning what research in communication prosthetics and agency should be.

Already in California were several groups of computer engineers who saw computer technology as a transformative force for society. McLuhan-like, they believed that the technology itself was already producing deep changes in consciousness, but that belief would not stop them from hurrying things along. They developed programs to make dial-up bulletin boards into social forums within which certain people with access to computers and modems could quickly form new kinds of communities unrestricted by barriers of distance or, perhaps more significantly, of physical appearance. (They were not unrestricted by other means—such as ethnicity, class, income, and fluency in the English language.) Wide dissemination of the telephone numbers of these early virtual communities led to unexpected clashes as different and mutually incomprehensible cultures faced off within the electronic environment. As the survivors built new communities out of the ashes of the old, they unconsciously built much older Western theories of social life into their systems . . . such as defensive countermeasures, surveillance, and control. This might be called the Great Wall moment in the history of virtual systems, in retrospect a rather primitive beginning, but more sophisticated techniques were not long in arriving.

Working in tandem with the early researchers were a few people who were directly concerned with designing and implementing environments for virtual social interaction. They designed their environments as games. The earliest of these were the multiple-user dungeons, or MUDs. The MUDs were direct descendants of a species of role-playing game known as Dungeons and Dragons, and they later changed (or rather attempted to change) the full name of their environments to multiple-user *domains* in an effort to attain a modicum of respectability.

The concept of MUDs was taken up as a research tool at the Xerox Corporation's Palo Alto Research Campus (PARC), where anthropologists have for a number of years observed social interactions within structured virtual environments with an eye toward eventual uses in the workplace. But the best known of the descendants of text-based MUDs was designed at LucasFilm as a pay-per-minute virtual game. Players entered *Habitat* via modem in the same way they would access any of the on-line services such as American Online or CompuServe. Once inside, they met other players, engaged in treasure hunts, apprehended (or became) criminals, published newspapers, married, divorced, and in general replicated in the virtual environment many of the pleasures and annoyances of life in the physical world. But the experiment never quite caught on in the United States. Almost unknown here, *Habitat* was acquired by Fujitsu, a Japanese company, and moved to a mainframe in Tokyo. There it became extremely successful, and attracted approximately 1.5 million inhabitants—an astonishing number even in light of LucasFilm's ambitious predictions for their American version.[12]

Habitat is a useful early example of how economies evolve in virtual spaces, in particular because sex work is a common form of employment in the simulation. Since Chip Morningstar and Randall Farmer—*Habitat*'s designers—were running nine months behind on the code's ship date, they didn't have time to provide ways for inhabitants to assume any of the vanilla sexual positions. For example, there is no code to describe characters lying on top of each other, and a fortiori dog positions are unknown.[13] Thus in order to engage in sex in *Habitat* people must be inventive, and so they have been. Also, because Fujitsu keeps good records, it is possible to get some idea of how gender works in the space. An item of particular interest to me is that at any given time approximately 15 percent of the *Habitat* population is actively engaging in cross-dressing or cross-gender behavior.

A subset of multiple-user domains consists of multiple-player games set in virtual environments. The earliest of these was a multiple-user environment called RBT, for Reality Built for Two, constructed at VPL Research in Palo Alto by one of the best-known of the first-generation virtual hackers, Jaron Lanier. Lanier's trademark dreadlocks became widely recognized in the world's business community when his picture (or, more correctly, an engraving of his face—*WSJ* never uses photographs) appeared on the front page of the *Wall Street Journal* on January 23, 1990. Lanier's steamy hyperbole on the subject of virtual reality is legendary in the virtual communities, and if his entry into the world of international finance did not precisely signal the arrival of the young virtuality industry, at least it indicated that industry's vigor and sent a message that it should be taken seriously. In 1991, VPL was acquired by Thomson CSF. Thomson, which VPL modestly described in its investors' brochure as a "French electronics firm," is the largest defense contractor in Europe and was a major embarrassment to the largely Greenpeace-oriented VPI. Thomson's subsequent gutting of VPI in 1992 is by now well known.

Currently there is only one other player in the high-stakes field of virtual games entrepreneurship. That is John Waldern of W Industries, in Leicester, England. The informing philosophy behind W's games is the essential attraction of exotic total-immersion visual environments coupled with the proven thrill of bang-bang-shoot-'em-up action. While this use of the technology holds all the thrill for me of chopping up Abel Gance's *Napoleon* to insert commercials for television viewing, there is no question that arcade games will represent a significant drive behind technological innovation in virtual-worlds equipment for the public sector. In addition, there also seems no question that a significant proportion of young people will spend a significant and increasing proportion of their waking hours playing computer-based games in one form or another, and so far the implications of this trend have yet to be fully addressed in academic forums. A major obstacle appears to be the feeling on the part of many academics that computer games are beneath serious notice, a situation perhaps best characterized as holding our cocktail party in a house that is already ablaze. Within a short time, the number of hours that a broad segment of children will spend playing computer-based games will exceed the number of hours that they spend watching television. It is entirely possible that computer-based games will turn out to be the major acknowledged source of socialization *and* education in industrialized societies before the 1990s have run their course.[14]

While the current generation of multiple-player games would seem to have no particular redeeming virtue, their designers are among the fiercest of the techno entrepreneurs. In addition, of all the possible commercial uses bruited about for virtual-worlds equipment, multiple-player games are the only commercial application that is currently returning a profit. Clearly they speak to some deep desires on the part of a significant number of consumers. Thus there will inevitably be more of them.

It is impossible to study the emergence of virtual systems without acknowledging the overwhelming influence exerted upon the entire field of virtual technologies research by the military. The earliest large-scale virtual environments were built for military purposes by engineers who were working for military organizations: Ivan Sutherland, who began his research in three-dimensional displays in 1966, became director of the Information Processing Techniques Office at the Defense Advanced Research Projects Agency (DARPA); Tom Furness, who started the "Supercockpit" project for the U.S. Air Force in 1965, became chief of the Visual Display System Branch, Human Engineering Division of the Armstrong Aerospace Research Laboratory at Wright-Patterson Air Force Base; Scott Fisher and Mike McGreevy did a good deal of their work at NASA Ames in the late 1980s.

Sooner or later we can trace any funding back to government sources; if we include MIT, which has always been heavily funded by military budgets, there is almost no one working in the field today whose original research is not or was not funded by military money. Still, some of the early and influential researchers were not directly involved with government funding. Myron Krueger is more of an artist than scientist; Jaron Lanier worked in the video game industry and wanted to become a composer; Fred Brooks is head of the computer science department at the University of North Carolina at Chapel Hill; Brenda Laurel took her M.F.A. at Ohio State in acting and directing, and later earned a Ph.D. in theater criticism. For better or worse, my focus in this essay is not on the military or even particularly on government; any number of my colleagues have done a far better and much more thorough job of studying military and government involvement in research and technological development than I could possibly do. In particularly, Bruce Sterling's account of the U.S. military's networked environmental battle simulation SIMNET is eminently worth reading (although it may be hard to find, since it appeared in the now-scarce first issue of *Wired* magazine).

Several critical events in the development of virtual systems theory do not seem to me to fit comfortably into a narrative description of the emergence of the technology of virtual systems. The first documented account of a virtual cross-dresser, for example—a person who caused considerable consternation and some misery among the women in the virtual community that he frequented—requires a different sort of description from that accorded the first tree-structured bulletin board. It is difficult to look back over such a brief period to 1985 and realize how naive most inhabitants of the virtual communities were at the time "Julie Graham" was giving her sensitive and helpful, albeit unquestionably deceptive, advice to unhappy women on the nets. From the first documented instance to 1992, when at any given time 15 percent of the population of *Habitat* (or, by one estimate, 150,000 people) was engaged in cross-gendered behavior, represents a span of seven years.

The trial of a man who was accused of raping a woman with multiple personalities by seducing one of her personae is important to this work. The trial became something of a spectacle, recalling in new surroundings the old power to fascinate that still inheres in freaks and monsters, the power of near-legibility; it raised issues of multiplicity and continuity, of what constitutes a single identity in social and legal terms. The trial itself and the media circus that surrounded it were similar in useful ways to the landmark suit brought by the Mashpee Wampanoag Tribal Council, Inc., against the town of Mashpee, on Cape Code, which James Clifford reported in his marvelous paper "Identity in Mashpee."

In his study Clifford identified three assumptions that he felt compromised the Mashpees' case against the government. He described these as (1) the idea of cultural wholeness and structure, (2) the hierarchical distinction between oral and literate forms of knowledge, and (3) the narrative continuity of history and identity. I found that Clifford's three points constituted a provocative background for the trial then under way in Oshkosh, Wisconsin, even in light of the fact that it was just such issues that the attorneys and judge in the Oshkosh case hoped would *not* intrude into an already complex debate.

I use the Oshkosh trial here to constitute a parallel narrative thread to my account of events in Silicon Valley in the 1970s and 1980s, and as a specimen of one kind of public response to a visible transgression of cultural norms of unitary subjectivity. That is, it is not simply the spectacle of difference in operation, but a voice from the shadows—a reflection of deeper conflicts and negotiations regarding the physical and conscious expression of a drive for closure, which is one of Western culture's most important ways of making meaning. Joseph Campbell refers to the social emergence of the drive for culture as the *yoke of individuality.* There is nothing ontological about this idea. What I call the drive for closure is itself an emergent manifestation of the interactions of complex events and forces. At this point I become daunted by complexity. Maybe I'm talking about what Foucault calls power, and maybe not; whatever it is, it is damnably hard to describe. Samuel Delany once described power as something like a thin mist in a valley, which is all but invisible when one is in it, and which only becomes visible when one can look back on it from a height or a distance.[15] Delany said it in the context of a work of fiction, which is a mode of representation with which I have great sympathy; and how to say just that in an academic context is for me a serious preoccupation.

Development of self-awareness takes place in a field that is already contoured by that invisible and impalpable structure called power. And while there is still plenty of mystery about how the self manages to emerge under these circumstances, there is an even deeper mystery about how self and power mutually constitute each other. A particle physics approach to psychology: We see selves, and we see power, and we talk about how they exchange forces in terms of discrete quantities or thoroughly muzzy qualities; but at bottom there are still problems in representing this exchange in satisfying terms. Describing the sort of entities that could move easily through such a field and produce satisfying accounts is a concrete example of Haraway's Three Aspects of Representation, as discussed *passim* in her preface to *Coyote's Sisters* (Tokyo: Routledge, 1999).[16] For those who have not yet received their copies of *Coyote's Sisters*, of course the three aspects are these: refuse closure; insist upon situation; and seek multiplicity.

I hope to reconsider the disparate accounts I have presented and attempt to point out their correspondences and divergences *without* trying to produce an overarching theoretical framework that appears to encompass them all. In so doing I am performing the activity for which I was trained and which embroiled me in such mischegas in the Science Studies Program at San Diego: attempting to hold these various discourses in productive tension without allowing them to collapse into a univocal account. My game is for the reader—that's you—to perform your own synthesis, if synthesis is your game.

I treat the strategy I have just outlined as a challenge—the challenge of how to best convey information to an imagined "reader." This could come under a broader heading along with a constellation of issues generally identified as part of what is sometimes referred to as a crisis of representation in the social sciences. How best to convey a complex description of a culture whose chief activity is complex description? Here it is useful to keep in mind that while I am attempting to describe cultures of sorts, even though they do not fit many of the customary definitions of cultures,[17] it is necessary to embed other information that is equally important and that becomes something else if it is extracted from the context of cultural quasi-description.

The choice among the three or perhaps four representational methods that would best serve the purposes here would be quite difficult. If I had to narrow the field to two, the choice would be between the method that Ursula Le Guin employed with such excellent results in *Always Coming Home,* and I hope that her success with it will rub off on me in the future. The first part of the book would consist of parallel narrative threads interspersed with descriptions of artifacts, information on cultural byways, "tribal" songs and poetry, and some interpretation of "tribal" philosophy and epistemology by members of the "tribe." I would have left it at that but for what I feel are certain academic constraints, and consequently the second section would consist of my own theoretical interpretations of what is going on. The other possible choice, as I indicated earlier, would be a cross between Traweek's *Beamtimes and Lifetimes* and Tolstoy's *War and*

Peace. Those who do not have much truck with anthropology of emergent or perhaps phantasmic cultures may find their purpose better served by ignoring these remarks.

Finally I do something that for lack of a better description might be called implications and consequences. Where does this stuff lead? What kind of world are these folks bringing into being? Or, perhaps to take up the questions raised by Deleuze, Guattari, and De Landa, what kind of system is using them to become realized? To address this question I frame my last few words in the context of a discussion with my daughter, who is eleven years old as of this writing and who is still trying to figure out just what it is that I do. And at the very end, my favorite person puts in an appearance—Anne Rice's fictional antihero the Vampire Lestat, who has mysteriously acquired a Ph.D. in anthropology. In the context of a work on cultural theory, Lestat may have a few pithy things to say about vision between the worlds.

NOTES

1. This is the first instance of the collapse of fiction and fact (whatever that is), narratization and description, that the style of this essay implies. Being a novelist at heart causes me to create krasis narratives, and it's hard to know to what depth I need to explicate them. Consequently let's try treating these two descriptive paragraphs as exemplary, to an extent that I will not carry on into the rest of the text. Both events are emblematic rather than specific. The bedroom is a combination of bedrooms from at least four different locales, including the young Kal-el's room on the farm in *Superman I.* There was no friend present; for reasons of my own I wanted to decenter the moment of discovery. Peering over a 24 × 24 console would have been impossible in 1955, since the state of the art was three-track recording, and at any rate I was too tall to have my nose at that level; the scene combines visits to various control rooms beginning in the late 1940s and extending into the 1970s, and when I was a preteen to pilgrimages to the transmitter of WOR itself, which, with its black bakelite monoliths and glowing 1920s-style power meters, resembled most closely some science-fiction author's depiction of mighty forces coiled to spring. There was little doubt in my mind that much of the early cinematic depictions of technology as mysterious and inexplicable power, as exemplified in *Metropolis,* came from the movie set designers' own visits to the few instantiations of technology at work that yet existed—that is, the control rooms of the few radio stations that maintained remote transmitter sites. These were invariably located far from populated areas, usually in swamps, thus specifically invoking the motif of lonely isolation in spooky circumstances.

2. All of this, of course, is about the interplay among communications technology, prosthetic community, the human body, and the uses of pleasure.

3. For some reason this sort of thing—having someone barge into my humdrum life and drag me off on some adventure—keeps happening, and I have gotten more good story material in such fashion than I like to admit.

4. This is perhaps the most egregious point of convergence of the two theses I have been pursuing. A more detailed description and analysis of the oddly interdependent issues of lesbian separatism and transgender can be found in my "other" book, *The Gaze of the Vampire: Tales from the Edges of Identity.*

5. My use of the world *talk* to refer to writing and reading is both playful and a considered position. Part of the work of this essay is to play in the boundaries between speech and writing as I discuss the play I observe in electronically prostheticized human interac-

tion. A typical example is a letter waiting in my e-mail box that begins, "Good to read from you."

6. Lippman had been developing these ideas in discussions at the MIT Media Lab over a period of time beginning in the early 1980s, but they were perhaps best captured by Stewart Brand in his recounting of talks with Lippman in *The Media Lab: Inventing the Future at MIT* (1987).

7. The invention of *Space War* is variously dated. Laurel (1991) puts it at 1962.

8. In particular I am referring to Deleuze and Guattari's discussions of deterritorialization and multiplicity in *Anti-Oedipus* (1983) and *A Thousand Plateaus* (1987), Virilio in *The Aesthetics of Disappearance* (1991a) and *The Light of Speed* (1991b), and Manuel De Landa in *War in the Age of Intelligent Machines* (1991).

9. I refer to Barker as "she" here without really being sure what s/he is. Some colleagues have assured me that Barker is a woman. Others claim that because of the spelling of the first name, Barker must be a man. Certainly this is something that should have been cleared up before publication, but I rather enjoy the confusion and the debates thus precipitated.

10. At the time I was first thrashing this out I had recently read Bruno Latour's *The Pasteurization of France* (1988) in its French incarnation, *Les Microbes: Guerere et paix suivi les irreductions.* As I struggled to regain whatever fluency in French I may have previously possessed, I completely missed Latour's pun on Tolstoy in the title of the book. It must have stuck around in the background.

11. Few of these adjectives, like *bloodthirsty* and *infamous,* are accidental, and I am fully aware of their tendentious character. These stories are experimental, and part of the experiment is to see how much (if any) of what I might term "dangerous" story forms can be recuperated into a different discourse without contaminating it (another deliberate word choice). The events at the Atari Lab are emblematic of what happened when the first generation of graduates—bright, dedicated, and to an extent thrillingly conscious of the liberatory potential of their creations—hit the buzz saw of commodification and then the street.

12. There are wide variations in the estimates of *Habitat*'s actual population, even within Fujitsu. The actual figure is probably lower, but how much lower is uncertain.

13. Morningstar and Farmer maintain that it never occurred to them that the characters might want to have sex; they were too preoccupied with just getting characters to look reasonably right and to be able to walk around. It is also possible that LucasFilm might have frowned on the idea, had it arisen at the time. It is not clear whether Fujitsu would have exercised the same scruples, nor that it would have made much difference vis-à-vis sexual activity within the simulation.

14. My use of the term *computer-based* is already becoming an anachronism, because the meanings of culturally defined objects such as television, telephone, cable, and computers, and the boundaries between them, are already in hot debate and increasing flux. Nicholas Negroponte had already pointed out in the late 1970s that there would soon come a time when there might be more MIPS (a measure of processor speed) in kitchen appliances than in the objects commonly called computers. This development prefigures, in part at least, the cultural redefinition, now under way, of these objects. It is partly driven by economics and partly by the effect of ubiquitous technology (technology so familiar as to be culturally invisible) on engineers' interpretations of the boundaries of their specialties, as well as ubiquitous technology's effect on the cultural paradigm of biological-machine binarism. An exhilarating and problematic time.

15. In Samuel R. Delany, *Tales of Neveryon* (New York: Bantam Books, 1979).

16. I want to emphasize that this sort of thing, that is, quoting from nonexistent work, is meant in a wholly humorous way. However, the case in point had its beginnings in a real and somewhat bizarre event. When I first came to the History of Consciousness program at the University of California at Santa Cruz, I had several dreams in which I was reading scholarly papers written by Haraway. Later, while writing a critique of some aspect of scientific research I absentmindedly quoted from one of the papers I'd dreamed about, confusing it in my fevered brain with a "real" Haraway paper. Some time later, while trying to attribute

page and line, I realized that the paper I'd cited was in physical terms nonexistent. The quote itself, however, whatever its true source in my memory, was exactly apt and productively useful. I mentioned this to Haraway, who jokingly requested coauthorship if I ever wrote out the papers I'd dreamed I read. In retrospect this was a rather novel instantiation of the mentoring relationship, neatly avoiding problems of age and experience, while evincing several of the advantages of virtuality as well as a considerable amount of tribality. Virtual mentoring, or shamanistic training? In regard to the quote, Haraway actually said, "Refuse closure and insist upon situation"; I have interpolated "Seek multiplicity."

17. I left this sentence in because it's fun. When I first wrote it, I had a lot of convincing to do, and a lot of forums to legitimize. Now academic panels on "virtual community" pop up like mushrooms, but times have changed so fast that the sentence makes me chuckle.

Consuming
Technoculture

"Cyberflesh girlmonster is even cannier. Now the user can click on witty little monsters and inviting lips that whisper: press here, press here, touch me, touch me."
–Sadie Plant, Zeros and Ones

"As we have seen, the information revolution that has today superseded the revolution in industrial manufacturing is not without danger, for the damage done by progress in interactivity may well be as harmful in the future as that done by radioactivity. . . . Unless social disintegration has already entered an irreversible phase, with the decline in the nuclear family and the boom in the population unit of the single parent."
–Paul Virilio, on the decline of family values, Open Sky

"Industrialism and privacy go hand in hand. Traditional mass-production companies had little reason to learn much about individual customers because they couldn't do anything with the knowledge anyway. Mass-customization creates a new dilemma: fear of companies knowing too much. The problem is that the act of gathering information about customers makes some of those customers nervous indeed. The (not unreasonable) worry is information gathered ostensibly to serve you better might someday be used against you."
–John Browning and Spencer Reiss

"Virtual cross-dressing is not as simple. Not only can it be technically challenging, it can be psychologically complicated. Playing in MUDs, whether as a man, a woman, or a neuter character, I quickly fell into the habit of orienting myself to new cyberspace acquaintances by checking out their gender. . . . On many MUDs, offering technical assistance has become a common way in which male characters 'purchase' female attention, analogous to picking up the check at an RL dinner."
–Sherry Turkle, Life on the Screen

Ien Ang

New Technologies, Audience Measurement, and the Tactics of Television Consumption

The Problem of the Audience

In February 1990, Walt Disney Studios decided to prohibit cinema theaters in the United States from airing commercials before screening Disney-produced movies. The decision was made because the company had received a great number of complaints from spectators who did not want to be bothered by advertising after having paid $7.50 for seeing a film, leading the company to conclude that commercials "are an unwelcome intrusion" into the filmgoing experience (Hammer 1990, p. 38). Of course, Disney's decision was informed by economic motives: it feared that commercials before films would have a negative effect on the number of people willing to go to the movies, and thus on its box-office revenues. As a result, the issue of in-theater advertising is now a controversial one in Hollywood.

This case clarifies a major contradiction in the institutional arrangement of the cultural industries. More precisely, the conflicting corporate interests represented by two types of consumption are at stake here: a conflict between media consumption, on the one hand, which is the profit base for media companies such as Disney, and the consumption of material goods, on the other, presumably to be enhanced by the showing of commercials. In this case, the conflict inheres in the very logic of cinema spectatorship as a consumer activity, both economic and cultural. Films are discrete media products, to be watched one at a time by consumers who pay a fixed entrance fee in advance in order to be able to see the film of their choice. In this exchange, commercials are not included in the bargain. On the contrary, it is suggestive of the controversial social meaning of advertising that commercials are seen to hurt rather than enrich the value of cinemagoing. In the cinema, the consumption of the film is to be

From *Living Room Wars* (New York: Routledge, 1996). Reprinted by permission of the author and Routledge.

clearly marked off from the selling of goods and services through advertising, both in the experience of the film consumer and in the economic logic of the industry.

The situation is altogether different with television. The very corporate foundation of commercial television rests on the idea of 'delivering audiences to advertisers'; that is, economically speaking, television programming is first and foremost a vehicle to attract audiences for the 'real' messages transmitted by television: the advertising spots inserted within and between the programs (e.g., Smythe 1981). The television business, in other words, is basically a 'consumer delivery enterprise' for advertisers. So, in the context of this structural interdependence of television broadcasters and advertisers, television consumption takes on a double meaning: it is consumption both of programs and of commercials; the two presuppose one another—at least, from the industry's point of view. Once a consumer has bought a television set, she or he has bought access to all broadcast television output, and in exchange for this wholesale bargain she or he is expected to expose herself or himself to as much output as possible, including most importantly the commercials which in fact make the financing of the programs possible. This merging of the two types of consumption is corroborated in the occurrence of one single activity, a presumably one-dimensional type of behavior: 'watching television.' This complex intermingling of economic conditions and cultural assumptions with regard to television consumption is a necessary precondition for the construction of an institutional agreement about the exchange value of the 'audience commodity' that is bought and sold. As is well known, this agreement is reached through the intermediary practice of audience measurement, producing ratings figures on the basis of the amount of 'watching television' done by the audience. These figures are considered to be the equivalent to box-office figures for cinema attendance (see, e.g., Meehan 1984; Ang 1991).

But this equivalence is fundamentally problematic, as I will try to show in this chapter. Undertaken by large research companies such as Nielsen and Arbitron in the United States and AGB in Britain and continental Europe, audience measurement is an entrenched research practice based upon the assumption that it is possible to determine the objective size of the 'television audience.' However, recent changes in the structure of television provision, as a result of the introduction of new television technologies such as cable, satellite, and the VCR, have thrown this assumption of measurability of the television audience into severe crisis. The problem is both structural and cultural: it is related to the fact that 'watching television' is generally a *domestic* consumer practice, and as such not at all the one-dimensional, and therefore measurable, type of behavior it is presumed to be.

The domestic has always been a contested terrain when it comes to the regulation of consumption. It is a terrain which, precisely because it is officially related to the 'private sphere,' is difficult to control from outside. Of course, it is true, as the young Jean Baudrillard once stated, that "[c]onsumption is not . . . an indeterminate marginal sector where an individual, elsewhere constrained by social rules, would finally recover, in the 'private' sphere, a margin of freedom and personal play when left on his [*sic*] own" (1988 [1970], p. 49). The development of the consumer society has implied the hypothetical construction of an ideal consuming subject through a whole range of strategic and ideological practices, resulting in very specific constraints, structural and cultural, within which people can indulge in the pleasures of leisurely consumption.

Indeed, it is important to note that the day-to-day, domestic practice of television consumption is accompanied by the implicit and explicit promotion of 'ideal' or 'proper' forms of consumer behavior, propelled by either ideological or economic motives and instigated by the social institutions responsible for television production and transmission.[1] More generally, the acceptance and integration of television within the domestic sphere did and does not take place 'spontaneously,' but was and is surrounded by continuous discursive practices which attempt to 'normalize' television-viewing habits.

For example, Lynn Spigel (1988) has shown how American women's magazines in the late 1940s and early 1950s responded to the introduction of television in the home with much ambivalence and hesitation, against the background of the necessity for housewives to integrate household chores with the attractions (and distractions) promised by the new domestic consumer technology. Through the advice and suggestions put forward in these magazines, they helped establish specific cultural rules for ways in which 'watching television' could be managed and regulated without disturbing the routines and requirements of family life.

However, precisely because the home has been designated as the primary location for television consumption, a 'right' way of watching television is very difficult to impose. As Roger Silverstone has put it," [t]he status of television as technology and as the transmitter of meaning is . . . vulnerable to the exigencies, the social structuring, the conflicts and the rituals of domestic daily life" (1990, p. 179). The domestic is a preeminent site of everyday life and the everyday is, according to Michel de Certeau, the terrain in which ordinary people often make use of infinite local tactics to "constantly manipulate events in order to turn them into 'opportunities'" (1984, p. xix). 'Watching television' can be seen as one everyday practice that is often tactical in character, articulated in the countless unpredictable and unruly ways of using television that elude and escape the strategies of the television industry to make people watch television in the 'right' way. And as we shall see, the home environment only reinforces the proliferation of such tactics in the age of new television technologies.

However, the fact that television consumption has been historically constructed as taking place within the private, domestic context has paradoxically also been quite *convenient* for the television industry.

Precisely because the activities of 'watching television' usually take place in sites unseen, behind the closed doors of private homes, the industry could luxuriate in a kind of calculated ignorance about the tactics by which consumers at home constantly subvert predetermined and imposed conceptions of 'watching television.'

Again, the cinema provides a suitable comparison. Because the cinema audience is gathered together in a public theater, spectators' reactions to the screen are immediately available and therefore not easily ignored. For example, Disney's decision to ban commercials in theaters was, at least in part, a response to observations that audiences had booed and hissed a Diet Coke commercial in which Elton John and Paula Abdul sing the soft drink's praises (Hammer 1990). Similar audience resistance in front of the television screen at home, however, remains largely invisible to the outsider. At the same time, it seems fair to suspect that television viewers are in a far better position to avoid messages they do not want to be subjected to than cinema spectators, who are trapped in their chairs in the darkened theater, enforced to keep their gaze directed to the large screen. After all, television viewers have the freedom to move around in their own homes when their television set is on; there is no obligation to keep looking and they can always divert their attention to something else whenever they want to. But it is precisely this relative freedom of television audiences to use television in ways they choose to which has been conveniently repressed in the industry's imaginings of its consumers.

This repression is reflected in the rather simplistic methods of information gathering used by ratings producers to measure the size of the television audience (or segments of it). Historically, two major audience measurement technologies have dominated the field: the diary and the set-meter. In the diary method, a sample of households is selected whose members are requested to keep a (generally, weekly) diary of their viewing behavior. At the end of the week the diaries must be mailed to the ratings firm. In the second case, an electronic meter is attached to the television sets of a sample of households. The meter gives a minute-by-minute automatic registration of the times that the television set is on or off, and of the channel it is turned on to. The data are transmitted to a home storage unit, where they are stored until they are accessed by the central office computer during the night. The meter data, which only indicate numbers of sets on, form the basis for what are called 'gross ratings,' while the diary data, which are more cumbersome to produce because they presuppose the active cooperation and discipline of viewers of sample homes in filling out their individual diaries, are used to compose demographic information about audiences for specific programs.[2]

It should be noted that these methods of measurement are grounded upon a straightforward behaviorist epistemology. 'Watching television' is implicitly defined as a simple, one-dimensional, and purely objective and isolatable act. As Todd Gitlin has rightly remarked in relation to the electronic setmeter, "The numbers only sample sets tuned in, not necessarily shows watched, let alone grasped, remembered, loved, learned from, deeply anticipated, or mildly tolerated" (1983, p. 54). In other words, what audience measurement information erases from its field of discernment is any specific consideration of the 'lived reality' behind the ratings. In the quantitative discourse of audience measurement television viewers are merely relevant for their bodies: strictly speaking, they appear in the logic of ratings only in so far as they are agents of the physical act of tuning-in. More generally, the statistical perspective of audience measurement inevitably leads to emphasizing averages, regularities, and generalizable patterns rather than particularities, idiosyncrasies, and surprising exceptions. What all this amounts to is the construction of a kind of streamlined map of the 'television audience,' on which individual viewers are readable in terms of their resemblance to a 'typical' consumer whose 'viewing behavior' can be objectively and unambiguously classified. In other words, in foregrounding the stable over the erratic, the likely over the fickle, and the consistent over the inconsistent, ratings discourse symbolically turns television consumption into a presumably well-organized, disciplined practice, consisting of dependable viewing habits and routines.

Imagining television consumption in this way is very handy for the industry indeed: it supplies both broadcasters and advertisers with neatly arranged and easily manageable information, which provides the agreed-upon basis for their economic negotiations. The *tactical* nature of television consumption is successfully disavowed, permitting the industry to build its operations upon an unproblematic notion of what 'watching television' is all about. This, at least, characterized the relatively felicitous conditions of existence for (American) commercial television for decades.

Technology and Measurement

Since the mid-1970s, however, an entirely different television landscape has unfolded before the viewer's eyes, one characterized by abundance rather than scarcity, as a result of the emergence of a great number of independent stations, cable, and satellite channels. This, at least, is the situation in the United States, but it also increasingly characterizes European television provisions. By 1987, 49 percent of American homes had been connected to a basic cable system, giving them access to cable channels such as MTV, ESPN, and CNN, while 27 percent had chosen to subscribe to one or more pay cable channels, such as Home Box Office. All in all, thirty or more channels can be received in 20 percent of American homes. Furthermore, after a slow start the number of homes with VCRs had grown exponentially in the early 1980s, reaching about 50 percent in 1987 (*TV World* 1987). This multiplication of consumer options has inevitably led to a fragmentation of television's audiences, which in turn has led to a perceived inadequacy of the figures provided by the existing ratings services. What's happening in the millions of living rooms now that people can choose from so many different offerings? Consequently, diverse branches of the industry began to call for more finely tuned audience information, to be acquired through better, that is, more accurate, measurement.

This call for better measurement was articulated by criticizing the prevailing techniques and methods of measuring the television audience: the diary and the setmeter. For example, the proliferation of channels has acutely dramatized the problems inherent in the diary technique. Suddenly, the built-in subjective (and thus 'unreliable') element of the diary technique was perceived as an unacceptable deficiency. David Poltrack, vice president of research for CBS, one of the three major U.S. networks, voiced the problem as follows:

> It used to be easy. You watched M*A*S*H on Monday night and you'd put that in the diary. Now, if you have thirty channels on cable you watch one channel, switch to a movie, watch a little MTV, then another program, and the next morning with all that switching all over the place you can't remember what you watched. (Quoted in Bedell Smith 1985, p. H23)

And officials of the pop music channel MTV complained that their target audience, young people between twelve and twenty-four, consistently comes off badly in the demographic data produced through diaries, because "younger viewers tend not to be as diligent in filling out diaries as older household members" (quoted in Livingston 1986, p. 130). In short, agreement grew within the industry that the possibilities of 'channel switching' and 'zapping' (swiftly 'grazing' through different channels by using the remote control device) had made the diary an obsolete measurement tool. Viewers could no longer be trusted to report their viewing accurately: they lack perfect memory, they may be too careless. In short, they behave in too capricious a manner! In this situation, calls for a 'better' method to obtain ratings data began to be raised; and better means more 'objective,' that is, less dependent on the 'fallibilities' of viewers in the sample. A method that erases all traces of wild subjectivity.

The videocassette recorder has also played a major destabilizing role in the measurability of the television audience. 'Time shifting' and 'zipping' (fastforwarding commercials when playing back a taped program) threatened to deregulate the carefully composed television schedules of the networks. This phenomenon has come to be called 'schedule cannibalization' (cf. Rosenthal 1987), a voracious metaphor that furtively indicates the apprehension, if not implicit regret, felt in network circles about the new freedoms viewers have acquired through the VCR. Through the VCR, the tactical nature of television consumption clearly begins to manifest itself. In response, the industry demanded the measurement of the VCR audience: it wanted answers to questions such as: how often is the VCR used by which segments of the audience? Which programs are recorded most? And when are they played back?

In the face of this growing demand for more accurate and more detailed information about television consumption, the ratings business has now come up with the 'people meter,' a new audience measurement technology that was introduced in the United States in 1987.[3] The people meter is supposed to combine the virtues of the traditional setmeter and the paper-and-pencil diary: it is an electronic monitoring device that can record individual viewing rather than just sets tuned in, as the traditional setmeter does. It works as follows.

When a viewer begins to watch a program, he or she must press a numbered button on a portable keypad, which looks like the well-known remote control device. When the viewer stops watching, the button must be pressed again. A monitor attached to the television set lights up regularly to remind the viewer of the button-pushing task. Every member of a sample family has her or his own individual button, while there are also some extra buttons for guests. Linked to the home by telephone lines, the system's central computer correlates each viewer's number with demographic data about them stored in its memory, such as age, gender, income, ethnicity, and education.

There is definitely something panoptic in the conceptual arrangement of this intricate measurement technology (Foucault 1979), in that it aims to put television viewers under constant scrutiny by securing their permanent visibility. This is attractive for the industry because it holds the promise of providing more detailed and accurate data on exactly when who is watching what. The people meter boosts the hope for better surveillance of the whole spectrum of television-viewing activities, including the use of the VCR. Smaller audience segments may now be detected and described, allowing advertisers and broadcasters to create more precise target groups. New sorts of information are made available; hitherto hidden and unknown minutiae of 'audience behavior' can now be detected through clever forms of number crunching (see, e.g., Beville 1986a and 1986b).

Still, the existing versions of the people meter are by no means considered perfect measurement instruments, as they still involve too much subjectivity: after all, they require viewer cooperation in the form of pushing buttons. A professional observer echoes the widespread feelings of doubt and distrust when he wonders:

> Will the families in the sample really take the trouble? Will they always press the buttons as they begin watching? Will they always remember to press their buttons when they leave the room—as when the telephone rings, or the baby cries? (Baker, 1986, p. 95)

It should come as no surprise, then, that furious attempts are being made to develop a so-called *passive* people meter—one with no buttons at all— that sense automatically who and how many viewers are in front of the screen. For example, Nielsen, the largest ratings company in the United States, has recently disclosed a plan for a rather sophisticated passive people meter system, consisting of an image-recognition technology capable of identifying the faces of those in the room. The system then decides first whether it is a face it recognizes, and then whether the face is directed toward the set (unfamiliar faces and even possibly the dog in the house will be recorded as 'visitors'). If tested successfully, this system could eventually replace the imperfect, push-button people meter, so Nielsen executives expect (*San Francisco Chronicle* 1989). In short, what seems to be desired within the television industry these days is a measurement technology that can wipe out all ambiguity and uncertainty about the precise size of the audience for any program and any commercial at any given time.

This recent utopian drive toward technological innovation in audience measurement can be interpreted as a desperate attempt to repair the broken consensus within the television industry as a whole as to the meaning of 'watching television.' Indeed, from the industry's perspective, a kind of 'revolt of the viewer' seems to have erupted with the emergence of the new television technologies: 'watching television' now appears to be a rather undisciplined and chaotic set of behavioral acts as viewers zip through commercials when playing back their taped shows on their VCRs, zap through channels with their remote controls, record programs so as to watch them at times to suit them, and so on. "After years of submitting passively to the tyranny of [network] television programmers, viewers are taking charge," comments American journalist Bedell Smith (1985, p. H21). This 'taking charge' can be seen as the return of the tactical nature of television consumption to the realm of visibility, shattering the fiction of 'watching television' as a simple, one-dimensional, and objectively measurable activity which has traditionally formed the basis for industry negotiations and operations.

In other words, what has become increasingly uncertain in the new television landscape is exactly what takes place in the homes of people when they watch television. Reduction of that uncertainty is sought in improvements in audience measurement technology, with its promise of delivering a continuous stream of precise data on who is watching what, every day, all year long. But beneath this pragmatic solution lurks an epistemological paradox.

For one thing, as the macroscopic technological 'gaze' of audience measurement becomes increasingly microscopic, the object it is presumed to measure becomes ever more elusive. The more 'watching television' is put under the investigative scrutiny of new measurements technology, the less unambiguous an activity it becomes. 'Zipping,' 'zapping,' 'time shifting,' and so on, are only the most obvious and most recognized tactical maneuvers viewers engage in in order to construct their own television experience. There are many other ways of doing so, ranging from doing other things while watching to churning out cynical comments on what's on the screen (see, e.g., Sepstrup 1986). As a result, it can no longer be conveniently assumed—as has been the foundational logic and the strategic pragmatics of traditional audience measurement—that having the television set on equals watching, that watching means paying attention to the screen, that watching a program implies watching the commercials inserted in it, that watching the commercials leads to actually buying the products being advertised.

To speak with de Certeau (1984), it is that which happens beneath technology and disturbs its operation which interests us here. The limits of technology are not a matter of lack of sophistication, but a matter of actual practices, of 'the murmuring of everyday practices' that quietly but unavoidably unsettle the functionalist rationality of the technological project. In other words, no matter how sophisticated the measurement technology, television consumption can never be completely 'domesticated' in the classificatory grid of ratings research, because television consumption is, despite its habitual character, dynamic rather than static, experiential rather than merely behavioral. It is a complex practice that is more than just an activity that can be broken down into simple and objectively measurable variables; it is full of casual, unforeseen, and indeterminate moments which inevitably make for the ultimate unmeasurability of *how* television is used in the context of everyday life.

The problem I refer to here has been foreshadowed by a classic study by Robert Bechtel et al. (1972), who in the early 1970s observed a small sample of families in their homes over a five-day period. Ironically, the method these researchers used is very similar to that of the passive people meter. The families were observed by video cameras whose operations, so the researchers state, was made as unobtrusive as possible: "There was no way to tell [for the family members] whether the camera was operating or not. The camera did not click or hum or in any way reveal whether it was functioning" (ibid., p. 277). More important, however, were the insights they gained from these naturalistic observations. Their findings were provocative and even put into question the very possibility of describing and delineating 'watching television' in any simple sense as 'a behavior in its own right': they asserted that their "data point to an inseparable mixture of watching and nonwatching as a general style of viewing behavior," and that "television viewing is a complex and various form of behavior intricately interwoven with many other kinds of behavior" (ibid., pp. 298–99).

Logically, this insight should have led to the far-reaching conclusion that having people fill out diaries or, for that matter, push buttons to demarcate the times that they watch television is in principle nonsensical because there seems to be no such thing as 'watching television' as a separate activity. If it is almost impossible to make an unambiguous distinction between viewers and nonviewers and if, as a consequence, the boundaries of 'television audience' are so blurred, how could it possibly be measured?

This study was certainly ahead of its time, and its radical consequences were left aside within the industry, because they were utterly unbearable in their impracticality.[4] Instead, technological innovations in audience measurement procedures are stubbornly seen as the best hope to get more accurate information about television consumption. Still, in advertising circles, in particular, growing skepticism can be observed as to the adequacy of ratings figures, no matter how detailed and accurate, as indicators for the reach and effectiveness of their commercial messages. For example, there is a growing interest in information about the relationship between television viewing and the purchase of products being advertised in commercials. After all, this is the bottom line of what advertisers care about: whether the audiences delivered to them are also 'productive' audiences (i.e., whether they are 'good' consumers). Thus, in more avant-garde commercial research circles the search for ever more precise demographic categories, such as the people meter provides, has already been losing its credibility. As one researcher put it:

> In many cases, lumping all 18–49 women together is ludicrous. . . . Narrow the age spread down and it still can be ludicrous. Take a 32½ year-old woman. She could be white or black, single or married, working or unemployed, professional or blue collar. And there's lots more. Is she a frequent flier? Does she use a lot of cosmetics? Cook a lot? Own a car? Then there's the bottom line. Do commercials get to her? These are the items the advertiser really needs to know, and demographic tonnage is not the answer. (Davis 1986, p. 51)

The kind of research that attempts to answer these questions, currently only in an experimental stage, is known as 'single-source' measurement: the same sample of households is subjected to measurement not only of its television-viewing behavior but also of its product-purchasing behavior (see, e.g., Gold 1988). Arbitron's ScanAmerica, for example, is such a system. In addition to measuring television viewing (using a push-button people meter device), it supplies sample members with another technological gadget: after a trip to the supermarket, household members (usually the housewife, of course) must remove a pencil-size electronic 'wand' attached to the meter and wave it above the universal product code that is stamped on most packaged goods. When the scanning wand is replaced in the meter, the central computer subsequently matches that information with the family's recent viewing patterns, thus producing data presumably revealing the effectiveness of commercials (Beville 1986b; *Broadcasting* 1988). Needless to say, this system is technically 'flawed' because it necessitates even more active cooperation than just button-pushing. But the tremendous excitement about the prospect of having such single-source, multivariable information, which is typically celebrated by researchers as an opportunity of "recapturing . . . intimacy with the consumer" (Gold 1988, p. 24) or getting in touch with 'real persons' (Davis 1986, p. 51), indicates the increasing discontent with ordinary ratings statistics alone as signifiers for the value of the audience commodity.

Similarly, one British advertising agency, Howell, Henry, Chalde-cott and Lury (HHCL), has recently caused outrage in more orthodox circles of the advertising industry by launching a strong attack on the common practice of selling and buying advertising time on the basis of people meter ratings statistics. In an advertisement in the *Financial Times* it showed a man and a woman making love in front of a television set while stating: "Current advertising research says these people are watching your ad. Who's really getting screwed?" (see Kelsey 1990).[5] HHCL's alternative of getting to know the 'real consumer,' however, is not the high-tech method of computerized single-source research, but more small-scale, qualitative, in-depth, focus group interviews with potential consumers of the goods to be advertised.

What we see in this foregrounding of qualitative methods of em-pirical research is a cautious acknowledgment that television consumption practices, performed as they are by specific individuals and groups in par-ticular social contexts, are not therefore generalizable in terms of isolated instances of behavior. If anything, this marks a tendency toward a recogni-tion of what could broadly be termed the 'ethnographic' in the industry's attempts to get to know consumers. This ethnographic move is in line with a wider recent trend in the advertising research community in the United States and elsewhere to hire cultural anthropologists to conduct 'observational research' into the minutiae of consumer behavior that are difficult to unearth through standard surveys (Groen 1990)—an interesting and perhaps thought-provoking development in the light of the growing popularity of ethnography among critical cultural researchers!

Conclusion

What are we to make of these developments? To round off this chapter, then, some concluding remarks. First of all, it is important to emphasize that a research practice such as audience measurement is constrained by strict institutional pressures and limits. We are dealing here with an in-dustry with vested interests of its own. Market research firms are for eco-nomic reasons bound to respond to changes in demand for types of research on the part of media and advertisers. Furthermore, it is important to stress the *strategic*, not analytic, role played by research in the organization and operations of the cultural industries. Research is supposed to deliver in-formational products that can serve as a shared symbolic foundation for industry negotiations and transactions, and epistemological considera-tions are by definition subservient to this necessity. Thus, innovations in audience measurement should be understood in this context: in the end, market-driven research will always have to aim at constructing a 'regime of truth' (Foucault 1979) that enables the industry to improve its strategies to attract, reach, and seduce the consumer. In this respect, recognition of some of the tactics by which viewers appropriate television in ways unintended and undesired by programmers and advertisers may under some circumstances be beneficial, even inevitable, as I have shown above. But the interests of the industry cannot and do not permit a complete acceptance of the tactical nature of television consumption. On the contrary, consumer tactics can be recognized only in so far as they can be incorporated in the strategic calculations of media and advertisers. In other words, despite its increasing attention to (ethnographically oriented) detail, market research must always stop short of acknowledging fully the permanent subversion inherent in the minuscule but intractable ways in which people resist being reduced to the imposed and presumed images of the 'ideal consumer.'

If we take full account of the inherently tactical nature of television consumption, however, we must come to the conclusion that any attempt to construct positive knowledge about the 'real consumer' will always be provisional, partial, fictional. This is not to postulate the total freedom of television viewers. Far from it. It is, however, to foreground and dramatize the continuing dialectic between the technologized strategies of the industry and the fleeting and dispersed tactics by which consumers, while confined by the range of offerings provided by the industry, surreptitiously seize moments to transform these offerings into 'opportunities' of their own, making 'watching television,' embedded as it is in the context of everyday life, not only into a multiple and heterogeneous cultural practice, but also, more fundamentally, into a mobile, indefinite, and ultimately ambiguous one, which is beyond prediction and measurement. But this idea, which if taken seriously would corroborate the adoption of a fully fledged ethnographic mode of understanding, is epistemologically unbearable for an industry whose very economic operation depends on some fixed and objectified description of the audience commodity. Therefore, it is likely that technological improvement of audience measurement will for the time being continue to be sought, stubbornly guided by the strategically necessary assumption that the elusive tactics of television consumption can in the end be recaptured in some clear-cut and hard measure of 'television audience,' if only the perfect measurement instrument could be found.[6]

De Certeau speaks of a 'strange chiasm':

> [T]heory moves in the direction of the indeterminate, while technology moves towards functionalist distinction and in that way transforms everything and transforms itself as well. As if the one sets out lucidly on the twisting paths of the *aleatory* and the metaphoric, while the other tries desperately to suppose that the utilitarian and *functionalist* law of its own mechanism is 'natural.' (de Certeau 1984, p. 199)

Meanwhile, American film producers worry that, as advertising in cinema theaters proliferates, more would-be moviegoers will stay at home and watch the film on video. Advertisers, unrelentingly in search of new ways to reach their potential consumers, have not been too keen on putting their commercials on videotapes, reportedly because they distrust one element missing from cinemas: the fast-forward button (Hammer 1990).

NOTES

1. This statement should not be interpreted monolithically. Discourses of audience produced by and within television institutions are certainly neither homogeneous nor without contradiction. In Ang (1991) I discuss in detail the differing assumptions about the 'television audience' and its sustaining 'viewing behavior' as operative in the historical practices of American commercial television and European public service television, respectively.

2. For a comprehensive overview of the audience measurement industry in the United States, see Beville Jr. (1985).

3. I discuss extensively the introduction of the people meter in the American television industry in part II of Ang (1991). The people meter is currently seen as the standard technology for television audience measurement in countries with developed (i.e., commercial, multichannel) television systems, including most West European countries and Australia.

4. The study was part of the huge Report to the Surgeon General's Scientific Advisory Committee on Television and Social Behavior, which was commissioned to establish facts about the effects of television violence. Even in that context, however, Bechtel et al.'s (1972) project was marginalized. As Willard Rowland has noted:

> [A]s provocative as this research was, its design violated so many of the normal science requirements for acceptable survey research that it had little impact on the major directions taken by the overall advisory committee program. Indeed this study was permitted only as a way of testing the validity of survey questionnaires. The somewhat radical theoretical implications of its findings were largely overlooked at all levels of review in the project. (Rowland 1983, p. 155)

5. The evidence dug up by HHCL was already substantiated in an earlier, Independent Broadcast Authority-funded research by Peter Collett and Roger Lamb (1986), similar in setup to Bechtel et al.'s (1972) study, in which they confirmed the widespread occurrence of 'inattentive viewing' in the home.

6. In this respect, it is interesting to note that the American national television networks, faced with declining viewing figures for their programs, are now insisting on the incorporation of television viewing *outside* the home (in bars, college dormitories, hotels, and so on) in Nielsen's audience measurement procedures (Huff 1990).

REFERENCES

Ang, I. (1991). *Desperately Seeking the Audience.* London and New York: Routledge.

Baker, W. F. (1986). "Viewpoints." *Television/Radio Age,* November 10.

Baudrillard, J. (1988). "Consumer Society" [1970]. In his *Selected Writings,* ed. M. Poster. Stanford, Calif.: Stanford University Press.

Bechtel, R. B., C. Achelpohl, and R. Akers. (1972). "Correlates between Observed Behavior and Questionnaire Responses on Television Viewing." In E. Rubinstein, G. Comstock, and J. Murray (eds.). *Television and Social Behavior,* vol. 4: *Television in Day-to-Day Life: Patterns of Use.* Washington, D.C.: U.S. Government Printing Office.

Bedell Smith, S. (1985). "Who's Watching TV? It's Getting Hard to Tell." *New York Times,* January 6.

Beville, H. M., Jr. (1985). *Audience Ratings: Radio, Television, Cable.* Hillsdale, N.J.: Lawrence Erlbaum.

Beville, M. (1986a). "People Meter Will Impact All Segments of TV Industry." *Television/Radio Age,* October 27.

———. (1986b). "Industry is Only Dimly Aware of People Meter Differences." *Television/Radio Age,* November 10.

Broadcasting. (1988). "Arbitron to Go With Peoplemeter." June 27.

Collett, P., and R. Lamb. (1986). *Watching People Watching Television.* London: Independent Broadcasting Authority.

Davis, B. (1986). "Single Source Seen as 'New Kid on Block' in TV Audience Data." *Television/Radio Age,* September 29.

de Certeau, M. (1984). *The Practice of Everyday Life.* Trans. S. Randall. Berkeley: University of California Press.

Foucault, M. (1979). *Discipline and Punishment.* Tr. A. Sheridan. Harmondsworth: Penguin.

Gitlin, T. (1983). *Inside Prime Time.* New York: Pantheon.

Gold, L. N. (1988). "The Evolution of Television Advertising-Sales Measurement: Past, Present and Future." *Journal of Advertising Research* 28(3): 18–24.

Groen, J. (1990). "Consument thuis per video begluurd." *De Volkskrant,* January 20.

Hammer, J. (1990). "Advertising in the Dark." *Newsweek,* April 9.

Huff, R. (1990). "New Nielsen Study Boosts Numbers." *Variety,* November 26.

Kelsey, T. (1990). "The Earth Moves for the Ratings Industry." *The Independent on Sunday,* February 18.

Livingston, V. (1986). "Statistical Skirmish: Nielsen Cable Stats Vex Cable Net Execs." *Television/Radio Age,* March 17.

Meehan, E. (1984). "Ratings and the Institutional Approach: A Third Answer to the Commodity Question." *Critical Studies in Mass Communication* 1(2): 216–25.

Rosenthal, E. M. (1987). "VCRs Having More Impact on Network Viewing, Negotiation." *Television/Radio Age,* May 25.

Rowland, W. (1983). *The Politics of TV Violence.* Beverly Hills, Calif.: Sage.

San Francisco Chronicle. (1989). "New 'People Meter' Device Spies on TV Ratings Families." June 1.

Sepstrup, P. (1986). "The Electronic Dilemma of Television Advertising." *European Journal of Communication* 1(4): 383–405.

Silverstone, R. (1990). "Television and Everyday Life: Towards an Anthropology of the Television Audience." In M. Ferguson (ed.), *Public Communication: The New Imperatives.* London: Sage.

Smythe, D. (1981). *Dependency Road.* Norwood, N.J.: Ablex.

Spigel, L. (1988). "Installing the Television Set: Popular Discourses on Television and Domestic Space: 1948–1955." *Camera Obscura,* 16:11–48.

TV World. (1987). "The US is Watching." September.

Cynthia Cockburn

The Circuit of Technology: Gender, Identity, and Power

The sociological understanding of technology has made great strides in the past two decades. One sign of this is the fact that today there is no need to begin an article about technology with a disclaimer of 'technological determinism.' It is taken as given that technology is not a prime mover, that it is socially shaped, a suitable case for treatment by social science. Another claim, however, cogently argued over the same period, has not yet become an accepted proposition in mainstream theories of technology. This is feminists' claim that the social relations of technology are gendered relations, that technology enters into gender identity, and (more difficult for many to accept) that technology itself cannot be fully understood without reference to gender.

Despite agreement that technology is social, there remain unanswered questions that are especially insistent when one is exploring technology from a woman's viewpoint, or more particularly when one is making a feminist analysis. What do we mean by social when we speak of technology? What do we mean by technology when we speak of gender? What is the connection between technology and power? What kind of power are we talking about? In this chapter I will explore, in the process of developing the logic of a particular current research project, the relations of gender and technology and the space between some mainstream theories and feminist theories of technology.

At the outset a brief review of some of the representations of technology-as-social that have found expression in the past decade may be helpful. It is possible to distinguish simpler and more complex visualizations of the link between the notion of 'technology' and that of 'the social.' A relatively simple view is that which has become known as the 'social shaping' approach. Donald MacKenzie and Judy Wajcman and the authors they assembled in *The Social Shaping of Technology* (1985) firmly dismissed the case for technological determinism. They illustrated the

This chapter originally appeared in *Consuming Technologies*, ed. R. Silverstone and E. Hirsch (New York: Routledge, 1992). Reprinted by permission of the author and Routledge.

social shaping of technology by economic interests, "the choices arising from a need to reduce costs and increase revenues," and showed these to differ in different social contexts. "The way a society is organized, and its overall circumstances, affect its typical pattern of costs, and thus the nature of technological change within it" (MacKenzie and Wajcman 1985, p. 17). They also illustrated the shaping of technological choices by the state, particularly through military purchasing policy (and see Noble 1977 on the emergence of modern technology and the rise of corporate capitalism).

A somewhat more complex rendering of the relationship of the concepts 'technology' and 'society' was that of Trevor Pinch and Wieber Bijker who, in a path-breaking study of bicycle design published in 1984, attempted a unification of ideas arising in the study of science and those current in the sociology of technology. They "set out the constitutive questions that such a unified social constructivist approach must address analytically and empirically" (Pinch and Bijker 1984, p. 399). They began by adopting from the sociology of science the 'empirical programme of relativism' (EPOR), an approach to understanding the content of the natural sciences in terms of social construction. EPOR emphasized the fact that scientific findings are open to more than one interpretation ('interpretive flexibility'), that it is social mechanisms that limit this flexibility, and that 'closure' mechanisms are related to the wider social-cultural milieu.

Applying EPOR to an empirical study, Pinch and Bijker evolve an approach they call SCOT ('social construction of technology'). They demonstrate that the development of the bicycle was a process of alternating variation and selection of designs, that closure and stabilization of a model occurred not in accordance with some essential technological logic but through effective rhetoric on the part of some 'relevant social group' (making claims that stick), or by redefinition of the problem.

Also intent on this convergence of the sociologies of science and technology was Bruno Latour: indeed, he adopted the term 'technoscience' to emphasize this closeness. In Latour's work, and that of Michel Callon and John Law, his colleagues at the Ecole des Mines in Paris, we see the construction of facts (science) and machines (technology) represented as a *collective* process. The collectivity is a network of actors, each playing a part in the unfolding of events.

> By themselves, a statement, a piece of machinery, a process are lost. By looking only at them and at their internal properties, you cannot decide if they are true or false, efficient or wasteful, costly or cheap, strong or frail. These characteristics are only gained through incorporation into other statements, processes and pieces of machinery. (Latour 1987, p. 29)

Rather, "the fate of facts and machines is in later users' hands; their qualities are thus a consequence, not a cause, of collective action." Thus, "we are never confronted with science, technology and society, but with a gamut of weaker and stronger *associations*; thus understanding *what* facts and machines are is the same task as understanding *who* the people are" (Latour 1987, p. 259). Nature, 'the facts,' are not a sufficient explanation for the closure of controversies in science or technology.

Curiously, this understanding of technological change is both more and less 'social' than those it challenges. We see that 'black boxes' (closed controversies, settled artifacts) only occur when *people* ally to close them and others forbear (temporarily no doubt) to reopen them. On the other hand, the risk of voluntarism in the simpler social approaches is averted. Material, nonhuman actors are recovered and firmly identified in the process. As John Law summarizes the difference between 'social constructivism' and the actor-network concept, "social constructivism works on the assumption that the social lies *behind* and directs the growth and stabilization of artifacts. Specifically, it assumes that the detection of relatively stable directing *social interests* offers a satisfying explanation for the growth of technology." By contrast, he suggests that other kinds of factors also enter into both the shaping of artifacts and the social structure that results. "The stability and form of artifacts should be seen as a function of the interaction of heterogeneous elements as these are shaped and assimilated into a network" (Law 1987, p. 113). In a study of Portuguese imperial expansion he shows that the actors include not only the monarch, boat-builders, explorers, and natives of distant lands but also boat-building materials, winds, currents, and reefs.

One further useful concept introduced by the 'actor-network' school should be mentioned here, that of 'translation.' It had been conventional to represent the invention of a new technology as followed by a process of 'diffusion' and practical innovation in production or everyday life. Latour makes the point that a project continually changes shape and content as alliances are stitched together to achieve it. To build a' black box,' whether this is a theory or a machine, it is necessary to enroll others so that they believe it, take it up, spread it. The control of the builder is therefore seldom absolute. The new allies shape the idea or the artifact to their own will—they do not so much transmit as *translate* it (Latour 1987, p. 108).

Women, Technology, and the Uses of Theory

A difference between the sexes in relation to 'technology' was a lived reality throughout the industrializing years of the nineteenth century in Britain, as it had been for centuries, perhaps millennia, before. Men were the technologists and technicians of the Industrial Revolution, women were the factory hands that operated the new machines. Nor was this only the preference of employers. In earlier work I have shown for instance that some machines were indeed held by employers to be suitable for operation by women and others not; and also how men in the printing industry re-sented and resisted women seeking to learn their skills and use their tools (Cockburn 1983). There are many such examples. In the twentieth cen-tury the struggle of women to be admitted to the engineering union was not won until 1942, despite two World Wars in which women were used to fill the engineering jobs of men drafted to the military. The advent of microchip technology does not, as some believed it might, break the tech-nical sexual division of labor and give women the knowledge and know-how to design, produce, and control, as well as merely supply parts to or press buttons on, electronic equipment (Cockburn, 1985).

In the new-wave feminism of the 1970s the distance of women from 'men's' technologies was identified as an aspect of women's disadvantage. Among the liberatory projects of the contemporary women's movement have been attempts to help women overcome the barriers to gaining tech-nical knowledge and know-how. Women have organized training workshops in which women, taught by women, can get a grounding in carpentry, ma-sonry, electronic engineering, and other skills that in the economy at large are almost exclusively 'masculine.'

The feminist technology project has not ended here, however. Many women pointed out that the masculine trades and technologies were pre-eminent in society only because men themselves were dominant. New-wave feminism therefore also comprised a project to recover and recognize women's traditional technologies. Midwives, for instance, became a focus for a feminist countermovement against the dominance of a masculine medical profession.

Third, new-wave feminism included a clear critique of the ex-ploitative and destructive uses of technology. It was expressed, for instance, in women's activism against nuclear weapons, in women's environmen-tal projects, and in solidarity work for women disadvantaged in processes of development in Third World countries.[1]

These several approaches to understanding technology-as-social have been highly relevant to women working for an analysis of technol-ogy that would help sustain these liberatory projects; and feminist social science has contributed to them.

Donald MacKenzie and Judy Wajcman, for instance, had clarified that 'technology' should be seen not only as a piece of hardware but also as a process of work and a kind of knowledge. Such a definition fitted well with women's redefinition as 'technological' of many of the things that women traditionally know and do. Cooking, whether with a wooden spoon or an electric liquidizer, is technological. Who can weigh one expert knowl-edge against another, and who can say when knowledge becomes or ceases to 'technical' (cf. McNeil 1987)?

In a classic 'social shaping' essay Langdon Winner had demon-strated that artifacts have politics.

The things we call 'technologies' are ways of building order in our world. Many technical devices and systems important in everyday life contain possibilities for many different ways of ordering human activity. Consciously or not, deliberately or inadvertently, societies choose structures for technologies that influence how people are going to work, communicate, travel, consume, and so forth over a very long time. In the processes by which structuring decisions are made, different people are differently situated and possess unequal degrees of power as well as unequal levels of awareness.

Winner was not speaking of women, but he was talking the same language as feminists. Women were saying, for instance, that nothing could be clearer than the different impact on women and men of the invention and dissemination of high-rise housing or the ovulation-inhibiting contraceptive pill (Stanworth 1987).

It was, of course, apparent that women as a sex were distant from the laboratories, drawing offices, and boardrooms from which decisions about new technologies were emerging. The social shaping literature fitted well with women's beliefs that this might prove a source as well as a sign of women's disadvantage. Work such as that of David Noble on numerically controlled tools (1984), of William Lazonick (1979) on the self-acting mule, of David Albury and Joseph Schwartz (1982) on the miner's safety lamp, showed variously that class, military , national, or other interests might shape the design of tools and artifacts. Women were showing that male interests too were a factor influencing technological outcomes. Linotype composing machinery invented in the late nineteenth century was shaped according to a compromise among manufacturers, printing employers, and the interests of a body of male printers organized in a craft union (Cockburn 1983). Often the choice of new technologies reflects calculations by (usually male) employers as to the most profitable exploitation of female labor, whose position in the labor market is systemically weakened by male-dominant social relations (Glucksmann 1990). Finally, if social contexts mold technological choices, what of family? Feminists were showing how gender relations in the Western household and its characteristic sexual division of labor powerfully shape the way domestic technologies are taken up and used (Cowan 1983; Bose et al. 1984).

Women as Actors in Technology Networks

It was within a feminist theoretical context that a research project was initiated in 1988 by the European Centre for Coordination of Research and Documentation in the Social Sciences (The Vienna Centre). The Centre brought together women from eleven countries in the hope of stimulating some similar and simultaneous research, in countries with contrasting political economies, on the relationship of technological change to changes occurring in relations between the sexes. In this framework, several such empirical investigations are now under way. From Britain we are contributing to the Vienna Centre project a case study of the design and development, production, marketing, and use of the microwave oven.[2]

We approached the empirical work feeling that the actor-network approach would be a productive way into a theme of this kind. And so, up to a point, it has proved. We find, for instance, the anticipated links between the development and use of the magnetron for military and industrial purposes and the application of microwaves to cooking. We can trace a history of the mobilization of support for the launching of the microwave oven as a consumer durable in the white goods range. We can see that what microwave cooking becomes depends crucially on the association with the process of millions of women (and men) in as many households. When a new generation of microwave ovens, with the addition of grill, rotisserie, and convection heating facilities, adopts the attributes of old conventional ovens, we see that enough women wishing to roast their chickens to crisp, brown perfection the way their mothers did can constitute a collective actor with the power to influence technological development.

We can see actors playing clear alliance-building roles: the manufacturer's home economist (female) who invents the recipes that bind manufacturer, artifact, and housewife together; the advertiser who projects an image of the artifact and its associated cooking process that can be 'translated' into a sales pitch by a manufacturer's team of demonstrators and the sales staff of the distributive stores. We can see the possible forging of a collectivity of interest between the manufacturers and distributors of microwave ovens, on the one side, and, on the other, an emergent industry, the manufacturers of cook-chill foods.

Finally, we can see that the actors in this network are truly heterogeneous. Take the British manufacturing operation. Some of the actors are human. There is a corps of Japanese managers, complete with Japanese traditions of management, newly arrived on the Celtic fringe of Britain to set up a manufacturing facility for microwave ovens and other products. There is the local Labour mayor who has the connections to deliver a no-strike one-union agreement in exchange for several hundred new technology jobs. There are the young women made redundant by the closure of sewing factories, with work discipline and plentiful skills that may be exploited at low cost. Their existence is not without influence on the design of the assembly process: it bears on which jobs are automated and which are not.

There are nonhuman, material actors in the network. There is the geographical location, the distance of the site from European markets, that is a factor in calculating production costs and retail prices. There is the conductivity and weight of certain materials deployed in manufacture; the energy of microwaves that some actors declare damaging to human health, others claim are harmless. At one moment of crisis, when demand for microwave ovens collapses, production must be cut back and jobs are threatened, it is because a new actor has emerged onto the scene: the *Listeria* bacterium that infects cook-chill foods and frightens off the nation's homemakers, who decide microwaves are risky and go back to their old habits.

The drama of the evolution and, from moment to moment, the closure, the 'black boxing,' of any one model of the microwave oven can readily be told in these terms. We have come to feel, however, that the actor-network approach falls short in three ways from being a fully adequate tool for a feminist analysis.

The first problem is the easiest to remedy. It is the invisibility of women. The actor-network approach focuses mainly on the design and development phases of an artifact's life. The actors, being mainly design and development engineers, entrepreneurs, and financiers, are just that: not actresses. To take the microwave case, no women are visible among the top management of the Japanese multinational, nor are they present in the engineering phases of the microwave development. Ironically, women's invisibility has been increased by the shift we have made from technological impact studies to social shaping studies. For a hard fact remains that, in matters of technological change, women are more impacted upon than impacting.

Women can, however, be introduced as actors by the expedient of ensuring that the net of the 'network' is cast wide enough, that it includes the lowest-paid workers producing the artifact, the individuals who use or refuse it. This is why our Vienna Centre research projects selected as a focus domestic technologies, artifacts that are applied in the home. It is also the reason for our decision not to halt the research at the point of design closure, when the prototype becomes an unproblematic 'black box' and enters production as this year's model of washing machine, vacuum cleaner, or food processor. Instead we are following the trajectory and tracing the relations of the artifacts from design along the spiral of development, manufacture, marketing, sales, purchase, use, service, and repair—not forgetting the feedback to design as one model supersedes another.

Besides, we are supposing that two sets of probabilities are inscribed in the artifact's design: first, the social relations of its use, but also, second, the social relations of its production. For, as some of us are finding, the manufacturer-designer of the artifact is sometimes frankly uninterested in who will buy it and what their needs are. (He will rely on an advertising campaign to create a market.) His main concern is how cheaply the artifact can be mass-produced. In the wide 'network' enlisted in birth-to-death biography of our artifacts we can be sure to find women in many phases of their many lives. They will be visible as manual workers, as marketing consultants, as home economists, trainee engineers, unpaid domestics, as shoppers, wives, mothers, daughters.

Although women as a sex can be scripted in, however, their presence does not guarantee a gendered analysis. And the language and concepts provided by the actor-network discourse seem not quite adequate for generating one. The other two shortcomings in the approach can, I believe, be encapsulated as follows. First, there is a lack of concern with subjectivity, which leads to a neglect of the way technology (particularly as knowledge and process) enters into our *gender identity*. Second, there is an incomplete representation of the historical dimensions of *power*, so that we cannot explain what we see with our eyes every day: that men as a sex dominate women as a sex, a relation in which technology is implicated. Indeed, it is only by tackling these two issues that we can explain the first—women's relative absence from the processes of design and development of technology. I will expand a little on these points.

Gender Identity and the Relations of Technology

Gender is a term increasingly used to convey a differentiation, by cultural processes involving appearance and behavior, action, thought, and language, of two ideal social categories, the masculine and the feminine, and their normative mapping onto male and female bodies respectively. By extension, gender is used to symbolize other differences (of size or importance for instance) and other phenomena (such as colors) are enlisted as symbols of gender. Think of the iteration between hierarchical dualities such as pink/blue, weak/strong, body/mind, woman/man. There is observed to be a strong tendency within gender processes to complementarity: what is masculine is not feminine, and vice versa. Together masculine and feminine are conceived as making a whole human being—the couple or dyad.

Just as there has been a shift in the past decade away from technological determinism, so there has been a shift from biological determinism. It has been a significant conceptual breakthrough to acknowledge that gender is not predestined by biological sex difference. What we perceive as 'facts of nature'—'man,' 'woman,' 'sexuality'—are largely cultural constructs. If gender differentiation has a source in human social history, the best claim to it may well be a deeply rooted cognitive practice of generating dichotomies through which to order our world—up/down, day/night, white/black, good/bad—dichotomies that are simultaneously hierarchies (Goodson 1990). Like Latour's scientific 'fact' or technological 'black box,' gender is more of a doing than a being. Gender is a social achievement. Technology too.

Yet, in the social processes that comprise gender, as in the social processes of technological development, are caught up material actors. Much as climate and geomorphology, electrons and magnets, may play a part in the networks that enable a certain technological outcome, so physique (hormones, average height, menstruation and ovulation) enter into the social representation and the relational play that is gender. Gender identity is what people do, think, and say about material and immaterial things *in relation to* other people conceived of as sexed. It is necessarily relational (Connell 1987, p. 79).

Technology too, as we have seen, is increasingly understood as relational. As deployed in production, in everyday life, in the household, technological artifacts *entail* relations. They embody some (those that went into their making). They prefigure others (those implied in their use, abuse, or neglect). But they also enter into and may change relations they encounter. There is yet nothing gendered about this perception, but gendering is inevitably present.

Take the microwave oven. It embodies its relations of design and production. These relations are, among class and other things, a matter of ethnicity and gender. The relationship between head office designers in Japan and local engineers in Britain (a relationship not without its tensions) has shaped certain variations in the model we buy. The assumptions made by an entirely male design team about the households in which their artifact is to find a role will be based on a masculine experience of domesticity (mediated of course by class and race positioning). The punishing relations of the assembly line, where women (in the main) spend their days in a relentless thirty-five-second cycle of lifting, inserting, screwing, clipping, while men (in the main) hunk heavy goods about, maintain machinery, and control the production process, have contributed the microwave oven's internal structure. The negotiations between cooks (mainly women) and those for whom they cook (mainly men and children) concerning what is 'good cooking' have given a certain product its combination of microwave and convention heat sources.

The microwave brings forward from the factory into the kitchen certain relational possibilities. It opens up, for instance, a new casualness in the ease with which individual members of a household, instead of taking a meal together, eat whenever it suits them. It enables unskilled cooks to 'cook' things. It calls for a negotiation of unaccustomed safety rules between adults and children, governing the various dangers of unexpected heat, unseen radiation, inadequately sterilized food.

On the other hand, the microwave oven enters a household where certain relations are well embedded. Perhaps here the wife works, the man is unemployed. There will be a pattern of cooking practices: who cooks for whom. Perhaps the teenage daughter is anorexic (a relational illness) so that food is a highly political topic in the family. There will be compromises concerning who makes decisions about purchases and does the shopping. The practice may differ as regards food and as regards consumer durables such as a microwave oven.

Clearly it is difficult, if not obtuse, to attempt an understanding of technology in such a context without taking account of gender identities.

It is not, however, as is sometimes supposed, only when women are present that gender exists. Studies of the introduction of new technology may sometimes seem to enter a world peopled entirely by men. There are, for instance, few women actors in the world in which David Noble (1984) saw computerized, numerically controlled machine tools taking shape and getting used. Yet the relations of this technology are relations of masculinity—and what is that if not gender?

As we are identified by others, and constitute our own identities in the course of a lifetime's interactions, part of the process is invariably gendering. Ineluctably, technology enters into our gendered identities. Since their inception in the Renaissance, modern science and technology have been deployed symbolically by men, the active sex in this project, as masculine. This holds true for many periods and many technologies. Francis Bacon called his fellow sixteenth-century scientists to "penetrate the secrets of Nature" in inaugurating with him "the truly masculine birth of time" (Farrington 1970). The culture and imagery of the scientists responsible for the atom bomb in the Los Alamos laboratories during the Second World War were strikingly masculine (Easlea 1983). Tracy Kidder's novel, *The Soul of a New Machine,* is a late-twentieth-century celebration of hi-tech machismo in a computer design project (Kidder 1981). The heroic age of mechanization had shaped and been shaped by muscular male identities. There was a hope (or was it a fear?) that the age of electronics, of informatics, would weaken the masculine identification of and with technology. Recent studies, however, have shown that IT, mediated by different symbols, has in its turn been appropriated for masculinity (Gray 1987; Haddon 1988).

The development of modern technoscience and of modern Western masculinity have been performed simultaneously, through an overlapping set of relations. What we think of as 'technology' is the doings of men (women's doings do not count); what we think of as a man is a 'hands-on' type. Similarly, contemporary Western femininity has involved the constitution of identities organized around technological incompetence. In the extreme patriarchal relations of Fascist Italy, Mussolini wrote of an incompatibility between women and machines and banned women from the operation of machinery in production—until he found his war economy could not do without them (Macciocchi 1979).

From the microwave oven study we could draw an instance of the explanatory value of a perception of identity. The senior home economist in the manufacturing company was a woman. It is not incidental that her daughter was employed in her department. Her report of her work in interview showed her to identify as a woman and homemaker, as well as an employee, and to identify with the housewife-customer using her company's microwaves. She was also aware of (and angry about) the high evaluation of men's engineering skills relative to women's domestic science qualifications; that there were few women managers in the firm; and that she herself was denied management status. This gender set—identity/ identification/disidentification—was significant in the position, responsibility, and activity she negotiated for herself in the firm. It colored the way she guided her team of women consultants, published recipe books, advocated features of microwave design, and represented her department vis-à-vis the rest of the firm. She had insisted that the word 'Test' be prefixed to the sign 'Kitchen' above their door.

Technology, then, can tell us something we need to know about gender identity. Gender identity can tell us something we need to know about technology.

The Peculiarities of Power

Among the things gender may be able to clarify for us about technology is the nature of its implication in control, exploitation, and domination. For gender is not merely a relation of difference, it is one of asymmetry. Women commonly experience the masculine relations of technology as relations in which they are dominated and controlled. Some feel there is a connection between their own experience and what they observe to be the damaging relation of industrial technoscience with the natural environment. A concept of power is by no means absent from the actor-network theory of technology. But it is a representation of power as capacity, whereas in the lived world power is often experienced as domination. A representation of power as capacity, or effectiveness, does not readily accommodate what feminist theory has come to understand by 'male power'—that is, a persistent pattern in which men as a sex dominate women as a sex, exploiting and controlling women's sexuality and reproductive capacities and benefiting from their labor.

An important insight of feminism has been a concept of systemic male power, sometimes designated 'patriarchy.' An extensive literature on this theme unfolded in the 1970s and 1980s (Rubin 1975; Kuhn and Wolpe 1978; Eisenstein 1979; Sargent 1981; Walby 1990). In brief, the notion supposed the existence in all societies of some kind of normative arrangements governing reproductive/sex/gender—just as there are some kind of systemic arrangements governing production/labor/consumption. In the way that a mode of production might in one culture and at one historical moment be feudal, at another capitalist, so a sex/gender system might in theory at one moment and in one society be matriarchal, in another patriarchal, or indeed involve sex equality. A focus of feminist study was why almost all societies of which we have knowledge through anthropology and history evince a pattern (though in differing forms) of male power, female subordination.

Into the social shaping approach, which was often in a Neo-Marxist framework and carried a subtext of partisan opposition to class domination, feminists were relatively easily able to insert considerations of male domination and by discussing its possible connections with class exploitation. It has proved more difficult in the case of the actor-network approach, which is characterized less by liberatory politics than by an enthusiasm for the minutiae of technical decision making as intellectual puzzle and human drama.

For here the notion of power is not derived from class. Such structures are eschewed, not perhaps as erroneous so much as, for these purposes, irrelevant. In the actor-network account of the emergence of theories and artifacts, power expresses itself as the *influence* that can mobilize opinion for one's arguments, the constraints limiting one's *capability* to mold materials or sway opinion in accordance with one's individual or corporate will. In case studies of the generation of new technologies (the bicycle, 'bakelite,' the electric car) power takes the low-profile form of 'reduced discretion' among the network of actors (Law 1986b, p. 17).

Nonetheless, some of these same authors have usefully amplified their concept of power in other essays. Latour, for instance, suggests that power "is not something one can possess—indeed it must be treated as a consequence rather than a cause of action" (Latour 1986, p. 264). Power over something or someone is, he says, a composition made by many people and attributed to one of them. "The amount of power exercised varies not according to the power someone has, but to the number of other people who enter into the composition" (Latour 1986, p. 265). It is a logical derivative of the actor-network/translation model of technoscience. In another essay he and Michel Callon sketch the process in which those entities we see as powerful acquire their scale and reach. There is no inherent difference, they say, between a small weak actor and a massive strong one. The first grows into the second by a process of associating other actors in its projects. Organization, in effect, is no different from any other kind of 'black box': it is a negotiated achievement, of dubious stability (Callon and Latour 1981).

Latour, then, is not saying that class power is not 'real,' or that male domination does not 'exist.' Perhaps it is accurate to say he is agnostic about such things, saying that we can learn more about power by putting such concepts aside and looking at the mechanisms by which power can be shown to be effective. The approach is not dissimilar to that of Michel Foucault whose cumulate essays on medicine, the prison, and sexuality portray power as operating not by repressing subjectivity but by deploying and regulating it. For Foucault power has to be understood in 'an ascending analysis,' starting from its 'infinitesimal mechanisms' in the individual. This individual is not (as in feminism and socialism) the subject of a liberatory project. Rather subjectivity, even the body of the subjugated, is the effect and vehicle of power. In a celebrated lecture he put it as follows:

[P]ower, if we do not take too distant a view of it, is not that which makes the difference between those who exclusively possess and retain it, and those who do not have it and submit to it. Power must be analyzed as something which circulates. . . . Power is employed and exercised through a net-like organization [cf. actor-networks]. And not only do individuals circulate between its threads; they are always in the position of simultaneously undergoing and exercising power. . . . Individuals are the vehicles of power, not its points of application.

(Foucault in Gordon 1980, p. 98)

The similarity with Latour's 'power as the consequence, not the cause, of action' will be immediately apparent.

Let us return to the microwave oven. Actor-network language is capable of expressing much of what we see. Women at least are not cast as passive victims—they too can be actors. The women who may try and fail to acquire engineering skills, or, having obtained them, to get jobs in the white goods industry; the women in the office and assembly line who may defer to men's 'natural' ability with technology; the housewife who 'prefers' to be the one to do the cooking even though that means effectively being a personal servant to her husband—we need not assume they are unwitting dupes. We can wonder whether, from their position of subordination, they strike some kind of bargain. As Foucault says, "what makes power hold good, what makes it accepted, is simply the fact that it doesn't only weigh on us as a force that says no, but that it traverses and produces things, it induces pleasure, forms knowledge, produces discourse" (Foucault in Gordon 1980, p. 119). Women can recognize themselves in this.

There is, however, difficulty in responding with the actor-network approach to feminist questions that continue to call for an answer. Perhaps it is that in mainstream technology theory the key question has been how to explain change, while for feminists it seems more urgent to explain continuity. Why do gender relations survive so little changed through successive waves of technological innovation?What is the connection between organization and gender identity? Can men as a sex be said to organize their power over women? Are technology relations a medium through which they do it? What part do women play in perpetuating this? Is there an erotic dimension linking technological and sexual control and abandon (Hacker 1989)?

In the actor-network approach to technoscience power is visible only as capacity. In life it often manifests itself as domination. In actor-networks, the performative approach, the social can have no predisposing patterns. For just as 'Nature' is not taken as preexisting the performance of the drama of actor and network, neither is 'Society.' When in life, time and again, men are cast for the parts that forge the alliances that influence technological outcomes, nothing explains why men are chosen. The theory permits actors to bring to the struggle certain 'resources.' Presumably

Cynthia Cockburn

women, being absent or in relatively powerless relationships to technological trajectories, lack these 'resources.' Hidden within the innocent concept of resources, however, is packed something of enormous scope that it
is barely competent to carry and which continually invites unpacking: the
historical patterning of the social by class, race, and gender. We know that
such patterns or structures are not unchanging, that performance or practice can alter them. In dwelling on them we do not need to be determinist
or functionalist. But we also know that in 'the social' there is a loading of
the dice that produces a probability that the collective associations and
translations that result in effective power in these technology networks
will continue to afford one sex domination over the other—unless it is
somehow interrupted through an active politics of sex/gender.

NOTES

1. For a thorough review and analysis of two decades of feminist interaction with technology, see Judy Wajcman's forthcoming book, *Feminism Confronts Technology.*

2. The British component of this research, described here, is funded by the Economic
and Social Research Council and being carried out at The City University, where the author
is a research fellow in the Centre for Research in Gender, Ethnicity and Social Change.

REFERENCES

Albury, D., and J. Schwartz. (1982). *Partial Progress: The Politics of Science and Technology.*
London: Pluto.

Baehr, H., and G. Dyer (eds.). (1987). *Boxed-in: Women on and in TV.* London: Routledge.

Bijker W., T. P. Hughes, and T. Pinch (eds.). (1987). *The Social Construction of Technological Systems.* Cambridge, Mass.: MIT Press.

Bose, C., P. Bereano, and M. Malloy. (1984). "Household technology and the social construction of housework." *Technology and Culture* 25: 53–82.

Callon, M. (1987). "Society in the making: The study of technology as a tool for social analysis." In W. Bijker et al. (eds.), *The Social Construction of Technological Systems.* Cambridge, Mass.: MIT Press.

Callon, M., and B. Latour. (1981). "Unscrewing the big Leviathan: How actors macro-structure
reality and how sociologists help them do so." In K. Knorr-Cetina and A. V. Cicourel
(eds.), *Advances in Social Theory and Methodology.* London: Routledge & Kegan Paul,
277–303.

Cockburn, C. (1983). *Brothers: Male Dominance and Technological Change.* London: Pluto.

———. (1985). *Machinery of Dominance: Women, Men and Technical Knowhow.* London:
Pluto.

Connell, R. W. (1987). *Gender and Power.* Cambridge: Polity.

Cowan, R. Schwartz. (1983). *More Work for Mother.* New York: Basic Books.

Easlea, B. (1983). *Fathering the Unthinkable: Masculinity, Science and the Nuclear Arms
Race.* London: Pluto.

Eisenstein, Z. R. (ed.). (1979). *Capitalist Patriarchy and the Case for Socialist Feminism.*
New York: Monthly Review Press.

Farrington, B. (1970). *The Philosophy of Francis Bacon.* Liverpool: Liverpool University Press.

Glucksmann, M. (1990). *Women Assemble: Women Workers and the New Industries in In-
terwar Britain.* London: Routledge.

Goodison, L. (1990). *Moving Heaven and Earth: Sexuality, Spirituality and Social Change.* London: The Women's Press.

Gordon, C. (ed.). (1980). *Michel Foucault: Power/Knowledge.* Brighton: Harvester Press.

Gray, A. (1987). "Behind closed doors: Women and video." In H. Baehr and G. Dyer (eds.), *Boxed-in: Women on and in TV.* London: Routledge, 38–54.

Hacker, S. (1989). *Pleasure, Power and Technology: Some Tales of Gender, Engineering and the Cooperative Workplace.* Boston: Unwin Hyman.

Haddon, L. (1988). "The roots and early history of the British home computer market: Origins of the masculine micro." Ph.D. thesis, Imperial College, London University.

Hughes, T. P. (1987). "The evolution of large technological systems." In W. Bijker et al. (eds.), *The Social Construction of Technological Systems.* Cambridge, Mass.: MIT Press, 51–82.

Kidder, T. (1981). *The Soul of a New Machine.* Harmondsworth: Penguin.

Knorr-Cetina, K., and A. V. Cicourel (eds.). (1981). *Advances in Social Theory and Methodology.* London: Routledge & Kegan Paul.

Kuhn, A., and A. M. Wolpe. (eds.) (1978). *Feminism and Materialism.* London: Routledge & Kegan Paul.

Latour, B. (1986). "The powers of association." In J. Law (ed.), *Power, Action and Belief: A New Sociology of Knowledge?* London: Routledge & Kegan Paul, 264–80.

———. (1987). *Science in Action.* Milton Keynes: Open University Press.

Law, J. (ed.). (1986a). *Power, Action and Belief: A New Sociology of Knowledge?* London: Routledge & Kegan Paul.

———. (1986b). "Power/knowledge and the dissolution of the sociology of knowledge." Editorial introduction to J. Law (ed.), *Power, Action and Belief: A New Sociology of Knowledge?* London: Routledge & Kegan Paul, 1–19.

———. (1987). "Technology and heterogeneous engineering: The case of Portuguese expansion." In W. Bijker et al. (eds.), *The Social Construction of Technological Systems.* Cambridge, Mass.: MIT Press, 111–34.

Lazonick, W. H. (1979). "Industrial relations and technological change: The case of the self-acting mule." *Cambridge Journal of Economics* 3:231–62.

Macciocchi, M. A. (1979). "Female sexuality in Fascist ideology." *Feminist Review* 1: 67–82.

MacKenzie, D., and J. Wajcman (eds.). (1985). *The Social Shaping of Technology.* Milton Keynes: Open University Press.

McNeil, M. (ed.). (1987). *Gender and Expertise.* London: Free Association Books.

Noble, D. F. (1977). *America by Design.* New York: Alfred A. Knopf.

———. (1984). *Forces of Production: A Social History of Industrial Automation.* New York: Alfred A. Knopf.

Pinch, T., and W. Bijker. (1984). "The social construction of facts and artifacts: or how the sociology of science and technology might benefit each other." *Social Studies of Science* 14:299–341.

Reiter, R. (ed.). (1975). *Toward an Anthropology of Women.* New York: Monthly Review Press.

Rubin, Gayle. (1975). "The traffic in women: Notes on the 'political economy' of sex." In R. Reiter (ed.), *Toward an Anthropology of Women.* New York: Monthly Review Press, 157–210.

Sargent, L. (ed.). (1981). *Women and Revolution: A Debate on Class and Patriarchy.* London: Pluto.

Stanworth, M. (1987). *Reproductive Technologies: Gender, Motherhood and Medicine.* Cambridge: Polity.

Wajcman, J. (1991). *Feminism Confronts Technology.* Cambridge: Polity.

Walby, S. (1990). *Theorizing Patriarchy.* Oxford: Blackwell.

Winner, L. (1985). "Do artifacts have politics?" In D. MacKenzie and J. Wajcman (eds.), *The Social Shaping of Technology.* Milton Keynes: Open University Press, 26–38.

Cynthia Cockburn

ACKNOWLEDGMENT

My thanks to Judy Wajcman, to Susan Ormrod (my co-researcher on the microwave oven study), and to other members of the Vienna Centre group, for reading and commenting on this chapter. It has benefited greatly from their advice, but remains a personal view.

Helen Cunningham

Moral Kombat and Computer Game Girls

Today, perhaps the most important question that needs to be addressed in relation to children's audiovisual media is not *what* children are watching on television but what they are *doing* with the television set. Television is no longer (if it ever was) a passive medium.[1] The age of interactive television is upon us.

In the early 1980s, when the video game was just beginning to emerge as an adjunct of the television entertainment system, most of the search undertaken suggested that playing video games was a predominantly male leisure pursuit.[2] In the late 1970s and early 1980s, when this research was carried out, video game-playing was not a home-based activity. Arcades were the place where game-playing took place and these arcades were populated by adolescent males.[3] A decade later, in the early 1990s, the second wave of game-playing has occurred primarily through the purchase of dedicated game consoles which are plugged into the home television set. The main companies in the console industry are Sega and Nintendo. First, access has opened up for a lower age group who previously could not play video games in arcades,[4] and second, as the context of games-playing has changed, female participation in games culture has increased. Young girls can now be found huddled around 'Gameboy' handhelds in the playground, and girls as well as boys now talk of 'Sonic' (Sega's trademark) and 'Super Mario' (Nintendo).

There are over 9.4 million computer game consoles in use in the United Kingdom[5] and the vast majority of these are used by children aged between seven and fifteen. Computer games are the new children's medium of the 1990s—and have sparked a moral panic within the United Kingdom. Schoolteachers and psychologists have all spoken up against prolonged childhood exposure to this new medium.[6] The aim of this chapter is to question the recent changes within the computer games industry, the moral

This chapter originally appeared in *In Front of the Children,* ed. Cary Bazalgette and David Buckingham (London: BFI Publishing, 1995). Reprinted by permission of BFI Publishing and the author.

panic surrounding computer games culture, and the extent of the gendered nature of computer game-playing.

Methodology

My own interest in computer games and the panic surrounding them began through observations of my ten-year-old sister and her involvement in computer games culture. The pleasure she found both in playing the games and in the wider culture of reading the gaming magazines, watching computer game programs on television, and identifying herself as a games player (her 'Sonic' T-shirt became her favorite item of clothing), was not explained by either the media panic or academic research, both of which portrayed game-playing as boys' culture.[7] This chapter was written at the beginning of my research into computer games and children's cultures, and is consequently exploratory and speculative. This research is based on informal interviews with my sister Julie and a group of her 'best friends,' all of whom I knew shared Julie's enthusiasm for and enjoyment of computer games. The interviews took place in Summer 1993. Around this time the media hysteria over the effects of computer games on children's development was at its peak and most of the girls were aware of the debates. I told the girls that I was interested in computer games and wanted to talk to them about what they thought about game-playing. My position within this informal interview situation was shaped by my position as Julie's older sister. Most of the girls had known me for as long as they have known Julie, and they were aware that I also often play computer games. Although my status as adult was apparent, since I am Julie's sister (rather than a parent or teacher) my questioning and listening were not seen as critical or authoritative. This research cannot be seen as representative of young girls' involvement with computer games in general, but as documenting the culture of a group of lower-middle-class ten-year-old girls. The parents of some of these girls were also interviewed, to ascertain their views on their daughters' participation in computer game cultures.

The Computer Games Industry

Leslie Haddon[8] provides a detailed history of the development of the computer games industry; but this history ends in 1991, which is the year when Sega and Nintendo began their big push for computer game console sales to children in Britain. Prior to 1990 these two Japanese firms were unheard of in British homes, but three years on 60 percent of children own a computer games machine and 80 percent are said to play computer games regularly.[9]

Sega and Nintendo are the brand leaders who dominate the market internationally. Nintendo is the largest of the two companies worldwide, but in the United Kingdom Sega holds the largest share of the market. The Japanese and American markets for computer games are fast approaching saturation. The battle for supremacy in the computer game markets is now taking place in Europe, where profit margins are much higher because of the higher cost of game cartridges.[10] Once a console has been purchased the game cartridges are incompatible—the console manufacturers have a monopoly. In Britain alone the market for the software (cartridges and discs) is worth about £755 million and is still growing.[11] In 1989 the computer games market was worth £78 million, but by 1993 sales of games, consoles, magazines, and accessories totaled £1.1 billion.[12]

The games industry and other areas of the entertainment industry are joining forces. Computer games are often more profitable than movies. One game, *Super Mario Bros. 3*, has sold 14 million copies and has generated more money than the movie *ET*.[13] Nintendo now makes greater profits than all of the American movie studios combined.[14] Sega and Nintendo have licensing deals with film companies in order to use animated characters—including Disney cartoon characters—and many computer games are based on successful films: *Robocop, The Terminator, Jurassic Park*, and *Aladdin* are just some of the films which have transferred to the games console. This cross-media licensing has not all been one way. The film *Super Mario Brothers* was released in 1993 and both *Super Mario* and *Sonic the Hedgehog* have their own cartoons broadcast on Sunday morning children's television in the United Kingdom. The advertising industry has also expressed an interest in computer games as a medium for product placement. In December 1993 the game *Snapperazzi* was launched, devised by the *Sun* newspaper and sponsored by the pizza delivery chain Domino's Pizza and the Leaf U.K. brand Fizzy Chewits. In the United States sponsored games have featured 'Spot,' the character in 7 Up's advertising.[15] Sponsorship and product placement within computer game software will probably increase since in Britain there is as yet no regulation of such practices.

It is not only the entertainment industry that has been affected by the rise of the computer game. Over the past five years Nintendo has made bigger profits than many of the international computer giants.[16] The microprocessor within computer game consoles is similar to that used in many personal computers. When the chief executive officer of Apple computers was asked what computer company he feared most in the 1990s, his reply—'Nintendo'—was not altogether surprising.[17] These games consoles, at present used primarily for games-playing, are 8-bit and 16-bit computers, which with the addition of other pieces of hardware are capable of becoming more than just toys for playing games. Sega and Nintendo have had great success at getting children involved in computer technology, having put computers in the homes of 60 percent of households with children. This entertainment-led technology revolution has been achieved not by cable television or by any of the leading computer manufacturers, but by a blue hedgehog and a plumber in red dungarees.

Computer game culture was initially welcomed by many parents, any childhood contact with computers being seen as educational. If parents had worries over their children playing computer games, they were concerned about the cost of the cartridges more than about the content— *Sonic* and *Super Mario Brothers* appeared harmless.

As most children's first experience of computers is through computer games in the home, many of the parents I interviewed bought computer game consoles for their children believing that an active enthusiasm for computer games would be the first step toward computer literacy, and hoping that once the interest in game-playing waned the enthusiasm for computers and new technology would remain. Early familiarity with computers was seen as important for young children. This view was expressed particularly in relation to daughters. The parents I talked to were very aware of the need to encourage their daughters to be involved with new technologies. The occupations of these parents, many of whom worked in education, were perhaps a factor in shaping their views on the importance of children's access to technology. Many of them attempted to convince their children that a personal computer would be the most sensible and suitable piece of equipment to buy. These parents hoped that the computer would also be used for its 'real' purpose once the games had been mastered. However, when it came to purchasing the equipment, the power of Sega's and Nintendo's advertising campaigns had convinced most children I spoke to that a 'SNES' or 'Megadrive' were the only credible consoles on which to play games.

Nintendo increasingly markets itself as a console which can be used for both entertainment and educational functions. 'Edutainment' is the word used in their promotional leaflets to parents. The 'Mario Edutainment Series' of software is promoted as relevant to the National Curriculum, and it also has other titles which are promoted as educational for preschool children. The interactive and cooperative nature of computer game-playing provides the potential for computer games as a tool in children's early learning. This may be an optimistic view, however, because at present the main use of these machines are being put to is games-playing, and none of these 'edutainment' games appears near the top of the game charts. The computer games industry is driven by commercial imperatives and the genres of games currently being produced are very profitable.

The industry is not likely to attempt to push consoles or software in any other direction until the market for entertainment games in the United Kingdom is saturated. Computer games have become part of 'youth culture' and the iconography of the games has been used in advertising, in 'Youth TV' programs, and within club cultures. Computer games have been appropriated by young people and have been marketed to them as anarchic, rebellious, and anti-establishment. Neither young people nor the games manufacturers want to lose their street credibility by being seen as too educational. Nintendo's 'edutainment' series reflects their marketing strategy, which appears very different from that of Sega. Nintendo, which is the market leader in the United States, has a wholesome family entertainment image. The version of *Mortal Kombat* available for Nintendo consoles did not contain the 'gore' codes to operate the notorious 'death moves.' As well as the 'edutainment' series Nintendo has produced a variety of leaflets for parents informing them of the educational benefits of Nintendo software.

Haddon documents the history of game-playing as being a young male preserve and sees no reason for this male dominance of gaming to change.[18] However, the move of computer games from 'street culture' in the arcades to 'bedroom culture'[19] in the home (which has been achieved through the dedicated games console) has transformed the experience of games-playing for young girls. This domestication of computer games has fed into girls' existing 'bedroom culture,' and now both boys and girls spend hours in their bedrooms playing computer games with friends. As so often, when a new mass medium emerges it is accompanied by a moral panic.[20] The increasing popularity of computer games among young children has been accompanied by concern over the effects of this medium. These debates over the *effects* of computer games on young children have been very similar to past debates within television studies over the effects of television viewing.[21] The three main issues cited in the attacks on computer game-playing are firstly the addictive nature of game-playing, which is thought to hinder children in social interaction; second, the violent content of the games; and third, the gender stereotyping within the games. I will discuss each of these in turn.

Addiction to 'Kiddie Cocaine'

Early fears about computer games concerned the ability of children who spent many hours alone playing computer games to interact socially with other children. Fears over the addictive nature of computer games were expressed. Reports of children who play computer games for up to six hours at a time, and the analogy of computer games to 'Kiddie Cocaine,'[22] have contributed to these fears. The Professional Association of Teachers believes that computer games can discourage children from reading and writing and make them aggressive. Jackie Miller, deputy general secretary of the PAT, believes "children's social and educational development is being affected by the games."[23] Computer games are also accused of eroding traditional children's games, with fears of computers taking over children's lives.

These fears are not substantiated by my own research. The girls I interviewed claimed to play computer games for approximately one hour a day, and all talked of other hobbies and activities on which they spent far more time. Computer games were often regarded as being the 'last resort' when there was nothing else to do.

> SARAH: Games are good to play when you're bored.

> ALISON: It's good to play [computer games] when it's raining or late [at night] and you can't go out.

Julie pointed out that the amount of time spent playing computer games varied depending on whether a new game had recently been purchased.

> JULIE: I hardly play computer games at all now because I can do all my games and so it's boring. I only play about once a week now until I can afford a new game.

> EMMA: When I get a new game I play it loads.

This fascination with a newly acquired toy is not unique to computer games. Children can also be 'addicted' to other leisure pursuits.

> JULIE: Sometimes when I am reading I don't want to put it down and I just sit up and read for hours when my mum thinks I have turned the light out.

Fears over children's addiction to reading are not usually expressed. Perhaps fears over this addiction to a new technology have been voiced by an older generation who cannot understand the appeal of the games, preferring children to play traditional games. This generation gap, and parental ignorance of games, is often part of their appeal. Fears of this new technology are due to the hitherto unknown interactive nature of computer games; talk of virtual reality and cyberspace is often beyond the everyday knowledge of parents.

Computer games have been criticized for causing antisocial behavior and for causing children to become isolated and unhealthy through spending many hours in a darkened room staring at the bright, fast-moving images on the television screen. This removal of children from the streets by a modern-day Pied Piper in the form of a hedgehog or a plumber has come at a time when parental fears of letting children play in the streets have never been higher. This retreat into the bedroom has more to do with fear of strangers and fast cars than of any addiction to 'Sonic' or 'Super Mario.' The move to 'bedroom culture' was welcomed by many parents who were relieved that their children had taken up home-based activities rather than 'hanging around' in the streets.

The criticism that playing computer games leads to social isolation is a far cry from the fears expressed over video games in the early 1980s when video games were blamed for delinquency as arcades became the collective point for youths to hang around.[24] Ironically, now that computer games have increasingly moved into the individual player's bedroom, the games are seen as causing isolation. Yet among the children with whom I have carried out this research the majority played computer games with friends or family members. They all expressed the view that game-playing was more fun when it was a social activity. Even when children play games alone, knowledge of computer games leads to social interaction in the playground as children talk about the latest games and exchange tips and 'cheats.' Many children also write reviews of games for school magazines. Computer game-playing is a shared experience. The game player who has progressed furthest through the games' 'levels' and 'worlds' shares her tactics. The girls I interviewed talked about how they phoned each other when they came to a level they could not complete and pooled their knowledge. Computer games are a site of shared knowledge rather than of individual competition. This feature of games-playing could potentially be of great benefit in 'educational' games.

Violence and Aggression in Game Play

The second main focus of the criticism of computer games has concerned the content of the games being played. When the narratives of the games are analyzed they can be seen to fall into six main genres.[25] The two genres most popular with the children I interviewed were 'Platformers' and 'Beat 'em ups.' Platform games such as *Sonic* and *Super Mario* involve leaping from platform to platform, avoiding obstacles, moving on through the levels, and progressing through the different stages of the game. Beat 'em ups are the games which have caused concern over their violent content. These games involve fights between animated characters. In many ways this violence can be compared to violence within children's cartoons where a character is hit over the head or falls off a cliff but walks away unscathed.

Controversy has occurred in part because of the intensity of the game play, which is said to spill over into children's everyday lives. There are worries that children are becoming more violent and aggressive after prolonged exposure to these games. As a computer game-player myself and also through my observations of Julie and her friends' game play, I have no doubt that playing computer games does involve feelings of intense frustration and anger which often expresses itself in aggressive 'yells' at the screen. It is not only the 'Beat 'em up' games which produce this aggression; platform games are just as frustrating when the characters lose all their 'lives' and 'die' just before the end of the level is reached. Computer gaming relies upon intense concentration on the moving images on the screen and demands great hand-to-eye coordination. When the player loses and the words 'Game over' appear on the screen, there is annoyance and frustration at being beaten by the computer and at having made an error. This anger and aggression could perhaps be compared to the aggression felt when playing football and you take your eye off the ball and enable the opposition to score. Yet the aggression felt in computer game play is vented on the screen and does not necessarily turn players into more aggressive and violent children. The annoyance experienced when defeated at a computer game is what makes gaming 'addictive': the player is determined not to make the same mistake again and to have 'one last go' in the hope of doing better next time.

Indeed the aggression generated in game-playing and focused onto the machine could be seen as *relieving* stress, tension, and pent-up aggression. My own computer game-playing is my method of switching off from the day's pressures by focusing on 'Super Mario' and how to get past the boss character at the end of the level.[26] I am not alone in using computer games in this way. While I was researching the gendered image of computer games many female colleagues and friends have 'confessed' their enjoyment of game-playing as a method of switching off and so managing stress.

Some of the concern over the violence of computer games has been about children who are unable to tell the difference between fiction and reality and who act out the violent moves of the games in fights in the playground. But children have always played fighting games. I have childhood memories of Saturday afternoons spent watching wrestling on television and copying the moves of 'Giant Haystacks' and 'Big Daddy,' trying in vain to overpower my brothers. This concern over the violent nature of popular culture is not new,[27] but Haddon suggests that it is the combination of violence and the new and as yet unknown powers of the interactive technology of computer games that is the cause of contemporary concerns. Haddon summarizes the view put forward by many: "concerns about the 'violent' narratives of games, referred back to the underlying technology of the media . . . the experience of violence through mediating technology would de-sensitise users to aggression."[28] Psychologists have put forward similar concerns. Professor Cary Cooper has commented: "The problem with video games is that they involve children more than television or films and this means there are more implications for their social behaviour. Playing these games can lead to anti-social behaviour, make children aggressive and affect their emotional stability."[29]

A theme behind this moral panic is ideological: law and order are seen to be weaker now than ever before, as more and more people report that they live in fear of violence. Television was blamed for this rising tide of violence, and now video and computer games are seen as inciting our children to violence. Each new form of popular cultural entertainment is seen as being more harmful than the previous one.

The manufacturers play down the violent nature of the games, claiming that violence in games reflects violence in society. Simon Morris, Sega UK's marketing director, points out: "Violence is a problem that is part of our society and we are not to blame for that. Our games are produced as a result of consumer demand and we are just responding to what people want to buy."[30] The girls I interviewed were well aware of this media concern that computer games cause violence and were very keen to express their rejection of this view.

> ELIZABETH: People were violent before computer games were invented.

The girls were aware of the media's coverage of computer game violence as being a problem which centered on boys rather than girls. Even so the computer games were not seen as the cause of boys' violence.

> EMMA: It's not playing games that makes boys violent, boys are already violent.

> JULIE: Boys are born violent.

> SARAH: Yeah, boys are just boys.

Although the girls did regard boys as being more violent than girls, they did not see this as an effect of the 'violent' games. The girls were quick to claim the 'Beat 'em up' genre as part of *their* game culture too.

Gender and Computer Game Culture—Game Boys and Girls?

The main focus of the media panic over the popularity of computer games among children has been the feared addictive nature of game-playing and the violent narratives of the games. Within this discussion of computer game culture the computer games industry, journalists, and also many academics have portrayed game-playing as a predominantly male preserve. My research suggests that this is no longer true and that the gendered nature of game-playing is changing. A large number of girls appear to enjoy playing computer games. Female participation games culture has significantly increased since computer game-play moved out of the arcades and into the home. This increase in girls' participation with computer games is encouraging. Most children's first experience of computers is through computer games, and it is vital that girls are included in these arenas where familiarity with new technology is established.

Angela McRobbie has posed the question of why girls are invisible in many accounts of youth cultures. She suggests that female invisibility is due in part to the media's concentration on violent aspects of youth culture.[31] This can be seen to be true within games culture. The girls I spoke to all mentioned platform games as being their current favorite game, rather than the more violent genre of 'Beat 'em ups.' Media concern has concentrated on 'Beat 'em ups' and the aggressive nature of *boys'* play rather than that of girls. However, the girls did indicate that they also liked to play the more violent games.

> SARAH: I like violent games . . . especially the one where you kill and blood comes out.

The 'Beat 'em up' game *Mortal Kombat* to which I presumed Sarah was referring had only just been released amid a barrage of media publicity and all the girls were keen to play it. This popularity of 'gore' genres spread over into other media as every girl identified horror as her favorite type of book. The *Point Horror* series was mentioned by all as their current favorite reading.

Gender stereotyping within computer games is an issue that has been raised by many critics.[32] When human female characters are present they are often represented as the 'quest' to be saved by the male hero. In *Super Mario Brothers* the aim is to defeat 'Bowser' and free 'Princess Toadstool.' This gender stereotyping has to be acknowledged as a problem within the design of the games, even though it should be stressed that these games are still very popular among female games players. In the 'Beat 'em up' genre female characters are increasingly appearing and are not always weaker than their male rivals.[33] Patricia Marks Greenfield acknowledges the importance of involving girls in computer game-play as an "entry point into the world of computers," and she suggests that "there is an urgent need for widely available video games that make a firm contact with the fantasy life of the typical girl as with that of the typical boy."[34] The vast majority of games designers are male and perhaps game design reflects this. Nintendo has attempted to engage with 'the fantasy of life of young girls' and has marketed a Barbie game. This is a platform game in which Barbie travels through Mall World and Soda Shop World in her search for 'magical fantasy accessories' and ball gowns. This game was not well received by my younger sister and her friends:

> JULIE: I'd never buy that Barbie game, it's stupid.

> SARAH: No way . . . I'd rather play violent games any day.

But Julie has never played with Barbie dolls and is probably too old to start doing so. Hopefully, as more female games designers enter the industry, a wider range of genres will develop and more female characters will be available within the existing genres. The most popular games for Julie and her friends were *Sonic, Sonic 2, Super Mario Brothers, Super Mario Brothers 3, Mickey Mouse in Castle of Illusion, Streets of Rage,* and *Street Fighter 2*.[35] Although these games could all be described as action and adventure (and in some cases they are violent 'Beat 'em up' games), the playing of them should not be seen as gendered.

Although platform games appear to be the favorite choice among young children, both male and female, Julie and her friends enjoyed 'Beat 'em ups' and expressed pleasure in the violence of these games, enjoying being able to fight and win when playing against brothers or male peers. In most areas of society this violent and aggressive side of a girl/woman's nature has to be repressed in conformity to socially expected norms of what is acceptable 'feminine' behavior. Playing violent games gives female players the chance to express this aggression in a safe context.

When playing these games the girls often chose to operate the male characters because they had 'better moves.' This operation of male characters by female players, and female characters by male players, could be seen as 'transgender identification,'[36] but I do not believe that 'identification' adequately describes players' relationships with the characters on the screen. Computer game players choose characters to operate on the basis of the moves they can make or the special skills and functions they have.[37] Future research on computer games should perhaps concentrate on the experience of playing these games and how 'pleasures' are created.

Sega's and Nintendo's success at getting consoles into children's homes has changed the nature of female participation in games culture. Haddon outlines how girls' experience of computer games differs from that of boys, documenting how "girls would rely on brothers to inform them about the latest game" and how "few girls played games in the various public sites."[38] The explosion in computer game sales, not only to children but also to young adults, has increased access to computer games information. A number of children's television programs have computer game review spots, and programs such as *Bad Influence* on ITV and *Games Master* on Channel 4 are avidly watched by young children. Over fifty computer game magazines are currently available, many exclusively aimed at children, which provide reviews and ratings for new games and new consoles.

The public spaces in which games are available to be 'tested out' by young consumers have multiplied. These public spaces are quite often within shopping centers (department stores, electrical shops, record shops, and also specialist computer game shops) and as such they are much more accessible to appropriation by young girls than arcades have been. Girls' knowledge of what new games are available is now potentially much greater. Reliance on brothers for information has decreased as other public sources of computer games information have evolved.

Conclusion

Computer games as a medium have a great potential for involving girls in new technology. The moving of computer games play from the arcades into bedrooms and shopping centers has significantly changed the experience of game-playing. This changing context has to be recognized in any ac-count of computer game cultures. In expressing my thoughts on these is-sues I have generated more questions than I have answered. Computer games are a relatively new medium and it is difficult to predict how their popularity with my sister and her friends will be maintained as they grow older and develop new interests. In the absence of a larger study of patterns of computer game use among girls it is very difficult to assess how typical my sister and her friends are. A lot more research needs to be carried out on the appeal of games to females of all ages. The experience of playing games has been transformed since the days of *PacMan* and *Ms. PacMan.* Gillian Skirrow's claim that "the pleasure of computer games is gender specific . . . women do not play them"[39] is no longer valid.

Media interest in computer game-playing has ensured that game cultures are public cultures despite the domestication of the console. As games-playing has become an activity of public concern and interest, the continuing portrayal of computer games as an activity only of boys is very worrying. As in so many accounts of previous youth cultures, female participants have been rendered invisible.

NOTES

1. For an overview of the debates over 'passive' and 'active' audiences, see Dennis McQuail, *Mass Communication Theory: An Introduction* (London: Sage, 1987).

2. For accounts of video games which put forward this view, see Gillian Skirrow, "Hel-livision: An Analysis of Video Games," in Colin McCabe (ed.), *High Theory/Low Culture: Analysing Popular Television and Film* (Manchester: Manchester University Press, 1986), and Leslie Haddon, "Electronic and Computer Games: The History of an Interactive Medium," *Screen,* vol. 29, no. 2, 1988.

3. The most extensive research on the gendered nature of arcade game-playing is in S. Kaplan, "The Image of Amusement Arcades and the Difference in Male and Female Video Game Playing," *Journal of Popular Culture,* vol. 17, no. 1, 1983, pp. 93–98.

4. The minimum age to play legally in arcades in the United Kingdom is sixteen.

5. *Edge,* March 1994 (Future Publishing).

6. The Professional Association of Teachers has produced an information sheet out-lining the potentially harmful effects of computer games: see "Young People in the Firing Line," *Guardian Education,* April 13, 1993. For psychologist Dr. Mark Griffiths's comments, see "One in Five Children Addicted to Computer Games," *Guardian,* December 15, 1993. For Professor Cary Cooper's views, see "Children of an Electronic God," *Guardian,* April 27, 1993.

7. Both Gillian Skirrow in "Hellivision" and Leslie Haddon in "Electronic and Com-puter Games" document the appeal of computer games for young boys.

8. Leslie Haddon, "Interactive Games," in Philip Hayward and Tana Wollen (eds.), *Fu-ture Visions: New Technologies of the Screen* (London: BFI, 1993).

9. Catherine Bennett, "Game Boys and Girls," *Guardian,* December 2, 1993.

10. The Monopolies and Mergers Commission in the United Kingdom is currently in-vestigating the activities of Sega and Nintendo.

11. Sheila Hayman, "Control Freaks: How to Play God in Simcity," *Arena,* Winter 1993–94.

12. Nick Cohen, "This ain't no game," *GO*, October 1993.

13. Ibid.

14. David Sheff, "Game Wars," *Esquire*, October 1993.

15. Meg Carter, "Selling through Computer Games isn't Child's Play," *Independent on Sunday*, December 5, 1993.

16. Nintendo made bigger profits last year than Sony, Microsoft, and Apple Computers: see Sheff, "Game Wars."

17. Ibid.

18. Haddon, "Interactive Games," p. 144.

19. For a discussion of young girls' bedroom cultures, see Angela McRobbie, *Feminism and Youth Culture* (London: Macmillan, 1991).

20. For a fuller discussion of this, see Kirsten Drotner, "Modernity and Media Panics," in Michael Skovmand and Kim Christian Schroder (eds.), *Media Cultures: Reappraising Transnational Media* (London: Routledge, 1993), pp. 42–62.

21. See David Buckingham, *Children and Television: An Overview of Research* (London: British Film Institute, mimeo, 1987).

22. Catherine Bennett, in "Game Boys and Girls," suggested that "after crack-cocaine they are one of the most perfect forms of capitalism ever devised."

23. "Video Games: Young People in the Firing Line," *Guardian*, April 13, 1993.

24. See Desmond Ellis, "Video Arcades, Youth and Trouble," *Youth and Society*, vol. 16, no. 1, 1984.

25. These genres are: 1. Platformers: as the screen scrolls forward the characters have to leap from platform to platform (*Super Mario Brothers, Sonic the Hedgehog*); 2. Shoot 'em ups: hi-tech military games, the object of which is to defeat your opponent, often using military hardware; 3. Beat 'em ups: these involve fighting using hand-to-hand combat (*Streets of Rage, Streetfighter 2*); 4. Sports: these include football, golf, and Formula 1 racing simulations; 5. RPG/Strategy: these are role-playing and strategy games (*Sim City, Zelda 3*), 6. Puzzlers: these include computer game versions of board games and other puzzle games devised for the computer (*Tetris, Lemmings*).

26. A common feature of 'platform' games, where a boss character has to be defeated in order to pass through to the next level of the game.

27. See Geoffrey Pearson, "Falling Standards: A Short Sharp History of Moral Decline," in Martin Barker (ed.), *The Video Nasties* (London: Pluto, 1984).

28. Haddon, "Interactive Games."

29. See *Guardian*, April 13, 1993, and April 27, 1993.

30. "Video Games: Young People in the Firing Line."

31. McRobbie, *Feminism and Youth Culture*.

32. See Eugene F. Provenzo Jr., *Video Kids: Making Sense of Nintendo* (Cambridge, Mass.: Harvard University Press, 1991), and Marsha Kinder, "Playing with Power on Saturday Morning Television and on Home Video Games," *Quarterly Review of Film and Video*, vol. 14, 1992, pp. 29–59.

33. 'Blaze' in *Streets of Rage*, 'Chun Li' in *Streetfighter 2*, and 'Sonya Blade' in *Mortal Kombat* are all formidable females who are often chosen for their special moves.

34. Patricia Marks Greenfield, *Mind and Media: The Effects of Television, Video Games and Computers* (London: Fontana, 1984).

35. *Sonic* is a platform game featuring Sonic, a blue hedgehog whose aim is to free the Woodland Creatures who have been captured by Dr. Robotnik. *Sonic 2* is similar, but Sonic is now accompanied by his friend Tails the fox. In *Mickey Mouse in Castle of Illusion*, Mickey has to rescue Minnie from the witch Mizrabel. *Super Mario Brothers* and *Super Mario Brothers 3* feature Super Mario and his brother Luigi in their adventures through the Mushroom Kingdom in their search for Princess Toadstool, who has been captured by Bowser. *Streets of Rage 2* is an urban 'Beat 'em up' in which various characters attempt to rescue their buddies who have been captured by an evil gangster. *Street Fighter 2* is another urban 'Beat 'em up.'

36. Marsha Kinder writes of her son's transgender identification when he plays Princess Toadstool in *Super Mario Brothers 2.*

37. Princess Toadstool can float for short periods of time.

38. Haddon, "Interactive Games," p. 142.

39. Skirrow, "Hellivision," p. 115.

Ellen Seiter

Television and the Internet

Claims that the internet will revolutionize communications (as well as education, work life, and domestic leisure) are now commonplace. Yet there is a danger that computer communications—and by this I mean their uses, the discourses surrounding computers and the internet, and research about them—will substantially buttress hierarchies of class, race, and gender. One healthy corrective is to recognize the many parallels between television and the internet, and incorporate the insights of television audience research into the uses of technologies in the domestic sphere, the articulation of gender identities through popular genres, the complexity of individuals' motivations to seek out media, and the variety of possible interpretations of media technologies and media forms. Ethnography can offer a rich context of understanding the motivations and disincentives to using computers: an important research topic in a world in which nonusers are likely to be labeled recalcitrants, technophobes, or slackers.

Most academic research on digital technologies is currently being produced by departments of library and information science, schools of business and management, education schools, and computer science departments. This research tends to emphasize information-seeking and statistical patterns of usage, while ignoring perceptions about computers, the cultural contexts in which they are used, and the images, sounds, and words to be found on computer screens. We need a means of touching upon the form and content of the internet as well as the practices and motivations of computer users. We need to view the internet from the perspective of its many parallels with broadcast media, maintaining a healthy skepticism about its novel qualities as a communication medium. We need to be alert to the ways in which stereotyped notions of the audience are constructing a discourse around the internet that privileges white, middle-class males.

In this chapter, I begin by reviewing the ways in which familiar forms of television are migrating to computer screens, while television, for its part, is busily promoting use of the internet by television viewers. These connections between television and computers are taking place at

From *Television and New Media Audiences* (London: Oxford University Press, 1999). Reprinted by permission of the author and Oxford University Press. © Ellen Seiter 1999.

the level of corporations, as Microsoft attempts to enter the mass-market entertainment business by investing in media firms; at the level of technology, as computers with television tuners and video stream capabilities become more commonplace; and at the level of form and content, as familiar genres from television, radio, and newspapers are tried out on the internet, many of them sponsored by such giants of television advertising as Procter and Gamble and Nabisco. In the second section, I survey qualitative research on the use of computers in domestic settings among families, and trace the similarities between these findings and research on television viewing in the home. I argue that television audience research is well positioned to help us understand the heavily gendered use of computers in the domestic space. At the same time, research on computer use may help to push television audience research to a more thorough investigation of the connections between domestic and public uses of media, and to think more about television viewers as workers, not only as family members or individual consumers. In the third section, I look at the various work issues related to computer use, the increase of 'telecommuting,' and its implications for the study of communication technologies in the domestic sphere.

Television on Computers

Television sets and computer terminals will certainly merge, cohabit, and coexist in the next century. In 1996 computers with built-in television tuners became available, as did set-top boxes to allow internet access via television sets. Because of the proliferation of television sets throughout many homes, they are increasingly likely to share space with computers in the same room. Many people (most of my students, it seems) have become adept at watching television while using the computer. As the number of personal computers increases in middle-class homes, the boundaries between leisure and work time, public and private space, promise to become increasingly blurred (Kling 1996). As the internet develops from a research-oriented tool of elites to a commercial mass medium, resemblances between websites and television programming will increase.

The internet can be used to organize users around political matters in ways unimaginable through broadcast television or small format video—and fundamentalist Christians are one group that has already proven this potential. It seems increasingly likely, however, that commercialization of the Web will discourage activism in favor of consumerism and the duplication of familiar forms of popular mass media, such as magazines, newspapers, and television programs (Morris and Ogan 1996). The World Wide Web reproduces some popular genres from television (and radio) broadcasting: sports, science fiction, home shopping clubs, news magazines, even

cyber-soap operas with daily postings of the serialized lives of its characters. In fact, the most popular websites represent the same genres—science
fiction, soap operas, and talk shows—that form the topic of some of the
best television audience research, by, for example, Press (1991), Jenkins
(1992), Gillespie (1995), and Shattuc (1997). The much-publicized presence
of pornography on the internet also parallels the spectacular success of
that genre on home video.

   The prevalence of television material on the Web confirms the insight provided by media ethnographers of the importance of conversation
about television in everyday life, and suggests that television plays a central role as common currency, a lingua franca. Television fans are a formidable presence on the internet: in chatrooms, where fans can discuss
their favorite program or television stars; on websites, where fan fiction can
be posted; and as the presumed market for sales of television tie-in merchandise. The dissemination of knowledge of the programming language
for the creation of websites (or home pages) unleashed countless die-hard
television fans eager to display their television knowledge—and provides
free publicity for television producers. Hundreds of painstakingly crafted
home pages have been devoted to old television shows. For example, one
site devoted to the 1964–65 Hanna-Barbera cartoon *Johnny Quest*, provides
plot summaries and still frames of every episode ever made. In fact, the
Web is a jamboree of television material, with thousands of official and
unofficial sites constituting television publicity, histories, cable and broadcast schedules, and promotional contests. Search engines turn up roughly
three times as many references to television as they do for topics such as
architecture, chemistry, or feminism. Apparently, television was one of the
first topics people turned to when trying to think of something to interest
a large and anonymous group of potential readers—other internet users.

   It would be a mistake, however, to see the rise of television material on the internet exclusively from a fan or amateur perspective, because
the connections between television and computer firms are proliferating.
The association between television and the internet has been heavily promoted at the corporate level by access providers eager to lure as sponsors
companies that invest heavily in television advertising, and by others seeking sources and inspiration for the new internet 'programming' (Schiller
1997). The software giant Microsoft corporation has acquired stakes in
media entertainment companies, developed an interactive television network, and looked to television and film for the basis for entertainment
'software' with a more 'universal,' that is, mass-market, appeal. Microsoft's partnership with NBC to form the twenty-four-hour news cable network and on-line news and information service MSNBC is the most obvious
example. Microsoft also has joint-venture deals with the cable network
Black Entertainment Television; with Stephen Spielberg, Jeffrey Katzenberg,
and David Geffen's company Dream Works SKG; and with Paramount
Television Group—all of which is leading to speculation that Microsoft is

"morphing into a media company for the new millennium" (Caruso 1996). Disney, now the owner of the television network ABC, is also one of the biggest interactive media producers in the world. America Online, the commercial internet and e-mail access provider, has followed a vigorous commercial strategy, which includes extensive coverage of television in all its familiar publicity aspects as well as encouragement of fan activity, to build a broad base of subscribers and to court advertisers. The A. C. Nielsen company, the television industry leader in audience ratings, produces reports on internet users. Worldgate Communications is offering a set-top box and hand-held remote control device that allows a viewer to access websites tied to television programs currently being watched.

In 1996, Microsoft made its intention to break into broadcasting explicit when it released a revised version of its on-line service, whose browser interface sends users straight to an 'On Stage' section with six different channels, each hosting 'shows' (Helm 1996). The goal was to give users a better idea of what to expect from each program by standardizing its offerings, a strategy strikingly familiar from the history of early radio and television (Boddy 1990; McChesney 1996). Videostreaming, already a commonplace on the World Wide Web, has been implemented on websites such as Cable News Network's to replay the 'News Story of the Week.' Advertising industry analysts predict that animated advertisements on the Web will dominate in the years ahead.

Hardware and software manufacturers are scrambling to secure the market for sales to noncomputer owners of devices that will convert the plain old television set to an internet browser, or win the battle between the high-definition television sets favored by the electronics industry and the digital television/monitors favored by the computer industry. The computer position is that "consumers would rather have a cheaper box that would be either a computer monitor or a TV than have the less complicated, high-definition-TV set that the consumer electronics industry favors" (Auletta 1997, p. 77). Microsoft is exerting considerable muscle in political lobbying and industry influence over issues of digital television (Bank and Takahashi 1997), exemplified by Bill Gates's decision to purchase WEB TV for $425 million and to announce this decision at the annual convention of the National Association of Broadcasters.

The relatively high penetration rates of home computers among the professional classes (including writers about computer issues) often give the false impression that everyone has a computer. The majority of homes do not have a PC, but they do have a television. Therefore, the computer industry continues to eye television greedily as a future market. Microsoft has entered the cable television business, exploring set-top boxes and television programming, and vigorously campaigning to thwart the success of High Definition Television (HDTV) in favor of 'digital television,' which would use computer monitors. Television set manufacturers are gambling on a variety of designs that integrate Web access with television viewing through cable boxes, double windows, wireless keyboards, and television remote controls with data entry features. Digital television sets are being designed to maximize flexibility for future uses with satellite receivers, internet navigators, digital video disc players, and set-top boxes.

The extent to which Microsoft is explicitly using television and its mass appeal as a model for future endeavors was made explicit in a recent article by Ken Auletta about Nathan Myhrvold of Microsoft. Microsoft realizes that "the skills that made Microsoft successful in software—technical proficiency, rapid response—are not transferable to what the company calls the content business, which relies more on a bottom-up rather than a top-down model" (Auletta 1997, p. 76). The fact that personal computers are stuck at a penetration rate of about 36 percent has led computer industry people to eye television jealously, and to try to develop ways to link perceptions about computers to entertainment. At Interval Research Corporation, the think-tank started by Microsoft's cofounder Paul Allen, the Explorers market research group has adopted the strategy of using television—a universally accepted domestic technology—as a model for the development of future communication technologies (Ireland and Johnson 1995).

For its part, television plays a crucial role as publicizer of the Web and computer use. Television programs are already filled with references to computers and the internet that both dramatize the importance of the new technologies and attempt to play a major role in educating the public about new media. Television's appetite for novelty, as well as its fears about losing viewers to computer screens, make computers one of its predictable obsessions. Silverstone and Hirsch are right to point to the dual nature of communication technologies such as television sets and computers "as quintessentially novel objects, and therefore as the embodiment of our desires for the new," which simultaneously act as "transmitters of all the images and information that fuel those desires" (1992, p. 3).

Computer references have gone far beyond the television character staring into the computer screen—although this long-standing movie cliché has been solidly established as a convention of television drama. Television commercials refer viewers to websites; call-in programs now ask that e-mail be submitted, as do television shows from *Meet the Press* to Nickelodeon's children's lineup. Numerous cable shows include e-mails (reflecting the striking banality of much chatroom conversation) scrolling across the screen below the program material. Television news shows, especially docudramas and news magazines, have become so enamored of reproducing e-mail and internet communications (which are both easy to capture on camera and lend a feeling of novelty and a sense of connection to the real world of viewers), that the practice has become a copyright/ privacy concern among computer specialists (Lesch 1994).

On-line communications have been used both to support and to attack television shows and their sponsors. Television networks are exploiting e-mail and internet communications with audiences to gain feedback on script or character changes, to compile mailing lists for licensed products relating to shows, and to publicize tie-in merchandise. The creation of an internet home page for the *X-Files* was credited with saving the show from cancellation after its first season. The *X-Files* producers recognized the perfect synergy of its high demographic fans and the internet, and targeted its audience through the World Wide Web, a move that helped both to prove its audience share to executives contemplating axing the show, and to generate more publicity for the program. Protests against television animate the on-line communications as well; Christian Right organizations such as The American Family Association use the internet to organize protests against television sponsors of objectionable material (a list it calls 'The Dirty Dozen') and 'filth' (*NYPD Blue*).

Computer and internet research can benefit greatly from television research on television flow and the use of remote controls, the installment of the television as a domestic object, and conversation around television. For example, the internet poses problems similar to that of television 'flow' (Williams 1974), as Web 'programmers' (especially those with commercial sponsors) attempt to guide the user through a preplanned sequence of screens and links. While nearly every branch of the advertising industry is making moves to work on the internet, anxieties are already rife about ways to measure consumers. At first, the number of 'hits' a website received was enough to entice sponsors. Repeated anxieties have surfaced about the fact that banner advertisements are often ignored and software that attempts to track the behavior of Web users cannot detect when the transfer is stopped before the ad is delivered. Thus, talk of hits gave way to a preference for 'impressions'—a word more likely to carry weight with sponsors, with its desirable associations with lasting mental influence. While an estimated $1 billion was spent on internet advertising in 1997, and that figure is expected to double in 1998, advertisers remain skeptical about tracking software and accountability in measuring consumers. This concern for the meaningfulness of exposure to Web advertising closely parallels the anxieties of television advertisers about the attention span of television viewers, and their proclivity for 'zapping' commercials by switching channels, or 'zipping' past commercials on videotapes of prerecorded programs. From the advertiser's perspective, Web surfers can be just as fickle as television watchers, it seems.

In *Desperately Seeking the Audience* (1991), Ien Ang carefully deconstructed the fantasies of control over television viewers and the necessity of such fantasies to the daily functioning of the television industry. In the trade publication *Advertising Age*, these fantasies, and the battle over competing claims for accurate measurement of Web surfers, are now a major preoccupation. On the one hand, the internet is projected as a much better vehicle than television, because the Web user is presumably more attentive, more goal-directed—and wealthier. Yet, on the other hand, anxieties about measuring and controlling Web users are escalating, and energies are focusing around the development of measurement devices adequate to convince sponsors.

For its target market of the professional upper-middle class, advertisers are promoting the installation of the computer as a domestic object in a process similar to the guidelines for installing the television set in the home in the 1950s studied by Lynn Spigel (1992). Computers are advertised on utopian claims to enrich family life, enhance communications, strengthen friendship and kin networks, and, perhaps more important, make children smarter and give them a competitive advantage in the educational sphere. In advertising, in news broadcasts, in education journals, the computer is often defined against, and pitched as an improvement on the television set: where television viewing is passive, computer use is interactive; where television programs are entertaining in a stale, commercialized, violent way, computer software and the internet are educational, virtuous, new.

Negative feelings about television viewing (shame or defensiveness) affect what people are willing to say about television. Comparison between television viewing and use of the World Wide Web are inevitable. Like television programming, computer media—software, websites, and so on—serve as topics of conversation, but the latter hold more legitimacy among the educated middle classes. Among middle-class professionals, the group best positioned to parlay computer use into improved earning power, discussing a new website holds more cachet than talking over last night's sitcom. The negative association of being a computer nerd, or even a hacker, have abated considerably in the past decade (Turkle 1995), while computer magazines such as *Wired* have promoted fashionable postmodern associations with computer use. While some sanctions are associated with being a nerd, this stereotype has a higher gender, class, and intellectual standing than the couch potato. On the other hand, those with less disposable income and less familiarity with computers may reject computers for the values they represent (such as dehumanization), their emphasis on written rather than oral culture, their associations with white male culture (hackers and hobbyists), and their solitary, antisocial nature. The operations of 'distinction' will be especially important to bear in mind when doing empirical work on the social contexts of computer use.

Gendered Uses of Computers at Home

Television sets and computers introduce highly similar issues in terms of placement in domestic space, conflicts among family members over usage and control, and value in the household budget. We can expect these conflicts to be articulated within gender roles in the family. Some research on gendered conflicts over computers (Haddon 1992; Murdock et al. 1992; Giacquinta et al. 1993) reproduces themes of family-based studies about control of the television set. Already, researchers have noted a strong tendency for men and boys to have more access to computers in the home. Television studies such as those carried out by Ann Gray (1992), David Morley (1986), and myself (Seiter et al. 1989) suggest that women in nuclear families have difficulty watching a favorite television show (because of competition for control of the set from other family members, and because of shouldering the majority of childcare, housework, and cooking). If male family members gravitate toward the computer as hobbyists, the load of chores relegated to female family members will only increase, and make it more difficult for female members to get time on the home computer. Computers require hours of trial-and-error experimentation, a kind of extended play demanding excess leisure time. Fully exploring the internet demands time for lengthy downloading, and patience with connections that are busy, so much so that some have dubbed the World Wide Web the World Wide Wait.

In the family, computers can create anxiety, too: young children must be kept away from the keyboard because of potential damage to the machine. Mothers, who have traditionally been charged with securing the academic success of their children, would have a strong incentive to relinquish computer time to older children, who are thought to benefit greatly from all contact with the technology. When anxieties increase and moral panics are publicized about children's encounters with pornography through the computer, or the unhealthy effects of prolonged computer use, the brunt of responsibility for enforcing restrictions on computer use will fall to mothers and teachers.

Some qualitative research has already explored these areas, and some of the most valuable work has been informed by British cultural studies work. Silverstone et al. have offered a fascinating case study of a well-to-do London family whose home included a wide array of new technologies, and whose explicit ideology was one of encouraging children to use them. Yet the mother remained at a weary distance from the computer. She responded with irritation to researchers' questions about her feelings toward the communication technologies, claiming not to have feelings about technologies at all (1990, p. 35). Both parents desired that their children gain computer experience and preferred this to television watching, but expressed irritation with the boy's domination of their home computer. Measures were taken to try to secure computer time for the daughters in the home, but with mixed success. The mother felt alienated from the developing 'father and son' culture around the computer, and suffered arguments about the selection of computer games, but her frustration led her to take a course on computing. Tensions among highly motivated, well-educated females over computer technology deserve much more investigation.

Similarly, Giacquinta et al., in their qualitative study of white middle-class New York families, found marked differences between males and females, and between adults and children, including less use by females: "mothers were particularly estranged from the machines" (1993, p. 80). The study, conducted in 1984–87, included sixty-nine mothers, two-thirds of whom were employed full time outside the home. Mothers tended to use computers for word-processing and "did not engage in programming, tinkering, pirating, or game playing" (1993, p. 81). They found the violence in computer games objectionable. In general, they lacked "the interest, the need and the time to develop computing skills" (1993, p. 89). Daughters were more likely than mothers to use the computer, and had more resources to support them, such as classroom teaching, but they were not a focus of the girls' leisure-time activities.

Friendships, kin networks, and work relationships are crucial to the successful adoption of new technologies such as computers (Douglas 1988). Computer use often involves borrowing software, troubleshooting problems, trying out new programs, boasting or discussing successes, and cross-checking machines. Advice and encouragement are important components of this. The Giacquinta study found women rarely spoke with other women about their computers or assisted each other in learning. If women and girls tend not to talk about computers, they are at a sizable disadvantage over boys and men, especially those with considerable practice at hobby talk. In another study of home computing, in the English Midlands, Murdock et al. found, in a sample of one thousand households, that males outnumbered females as the primary computer user by a ratio of seven to one. They also found that those who did not talk to others about computers or borrow programs from friends or relatives were most likely to have stopped using their computers (1992, p. 150).

Jane Wheelock found, in a British study of thirty-nine families in "a peripheral region of the national economy," that there were three times as many sons as daughters interested in the home computer. Wheelock's study focuses on the ways that the household, operating as a complementary economy to the formal one, reproduces and produces labor power as related to computer use (1992, pp. 98–99). Daughters were more likely to be interested in computers if such interest was facilitated by parental and teacher encouragement, or if there were no sons in the family—or after the machine had been abandoned by the sons in the family (1992, p. 110).

Boys were much more likely than girls to use computers as a part of their social networks, something that, Wheelock notes, "increases boys' socializing, and shifts its locus towards the home; traditionally both are features of girls' experience" (1992, p. 111). Leslie Haddon's observations, based on time spent in a computer club, similarly suggested that girls may not use computers as a topic for school conversation even when they do use computers at home, being more likely to discuss rented videos (1992, p. 91). The girls in Haddon's study were also unlikely to play computers in public places, stores, arcades, or school clubs, but used them at home.

Most of these studies are somewhat dated, and do not provide any information about the use of electronic mail and the World Wide Web. These two uses of the computer, in their facilitation of personal communications, most closely resemble the telephone, a communications technology particularly valued by women (Rakow 1988; Livingstone 1992; Spender 1995). Reliable information about Web users is hard to come by, and most research relies upon self-selection of its sample, that is, people responding to various postings asking for internet users to fill out a survey form. Some of this recent survey information suggests that the internet may be attracting women, especially younger, white, middle-class women under thirty-five, at surprising rates. In 1995, the Georgia Tech World Wide Web users survey reported that 29 percent of their respondents are female—a number that has increased significantly over the past three years. There has been a substantial increase in female Web users between the ages of sixteen and twenty, and an increase in female users who teach kindergarten through twelfth grade—there was a 10 percent increase in female users in this category observed in one year, 1994–95. The importance of access through public education institutions for women is significant: 39 percent of women responding gained internet access that way, compared to 28 percent of men. Yet there are clear signs that women are less likely to use computers intensively in their free time. On the weekends—an indication of hobbyist users—the gap widens to 75 percent male and only 25 percent female: thus, three times as many men are weekend users of computers compared to women. Women users were less likely than men to spend time doing 'fun computing' or to use a computer for more than thirty-one hours a week. Although the Web is attracting men over the age of forty-six in increasing numbers—many of them as a retirement hobby—the relative numbers of women of that age who use the internet are declining.

Much has been written about the ways in which the internet can be used to explore personality and identity (for example, Turkle 1995; Star 1995). But women and the poor are going to be less advantageously positioned to engage in such activities for a complex set of reasons (Star 1995). As Roger Silverstone has explained:

> [T]he ability to use information and communication technology as a kind of extension of the personality in time and space . . . is also a matter of resources. The number of rooms in a household relative to the number of people, the amount of money that an individual can claim for his or her own personal use, the amount of control of his or her own time in the often intense atmosphere of family life, all these things are obviously of great relevance. (Silverstone 1991, p. 12)

Computerized Work

Are women more likely to use the internet if they use computers on the job? Working on a computer can mean very different things: if we are to understand the differential desire to use computers during leisure time, it is essential to make distinctions between kinds of computerized work. For example, huge numbers of female employees occupy clerical jobs that use computers for processing payroll, word-processing, conducting inventory, sales, and airline reservations—more than 16 million women held such positions in the United States in 1993 (Kling 1996). Women overwhelmingly outnumber men in the kinds of job where telephones and computers are used simultaneously: airline reservations, catalogue sales, telephone operators. In contrast, fewer than half a million women work as computer programmers or systems analysts.

The type of employment using a computer that is likely to be familiar to the largest number of women, then, is a kind of work where keystrokes might be counted, where supervisors may listen in on phone calls, where productivity is scrutinized on a daily and hourly basis, where conversation with co-workers is forbidden (Clement 1994; Iacono and Kling 1996). The stressful and unpleasant circumstances under which this kind of work is performed might explain women's alienation from computer technology and their tendency to stay away from it during their leisure time.

Some parallels exist between clerical work and teaching—one of the sole white-collar professions dominated by women. As Steven Hodas points out in his discussion of what he calls 'technology refusal' in schools, most teachers are women, and most educational technologists are men, who target their efforts at introducing computer technologies toward classroom teachers, not male administrators. Too often, technologists express the need for classroom computers in ways that derogate the work of teachers: in these discussions, "the terms used to describe the insufficiency of the classroom and to condescend to the folk-craft of teaching are the same terms used by an androgenized society to derogate women's values and women's work generally." When technologies fail in the classroom, the reaction is to "blame the stubborn backwardness of teachers or the inflexibility and insularity of the school culture" (Hodas 1993, p. 206). Hodas usefully reminds us that present-day arguments about the need for computers in schools mirror the same redemption through technology arguments that accompanied other media, most notably for our purposes, educational video:

> The violence that technologists have done to our only public children's space by reducing it to an 'instructional delivery vehicle' is enormous, and teachers know that. To abstract a narrow and impoverished concept of human sentience from the industrial laboratory and then inflict it on

children for the sake of 'efficiency' is a gratuitous, stunning stupidity and teachers know that, too. Many simply prefer not to collaborate with a process they experience as fundamentally disrespectful to kids and teachers alike. (1993, p. 213)

Telecommuting, working from the home through a modem or internet access to the office, is a different category of computerized work which is supposed to hold special appeal to women. Telecommuting is now officially sanctioned by the U.S. government, according to Rob Kling:

> A recent report developed under the auspices of the Clinton administration included a key section. 'Promoting Telecommuting,' that lists numerous advantages for telecommuting. These include major improvements in air quality from reduced travel; increased organizational productivity when people can be more alert during working hours; faster commercial transit times when few cars are on the roads; and improvements in quality of worklife. The benefits identified in this report seem so overwhelming that it may appear remarkable that most organizations have not already allowed all of their workers the options to work at home. (1996, p. 212)

While there are growing numbers of women doing pink-collar jobs such as clerical and sales work at home—that is, using their home computers at jobs such as credit-card verification and telephone solicitations—this form of telecommuting is also largely invisible in the mass media. Instead, we see images of female professionals using computers to work from home, perhaps while their one and only child conveniently naps in the next room. As Lynn Spigel has noted, computer publications suggest "a hybrid site of home and work, where it is possible to make tele-deals while sitting in your kitchen" (Spigel 1996, p. 11). In her discussion of a *Mac Home* magazine cover, Spigel suggests that "The computer and Net offer women a way to do two jobs at once—reproduce and produce, be a mother and hold down a high powered job. Even while the home work model of domestic space finds a place for women, it does not really break down the traditional distinctions between male and female" (Spigel 1996, p. 12). As more workers struggle to get their work done from home without the benefit of an office, questions about the use of communication technologies become especially interesting and well suited to ethnographic approaches.

Another interesting issue for audience research is the entertainment uses of computers in offices, as office workers have access to more entertainment and play functions through their computers. Sherry Turkle's book *Life on the Screen* (1995) examines a group comprised mainly of students or white-collar computer programmers who work on computers throughout the day. Turkle is especially interested in participation in multi-user domains, or MUDS: on-line fantasy games that can have dozens of players. Turkle describes this balance of play and work:

I have noted that committed players often work with computers all day at their regular jobs. As they play on MUDs, they periodically put their characters to sleep, remaining logged on to the game, but pursuing other activities. The MUD keeps running in a buried window. From time to time, they return to the game space. In this way, they break up their day and come to experience their lives as cycling through the real world and a series of virtual ones. (1995, p. 189)

While Turkle is primarily interested in the psychological dynamics of play with a variety of virtual selves, she rarely foregrounds the very specific class fraction which has the privilege to play on the job. In this world, white-color, upper-middle-class employers are finding that after a period of vigorous encouragement if not requirement of nearly constant computer and internet use in many occupations, employees are spending large parts of the day playing computer games, writing personal e-mails, and cruising the internet, and that it is increasingly difficult to confine white-collar employees to work-related rather than entertainment uses of these technologies—or in some cases to distinguish between the two. As one expert wryly put it in a discussion of the impact of digital video discs (DVDs), "The big application for DVD later this year will be desktop video playback, which will eliminate any remaining worker productivity that hasn't already been destroyed by Web surfing" (Hood 1997, p. 14).

Turkle gives many examples of women immersed in the play with identity and the capacity for writing one's own dramatic narratives that MUDs offer, and she is especially interested in the phenomenon of players impersonating the opposite sex in their virtual personae. Turkle's study fails to interrogate the particular class background of the MIT students and computer programmers she interviewed for her study, or to ask how particular class positions may predispose women to be attracted to the kind of play with virtual selves that she describes as postmodern. The luxury of such computer play is unimaginable in the context of most clerical computer work and unlikely in the home workstation when women race to get their work done before children come home from school, or hurry to turn their attention from the computer to housecleaning, shopping, or food preparation.

For writers such as Turkle, the internet offers an arena for exploring—and deconstructing—traditional gender roles that can be quite liberating. Such an analysis contrasts sharply with the concerns of Cheris Kramarae, who, in her analysis of the genres and kinds of information available on the World Wide Web, has noted the rapid proliferation of genres such as pornography, and modes of discourse, such as flaming, which may act as deterrents to women going on-line (Kramarae and Kramer 1995). Similarly, linguist Susan Herring has cast doubt on utopian claims for on-line communications in the workplace that have suggested these technologies may be advantageous for women getting ahead in professions.

This work suggests that even among white-collar professionals, computers may be more fun for men than for women. For example, in a study of academics using electronic discussion groups, Herring found that:

> [M]ale and female academic professionals do not participate equally in academic CMC [Computer Mediated Communication]. Rather, a small male minority dominates the discourse both in terms of amount of talk, and rhetorically, through self-promotional and adversarial strategies. Moreover, when women do attempt to participate on a more equal basis, they risk being actively censored by the reactions of men who either ignore them or attempt to delegitimize their contributions (1996, p. 225)

Herring concluded that there was nothing inherently wrong with the technology of computer bulletin boards; rather, the problem stemmed from the ways that old, familiar forms of gender discrimination—from the academic workplace and from society—dictated the ways that participants would communicate. Laura Miller has dismissed this type of complaint about the internet, warning that we need to be wary of the media's attempt to cast women as victims on the internet, as in discussions of flaming or the prevalence of pornography:

> The idea that women merit special protections in an environment as incorporeal as the Net is intimately bound up with the idea that women's minds are weak, fragile, and unsuited to the rough and tumble of public discourse. It's an argument that women should recognize with profound mistrust and resist, especially when we are used as rhetorical pawns in a battle to regulate a rate (if elite) space of gender ambiguity. (1995, p. 58)

What is needed in this discussion between two different camps of feminists—those who see the internet as rife with sexism (a group Miller unhelpfully calls the 'schoolmarms') and those who see such behavior as incidental problems, easily overcome by assertive behavior—is research that links women's reaction and attraction to the internet to other aspects of their lived experience, including education, class, sexual orientation, religion, and workplace culture, including how frequently they experience and how they deal with sexual harassment in their workplace.

As Cynthia Cockburn reminds us, "gender is a social achievement. Technology too" (1992, p. 39). The trick for media researchers will be to avoid a kind of static gender essentialism, where women, the elderly, and the poor are eternally, innately deficient vis-à-vis new technologies, while all men benefit from inborn technical know-how, an affinity to toying with machines, and an enthusiastic embrace of time-intensive forms of hobbyism. Ethnographic studies can help to provide materialist explanations for women's reluctance to immerse themselves in computer use and on-line communications, as well as explicating the circumstances under which women are attracted to and excel at computer skills.

The role of feminist researchers will include tempering enthusiasm for these new technologies, as Kramarae and Kramer argue:

> The so-called popular media treat the new electronic systems as sexy, but the media fail to deal seriously with most of the gender-related issues. As those who are working closely with its developments know, the Internet will not, contrary to what the media tout, ride the world of hostility, ignorance, racism, sexism, greed and undemocratic governments. The Internet has the potential for creating a cooperative collective international web; but, as with all technologies, the Internet system will be shaped by prevailing communication behaviors, economic policy, and legal decisions. (1995, p. 15)

One of the most important functions of ethnography, then, may be as an antidote to the hype about computers. Ethnography can describe, for example, the full context in which barriers to entrance onto the information highway exist—especially among people still worrying about keeping a car running on the real highway, much less having the extra time and money to maintain a computer at home.

When turning to the internet, it will be especially important to treat computer communications as deeply cultural, as employing fiction and fantasy as well as information, and as open to variable interpretations based on gender, race, and ethnicity. As the internet is championed as the ultimate in 'interactive' media use, it will be extremely important for media researchers to put forward appropriate skepticism about the laudatory uses of the term 'active' in discussions of media use, and to problematize the complex factors involved in attracting both television viewers and computer users to particular contents and genres.

Ethnography can offer an appreciation of some of television's advantages over computers (such as its accessibility by more than one person at a time, and its visual and aural modes of communication) and guard against the unnecessary pathologization of television viewing, which almost always acts to stigmatize the already social powerless. It will be important to look at the reasons why and the contexts in which television might be more appealing than the internet. We know that television—for all of its failings—has been cheap, easy to watch at home, enjoyable to talk about with others, and reasonably successful as an educational tool. It will be important to understand the ways in which such social features of broadcast technologies are missing or modified in new technologies. Finally, scholars should work to change the media and the structures of access to communication technologies, while politicizing the public discussion of media in ways that make explicit the gains and the losses at stake in promoting different representations of television audiences and computer users.

Ellen Seiter

REFERENCES

Ang, I. (1991). *Desperately Seeking the Audience*. London and New York: Routledge.

Auletta, K. (1997). "The Microsoft Provocateur." *New Yorker*, May 12: 66–77.

Bank, D., and D. Takahashi. (1997). "Microsoft Plans Big Digital TV Push, Stressing Hardware and Programming." *Wall Street Journal*, April 16: B6.

Boddy, W. (1990). *Fifties Television: The Industry and Its Critics*. Urbana: University of Illinois Press.

Caruso, D. (1996). "Microsoft Morphs into a Media Company." *Wired* 4/6: 126–29.

Clement, A. (1994). "Computing at Work: Empowering Action by 'Low-Level Users.'" *Communications of the ACM* 37/1: 52–65. Also reprinted in R. Kling, *Computerization and Controversy* (1996).

Cockburn, C. (1992). "The Circuit of Technology: Gender, Identity and Power." In R. Silverstone and E. Hirsch, *Consuming Technologies*, pp. 32–47.

Douglas, M. (1988). "Goods as a System of Communication." In *In the Active Voice*. London: Routledge & Kegan Paul, 20–29.

Giacquinta, J. B., J. A. Bauer, and J. E. Levin. (1993). *Beyond Technology's Promise: An Examination of Children's Educational Computing at Home*. Cambridge and New York: Cambridge University Press.

Gillespie, M. (1995). *Television, Ethnicity and Cultural Change*. London and New York: Routledge.

Gray, A. (1992). *Video Playtime: The Gendering of a Leisure Technology*. London and New York: Routledge.

Haddon, L. (1992). "Explaining ICT Consumption: The Case of the Home Computer." In R. Silverstone and E. Hirsch, *Consuming Technologies*, pp. 82–96.

Helm, L. (1996). "Microsoft Unveils Revamped Online Service." *Los Angeles Times*, October 11: D2.

Herring, S. C. (1996). "Gender and Democracy in Computer-Mediated Communication." *Electronic Journal of Communication* 3/2. Reprinted in S. C. Herring, *Computer-Mediated Communication: Linguistic and Cultural Perspectives*. Amsterdam and Philadelphia: J. Benjamin (1996), pp. 225–46.

Hodas, S. (1993). "Technology Refusal and the Organizational Culture of Schools." *Electronic Journal of Education Policy Analysis Archives*, 1/10. Reprinted in R. Kling, *Computerization and Controversy* (1996), pp. 197–218.

Hood, P. (1997). "The Wizard of Silicon Valley." *Newmedia* 6: 14.

Iacono, S., and R. Kling. (1996). "Computerization Movements and Tales of Technological Utopianism." In R. Kling, *Computerization and Controversy*, pp. 85–107.

Ireland, C., and B. Johnson. (1995). "Exploring the Future in the Present." *Design Management Journal* 6/2: 57–64.

Jenkins, H. (1992). *Textual Poachers: Television Fans and Participatory Culture*. London and New York: Routledge.

Kling, R. (ed.). (1996). *Computerization and Controversy: Value Conflicts and Social Choices*, 2nd ed. San Diego: Academic Press.

Kramarae, C. (ed.). (1988). *Technology and Women's Voices: Keeping in Touch*. New York: Routledge & Kegan Paul.

Kramarae, C., and J. Kramer. (1995). "Legal Snarls for Women in Cyberspace." *Internet Research: Electronic Networking Applications and Policy* 5: 14–24.

Lesch, S. G. (1994). "Your Work On Television? A View from the USA." *Computer Mediated Communication Magazine* 1/4: 5

Livingstone, S. M. (1992). "The Meaning of Domestic Technologies: A Personal Construct Analysis of Familial Gender Relations." In R. Silverstone and E. Hirsch, *Consuming Technologies*, pp. 113–30.

McChesney, R. W. (1996). "The Internet and U.S. Communication Policy-Making in Historical and Critical Perspective." *Journal of Communication* 46/1: 98–124.

Miller, L. (1995). "Women and Children First: Gender and the Settling of the Electronic Frontier." In J. Brook and I. A. Boal (eds.), *Resisting The Virtual Life.* San Francisco: City Lights.

Morley, D. (1986). *Family Television.* London: Comedia/Routledge.

Morris, M., and C. Ogan. (1996). "The Internet as a Mass Medium." *Journal of Communication* 46/1: 39–50.

Murdock, G., P. Hartmann, and P. Gray. (1992). "Contextualizing Home Computing: Resources and Practices." In R. Silverstone and E. Hirsch, *Consuming Technologies,* pp. 146–60.

Press, A. (1991). *Women Watching Television.* Philadelphia: University of Pennsylvania.

Rakow, L. (1988). "Women and the Telephone: The Gendering of a Communications Technology." In C. Kramarae (ed.), *Technology and Women's Voices,* pp. 207–28.

Schiller, D. (1997). "Les marchands à l'assaut de l'Internet" (Cornering the Market in Cyberspace). *Le Monde Diplomatique* 516/March: 1, 24, 25.

Seiter, E., H. Borchers, G. Kreutzner, and E. Warth (eds.). (1989). *Remote Control: Television, Audiences and Cultural Power.* London: Routledge.

Shattuc, J. (1997). *The Talking Cure: TV Talk Shows and Women.* New York: Routledge.

Silverstone, R. (1991). "Beneath the Bottom Line: Households and Information and Communication Technologies in an Age of the Consumer." London: PICT (Programme in Information and Communication Technologies), Policy Research Papers No. 17.

Silverstone, R., and E. Hirsch (eds.). (1992). *Consuming Technologies: Media and Information in Domestic Spaces.* London: Routledge

Silverstone, R., E. Hirsch, and D. Morley (1990). "Listening to a Long Conversation: An Ethnographic Approach to the Study of Information and Communication Technologies in the Home." CRICT Discussion Paper, Brunel University. Reprinted in *Cultural Studies* 5/2 (1991).

Spender, D. (1995). *Nattering on the Net: Women, Power and Cyberspace.* North Melbourne, Australia: Spinifex Press.

Spigel, L. (1992). *Make Room for TV: Television and the Family Ideal in Postwar America.* Chicago: University of Chicago Press.

———. (1996). "Portable TV: Studies in Domestic Space Travel." Paper delivered at Console-Ing Passions: The Annual Conference on Television, Video, and Feminism.

Star, S. L. (1995). "Introduction." *The Cultures of Computing.* Oxford: Blackwell, 1–28.

Turkle, S. (1995). *Life on the Screen.* New York: Simon & Schuster.

Wheelock, J. (1992). "Personal Computers, Gender and an Institutional Model of the Household." In R. Silverstone and E. Hirsch, *Consuming Technologies,* 97–112.

Williams, R. (1974). *Television: Technology and Cultural Form.* London: Fontana.

Boundaries,
Identities, Practice

"Armstrong pledges 'no redlining.' AT&T chief says broadband cable networks will not neglect poorer neighborhoods. AT&T expects to get FCC approval to merge with cable giant Tele-Communications, Inc. in the next few weeks."
                                                    –Broadcasting and Cable

"Given this hegemonic cartography, those of us (illegal aliens) living South of the digital border were forced to assume once again the unpleasant but necessary roles of webbacks. . . . In the barrios of resistance, . . . every block has a secret community center. There, the runaway youths called Robo-Raza III or 'floating greasers' publish anarchist laser-Xerox magazines, (and) edit experimental home videos on police brutality."
                                                    –Guillermo Gómez-Peña

"All hold-ups, hijacks and the like are now as it were simulation hold-ups, in the sense that they are inscribed in advance in the decoding and orchestration rituals of the media, anticipated in their mode of presentation and possible consequences. . . . The political sphere must keep secret the rule of the game that, in reality, power doesn't exist. Its strategy is, in fact, always creating a space of optical illusion, maintaining itself in total ambiguity, total duplicity in order to throw the others into this space."
                                                    –Jean Baudrillard

Andrew Ross

Hacking Away
at the Counterculture

Ever since the viral attack engineered in November 1988 by Cornell University hacker Robert Morris on the national network system internet, which includes the Pentagon's ARPAnet data exchange network, the nation's high-tech ideologues and spin doctors have been locked in debate, trying to make ethical and economic sense of the event. The virus rapidly infected an estimated six thousand computers around the country, creating a scare that crowned an open season of viral hysteria in the media, in the course of which, according to the Computer Virus Industry Association in Santa Clara, California, the number of known viruses jumped from seven to thirty during 1988, and from three thousand infections in the first two months of that year to thirty thousand in the past two months.[1] While it caused little in the way of data damage (some richly inflated initial estimates reckoned up to $100 million in downtime), the ramifications of the internet virus have helped to generate a moral panic that has all but transformed everyday "computer culture."

Following the lead of the Defence Advance Research Projects Agency (DARPA) Computer Emergency Response Team at Carnegie-Mellon University, antivirus response centers were hastily put in place by government and defense agencies at the National Science Foundation, the Energy Department, NASA, and other sites. Plans were made to introduce a bill in Congress (the Computer Virus Eradication Act, to replace the 1986 Computer Fraud and Abuse Act, which pertained solely to government information) that would call for prison sentences of up to ten years for the "crime" of sophisticated hacking, and numerous government agencies have been involved in a proprietary fight over the creation of a proposed Center for Virus Control—modeled, of course, on Atlanta's Centers for Disease Control, notorious for its failures to respond adequately to the AIDS crisis.

In fact, media commentary on the virus scare has run not so much

From *Strange Weather: Culture, Science, and Technology in the Age of Limits* (London: Verso, 1991). Reprinted by permission of Verso and the author. © 1991 by Andrew Ross.

tongue-in-cheek as hand-in-glove with the rhetoric of AIDS hysteria—for example, the common use of terms like *killer virus* and *epidemic;* the focus on high-risk personal contact (virus infection, for the most part, is spread through personal computers, not mainframes); the obsession with defense, security, and immunity; and the climate of suspicion generated around communitarian acts of sharing. The underlying moral imperative is this: you can't trust your best friend's software any more than you can trust his or her bodily fluids. Safe software or no software at all! Or, as Dennis Miller put it on *Saturday Night Live,* "Remember, when you connect with another computer, you're connecting to every computer that computer has ever connected to." This playful conceit struck a chord in the popular consciousness, even as it was perpetuated in such sober quarters as the Association of Computing Machinery—the president of which, in a controversial editorial titled "A Hygiene Lesson," drew comparisons not only with sexually transmitted diseases but also with a cholera epidemic, and urged attention to "personal systems hygiene."[2] Some computer scientists who studied the symptomatic path of Morris's virus across the internet have pointed to its uneven effects upon different computer types and operating systems, and concluded that "there is a direct analogy with biological genetic diversity to be made."[3] The epidemiology of biological virus (especially AIDS) research is being studied closely to help implement computer security plans. In these circles, the new witty discourse is laced with references to antigens, white blood cells, vaccinations, metabolic free radicals, and the like.

The form and content of more lurid articles like *Time's* infamous story, "Invasion of the Data Snatchers" (September 1988), fully displayed the continuity of the media scare with those historical fears about bodily invasion, individual and national, that are endemic to the paranoid style of American political culture.[4] Indeed, the rhetoric of computer culture, in common with the medical discourse of AIDS research, has fallen in line with the paranoid, strategic mode of Defense Department rhetoric established during the cold war. Each language repertoire is obsessed with hostile threats to bodily and technological immune systems; every event is a ballistic maneuver in the game of microbiological war, where the governing metaphors are indiscriminately drawn from cellular genetics and cybernetics alike. As a counterpoint to the tongue-in-cheek AI tradition of seeing humans as "information-exchanging environments," the imagined life of computers has taken on an organicist shape now that they too are subject to cybernetic "sickness" or disease. So too, the development of interrelated systems, such as the internet itself, has further added to the structural picture of an interdependent organism, whose component members, however autonomous, are all nonetheless affected by the "health" of each individual constituent. The growing interest among scientists in developing computer programs that will simulate the genetic behavior of living organisms (in which binary numbers act like genes) points to a future where the border between organic and artificial life is less and less distinct.

In keeping with the increasing use of the language of biology to describe mutations in systems theory, conscious attempts to link the AIDS crisis with the information security crisis have pointed out that both kinds of virus, biological and electronic, take over the host cell/program and clone their carrier genetic codes by instructing the hosts to make replicas of the viruses. Neither kind of virus, however, can replicate itself independently; both are pieces of code that attach themselves to other cells/programs—just as biological viruses need a host cell, computer viruses require a host program to activate them. The internet virus was not, in fact, a virus, but a worm, a program that can run independently and therefore appears to have a life of its own. The worm replicates a full version of itself in programs and systems as it moves from one to another, masquerading as a legitimate user by guessing the user passwords of locked accounts. Because of this autonomous existence, the worm can be seen to behave as if it were an organism with some purpose, or teleology, and yet it has none. Its only "purpose" is to reproduce and infect. If the worm has no inbuilt antireplication code, or if the code is faulty as was the case with the internet worm, it will make already infected computers repeatedly accept further replicas of itself until their memories are clogged. A much quieter worm than that engineered by Morris would have moved more slowly (as one supposes a "worm" should), protecting itself from detection by ever more subtle camouflage, and propagating its cumulative effect of operative systems inertia over a much longer period of time.

In offering such descriptions, however, we must be wary of attributing a teleology/intentionality to worms and viruses that can be ascribed only, and in most instances speculatively, to their authors. There is no reason why a cybernetic "worm" might be expected to behave like a biological worm in any fundamental way. So, too, the assumed intentionality of its author distinguishes the human-made cybernetic virus from the case of the biological virus, the effects of which are fated to be received and discussed in a language saturated with human-made structures and narratives of meaning and teleological purpose. Writing about the folkloric theologies of significance and explanatory justice (usually involving retribution) that have sprung up around the AIDS crisis, Judith Williamson has pointed to the implications of this collision between an intentionless virus and a meaning-filled culture:

> Nothing could be more meaningless than a virus. It has no point, no purpose, no plan; it is part of no scheme, carries no inherent significance. And yet nothing is harder for us to confront than the complete absence of meaning. By its very definition, meaninglessness cannot be articulated within our social language, which is a system *of* meaning: impossible to

include, as an absence, it is also impossible to exclude—for meaningless-ness isn't just the opposite of meaning, it is the end of meaning, and threatens the fragile structures by which we make sense of the world.[5]

No such judgment about meaninglessness applies to the computer security crisis. In contrast to HIV's lack of meaning or intentionality, cybernetic viruses are always replete with social significance. Their meaning is related, first of all, to the author's local intention or motivation—whether psychic or fully social, whether wrought out of a mood of vengeance, a show of bravado or technical expertise, a commitment to a political act, or in anticipation of the profits that often accrue from the victims' need to buy an antidote from the author. Beyond these local intentions, how-ever, which are usually obscure or, as in the Morris case, quite inscrutable, there is an entire set of social and historical narratives that surround and are part of the "meaning" of the virus: the coded anarchist history of the youth hacker subculture; the militaristic environments of search-and-destroy warfare (a virus has two components—a carrier and a "warhead"), which because of the historical development of computer technology, con-stitute the family values of information technoculture; the experimental research environments in which creative designers are encouraged to work; and the conflictual history of pure as against applied ethics in the science and technology communities, to name just a few. A similar list could be drawn up to explain the widespread and varied *response* to computer viruses, from the amused concern of the cognoscenti to the hysteria of the casual user, and from the research community and the manufacturing industry to the morally aroused legislature and the mediated culture at large. Every one of these explanations and narratives is the result of social and cultural processes and values; consequently, there is very little about the virus itself that is "meaningless." Viruses can no more be seen as an objective or necessary result of the "objective" development of techno-logical systems than technology in general can be seen as an objective, de-termining agent of social change.

For the sake of polemical economy, I would note that the cumu-lative effect of all the viral hysteria has been twofold. First, it has resulted in a windfall for software producers, now that users' blithe disregard for makers' copyright privileges has eroded in the face of the security panic. Used to fighting halfhearted rearguard actions against widespread piracy practices, or reluctantly acceding to buyers' desire for software unencum-bered by top-heavy security features, software vendors are now profiting from the new public distrust of program copies. So, too, the explosion in security consciousness has hyperstimulated the already fast-growing sec-tors of the security system industry and the data encryption industry. In line with the new imperative for everything from "vaccinated" worksta-tions to "sterilized" networks, it has created a brand-new market of viral vaccine vendors who will sell you the virus (a one-time only immunization

shot) along with its antidote—with names like Flu Shot +, ViruSafe, Vacci-
nate, Disk Defender, Certus, Viral Alarm, Antidote, Virus Buster, Gate-
keeper, Ongard, and Interferon. Few of the antidotes are very reliable,
however, especially since they pose an irresistible intellectual challenge
to hackers who can easily rewrite them in the form of ever more powerful viruses. Moreover, most corporate managers of computer systems and
networks know that the vast majority of their intentional security losses
are a result of insider sabotage and monkeywrenching.

In short, the effects of the viruses have been to profitably clamp
down on copyright delinquency, while generating the need for entirely new
industrial production of viral suppressors to contain the fallout. In this
respect, it is easy to see that the appearance of viruses could hardly, in
the long run, have benefited industry producers more. In the same vein,
the networks that have been hardest hit by the security squeeze are not
restricted-access military or corporate systems but networks like the internet, set up on trust to facilitate the open academic exchange of data, information, and research, and watched over by its sponsor, DARPA. It has
not escaped the notice of conspiracy theorists that the military intelligence
community, obsessed with "electronic warfare," actually stood to learn a
lot from the internet virus; the virus effectively "pulsed the system," ex-
posing the sociological behavior of the system in a crisis.[6]

The second effect of the virus crisis has been more overtly ideological. Virus-conscious fear and loathing have clearly fed into the paranoid
climate of privatization that increasingly defines social identities in the
new post-Fordist order. The result—a psychosocial closing of the ranks
around fortified private spheres—runs directly counter to the ethic that
we might think of as residing at the architectural heart of information
technology. In its basic assembly structure, information technology involves processing, copying, replication, and simulation, and therefore
does not recognize the concept of private information property. What is
now under threat is the rationality of a shareware culture, ushered in as
the achievement of the hacker counterculture that pioneered the personal
computer revolution in the early 1970s against the grain of corporate
planning.

There is another story to tell, however, about the emergence of the
virus scare as a profitable ideological moment, and it is the story of how
teenage hacking has come to be defined increasingly as a potential threat
to normative educational ethics and national security alike. The story of
the creation of this "social menace" is central to the ongoing attempts to
rewrite property law in order to contain the effects of the new information
technologies that, because of their blindness to the copyrighting of intellectual property, have transformed the way in which modern power is exercised and maintained. Consequently, a deviant social class or group has
been defined and categorized as "enemies of the state" in order to help rationalize a general law-and-order clampdown on free and open information

exchange. Teenage hackers' homes are now habitually raided by sheriffs and FBI agents using strong-arm tactics, and jail sentences are becoming a common punishment. Operation Sun Devil a nationwide Secret Service operation conducted in the spring of 1990, and involving hundreds of agents in fourteen cities, is the most recently publicized of the hacker raids that have resulted in several arrests and the seizure of thousands of disks and address lists in the past two years.[7]

In one of the many harshly punitive prosecutions against hackers in recent years, a judge went so far as to describe "bulletin boards" as "high-tech street gangs." The editors of *2600*, the magazine that publishes information about system entry and exploration indispensable to the hacking community, have pointed out that any single invasive act, such as trespass, that *involves* the use of computers is considered today to be infinitely more heinous than a similar act undertaken *without* computers.[8] To use computers to execute a prank, raid, fraud, or theft is to incur automatically the full repressive wrath of judges, urged on by the moral panic created around hacking feats over the past two decades. Indeed, a strong body of pressure groups is pushing for new criminal legislation that will define "crimes with computers" as a special category deserving "extraordinary" sentences and punitive measures. An increasingly criminal connotation today has displaced the more innocuous, amateur-mischief-maker-cum-media-star role reserved for hackers until a few years ago.

In response to the gathering vigor of this "war on hackers," the most common defenses of hacking can be presented on a spectrum that runs from the appeasement or accommodation of corporate interests to drawing up blueprints for cultural revolution: (a) Hacking performs a benign industrial service of uncovering security deficiencies and design flaws. (b) Hacking, as an experimental, free-form research activity, has been responsible for many of the most progressive developments in software development. (c) Hacking, when not purely recreational, is an elite educational practice that reflects the ways in which the development of high technology has outpaced orthodox forms of institutional education. (d) Hacking is an important form of watchdog counterresponse to the use of surveillance technology and data-gathering by the state, and to the increasingly monolithic communications power of giant corporations. (e) Hacking, as guerrilla know-how, is essential to the task of maintaining fronts of cultural resistance and stocks of oppositional knowledge as a hedge against a technofascist future. With all of these and other arguments in mind, it is easy to see how the social and cultural *management* of hacker activities has become a complex process that involves state policy and legislation at the highest levels. In this respect, the virus scare has become an especially convenient vehicle for obtaining public and popular consent for new legislative measures and new powers of investigation for the FBI.[9]

Consequently, certain celebrity hackers have been quick to play down the zeal with which they pursued their earlier hacking feats, while reinforcing the *deviant* category of "technological hooliganism" reserved by moralizing pundits for "dark-side" hacking. Hugo Cornwell, British author of the bestseller *Hacker's Handbook*, presents a Little England view of the hacker as a harmless fresh-air enthusiast who "visits advanced computers as a polite country rambler might walk across picturesque fields." The owners of these properties are like "farmers who don't mind careful ramblers." Cornwall notes that "lovers of fresh-air walks obey the Country Code, involving such items as closing gates behind one and avoiding damage to crops and livestock" and suggests that a similar code ought to "guide your rambles into other people's computers; the safest thing to do is simply browse, enjoy and learn." By contrast, any rambler who "ventured across a field guarded by barbed wire and dotted with notices warning about the Official Secrets Act would deserve most that happened thereafter."[10] Cornwall's quaint perspective on hacking has a certain "native charm," but some might think that this beguiling picture of patchwork-quilt fields and benign gentlemen farmers glosses over the long bloody history of power exercised through feudal and postfeudal land economy in England, while it is barely suggestive of the new fiefdoms, transnational estates, dependencies, and principalities carved out of today's global information order by vast corporations capable of bypassing the laws and territorial borders of sovereign nation-states. In general, this analogy with "trespass" laws, which compares hacking to breaking and entering other people's homes, restricts the debate to questions about privacy, property, possessive individualism, and, at best, the excesses of state surveillance, while it closes off any examination of corporate owners' and institutional sponsors' information technology activities (the most prized "target" of most hackers).[11]

Cornwall himself has joined the well-paid ranks of ex-hackers who either work for computer security firms or write books about security for the eyes of worried corporate managers.[12] A different, though related, genre is that of the penitent hacker's "confession," produced for an audience thrilled by tales of high-stakes adventure at the keyboard, but written in the form of a computer security handbook. The best examples of the "I Was a Teenage Hacker" genre is Bill (aka "The Cracker") Landreth's *Out of the Inner Circle: The True Story of a Computer Intruder Capable of Cracking the Nation's Most Secure Computer Systems*, a book about "people who can't 'just say no' to computers." In full complicity with the deviant picture of the hacker as "public enemy," Landreth recirculates every official and media cliché about subversive conspiratorial groups by recounting the putative exploits of a high-level hackers' guild called the Inner Circle. The author himself is presented in the book as a former keyboard junkie who now praises the law for having made a moral example of him:

> If you are wondering what I am like, I can tell you the same things I told the judge in federal court: Although it may not seem like it, I am pretty much a normal American teenager. I don't drink, smoke or take drugs. I don't steal, assault people, or vandalize property. The only way in which I am really different from most people is in my fascination with the ways and means of learning about computers that don't belong to me.[13]

Sentenced in 1984 to three years' probation, during which time he was obliged to finish his high school education and go to college, Landreth concludes: "I think the sentence is very fair, and I already know what my major will be. . . ." As an aberrant sequel to the book's contrite conclusion, however, Landreth vanished in 1986, violating his probation, only to later face a stiff five-year jail sentence—a sorry victim, no doubt, of the recent crackdown.

Cyber-Counterculture?

At the core of Steven Levy's 1984 bestseller, *Hackers,* is the argument that the hacker ethic, first articulated in the 1950s among the famous MIT students who developed multiple-access user systems, is libertarian and crypto-anarchist in its right-to-know principles and its advocacy of decentralized technology. This hacker ethic, which has remained the preserve of a youth culture for the most part, asserts the basic right of users to free access to all information. It is a principled attempt, in other words, to challenge the tendency to use technology to form information elites. Consequently, hacker activities were presented in the 1980s as a romantic countercultural tendency, celebrated by critical journalists like John Markoff of the *New York Times,* and Stewart Brand of *Whole Earth Catalog* fame, and by New Age gurus like Timothy Leary in the magazine *Reality Hackers.* Fueled by sensational stories about "phone phreaks" like Joe Egressia (the blind eight-year-old who discovered the phone company's tone signal by whistling) and Captain Crunch, groups like the Milwaukee 414s, the Los Angeles ARPAnet hackers, the SPAN Data Travelers, the Chaos Computer Club of Hamburg, the British Prestel hackers, *2600*'s BBS, "The Private Sector," and others, the dominant media representation of the hacker came to be that of the "rebel with a modem," to use Markoff's term, at least until the more recent "war on hackers" began to shape media coverage.

On the one hand, this popular folk hero persona offered the romantic high profile of a maverick though nerdy cowboy whose fearless raids upon an impersonal "system" were perceived as a welcome tonic in the gray age of technocratic routine. On the other hand, he was something of a juvenile technodelinquent who hadn't yet learned the difference between right and wrong; a wayward figure whose technical brilliance and proficiency differentiated him from, say, the maladjusted working-class J.D. street-corner boy of the 1950s (hacker mythology, for the most part, has been almost exclusively white, masculine, and middle class). One result of this media profile was a persistent infantilization of the hacker ethic— a way of trivializing its embryonic politics, however finally complicit with dominant technocratic imperatives or with entrepreneurial-libertarian ideology one perceives these politics to be. The second result was to reinforce, in the initial absence of coercive jail sentences, the high educational stakes of training the new technocratic elites to be responsible in their use of technology. Never, the given wisdom goes, has a creative elite of the future been so in need of the virtues of a liberal education steeped in Western ethics!

The full force of this lesson in computer ethics can be found *laid out* in the official Cornell University report on the Robert Morris affair. Members of the university commission set up to investigate the affair make it quite clear in their report that they recognize the student's academic brilliance. His hacking, moreover, is described as a "juvenile act" that had no "malicious intent" but that amounted, like plagiarism, the traditional academic heresy, to a dishonest transgression of other users' rights. (In recent years, the privacy movement within the information community— mounted by liberals to protect civil rights against state gathering of information—has actually been taken up and used as a means of criminalizing hacker activities.) As for the consequences of this juvenile act, the report proposes an analogy that, in comparison with Cornwall's *mature* English country rambler, is thoroughly American, suburban, middle-class, and *juvenile*. Unleashing the internet worm was like "the driving of a golf-cart on a rainy day through most houses in the neighborhood. The driver may have navigated carefully and broken no china, but it should have been obvious to the driver that the mud on the tires would soil the carpets and that the owners would later have to clean up the mess."[14]

In what stands out as a stiff reprimand for his alma mater, the report regrets that Morris was educated in an "ambivalent atmosphere" where he "received no clear guidance" about ethics from "his peers or mentors" (he went to Harvard!). But it reserves its loftiest academic contempt for the press, whose heroizing of hackers has been so irresponsible, in the commission's opinion, as to cause even further damage to the standards of the computing profession; media exaggerations of the courage and technical sophistication of hackers "obscures the far more accomplished work of students who complete their graduate studies without public fanfare," and "who subject their work to the close scrutiny and evaluation of their peers, and not to the interpretations of the popular press."[15] In other words, this was an inside affair, to be assessed and judged by fellow professionals within an institution that reinforces its authority by means of internally self-regulating codes of professional ethics, but rarely addresses its ethical relationship to society as a whole (acceptance of defense grants, and the like). Generally speaking, the report affirms the genteel liberal ideal that professionals should not need laws, rules, procedural guidelines, or fixed guarantees of safe and responsible conduct. Apprentice professionals ought to have acquired a good conscience by osmosis from a liberal education, rather than from some specially prescribed course in ethics and technology.

The widespread attention commanded by the Cornell report (attention from the Association of Computing Machinery, among others) demonstrates the industry's interest in how the academy invokes liberal ethics in order to assist in managing the organization of the new specialized knowledge about information technology. Despite, or perhaps because of, the report's steadfast pledge to the virtues and ideals of a liberal education, it bears all the marks of a legitimation crisis inside (and outside) the academy surrounding the new and all-important category of computer professionalism. The increasingly specialized design knowledge demanded of computer professionals means that codes going beyond the old professionalist separation of mental and practical skills are needed to manage the division that a hacker's functional talents call into question, between a purely mental pursuit and the pragmatic sphere of implementing knowledge in the real world. "Hacking" must then be designated as a strictly *amateur* practice; the tension in hacking between *interestedness* and *disinterestedness* is different from, and deficient in relation to, the proper balance demanded by professionalism. Alternatively, hacking can be seen as the amateur flipside of the professional ideal—a disinterested love in the service of interested parties and institutions. In either case, it serves as an example of professionalism gone wrong, if not very wrong.

In common with the two responses to the virus scare I described earlier—the profitable reaction of the computer industry, and the self-empowering response of the legislature—the Cornell report shows how the academy uses a case like the Morris affair to strengthen its own sense of moral and cultural authority in the sphere of professionalism, particularly through its scornful indifference to and aloofness from the codes and judgments exercised by the media, its diabolical competitor in the field of knowledge. Indeed, for all the trumpeting about excesses of power and disrespect for the law of the land, the revival of ethics in the business and science disciplines of the Ivy league and on Capitol Hill (both awash with ethical fervor in the post-Boesky/Milken and post-Reagan years) is little more than a weak liberal response to working flaws or adaptational lapses in the technocracy's social logic.

To complete the scenario of morality play example-making, however, we must also consider that Morris's father was chief scientist at the National Computer Security Center, the national Security Agency's public effort at safeguarding computer security. A brilliant programmer and code-breaker in his own right, he had testified in Washington in 1983 about the need to deglamorize teenage hacking, comparing it to "stealing a car for the purpose of joyriding." In a further Oedipal irony, Morris Sr.

may have been one of the inventors, while at Bell Labs in the 1950s, of a computer game involving self-perpetuating programs that were a prototype of today's worms and viruses. Called Darwin, its principles were incorporated in the 1980s into the popular hacker game Core War, in which autonomous "killer" programs fought each other to the death.[16]

With the appearance in the Morris affair of the Pentagon's guardian angel as patricidal object—an implicated if not victimized father—we now have many of the classic components of countercultural, cross-generational conflict. We might consider how and where this scenario differs from the definitive contours of such conflicts that we recognize as having been established in the 1960s; how the Cornell hacker Morris's relation to, say, campus "occupations" today is different from that evoked by the famous image of armed black students emerging from a sit-in on the Cornell campus; how the relation to technological ethics differs from Andre Kopkind's famous statement, "Morality begins at the end of a gun barrel," which accompanied the publication of the "do-it-yourself Molotov cocktail" design on the cover of a 1968 issue of the *New York Review of Books;* or how hackers' prized potential access to the networks of military systems warfare differs from the prodigious Yippie feat of levitating the Pentagon building. It may be that, like the J.D. rebel without a cause of the 1950s, the disaffiliated student dropout of the 1960s, and the negationist punk of the 1970s, the hacker of the 1980s has come to serve as a visible, public example of moral maladjustment, a hegemonic test case for redefining the dominant ethics in an advanced technocratic society.

What concerns me here, however, are the different conditions that exist today for recognizing countercultural expression and activism. Twenty years later, the technology of hacking and viral guerrilla warfare occupies a similar place in countercultural fantasy as the Molotov cocktail design once did. While such comparisons are not particularly sound, I do think that they conveniently mark a shift in the relation of countercultural activity to technology; a shift in which a software-based technoculture organized around outlawed libertarian principles about free access to information and communication has come to replace a dissenting culture organized around the demonizing of abject hardware structures. Much, though not all, of the 1960s counterculture was formed around what I have elsewhere called the *technology of folklore*—an expressive congeries of preindustrialist, agrarianist, Orientalist, and antitechnological ideas, values, and social structures. By contrast, the cybernetic countercultures of the 1990s are already being formed around the *folklore of technology*—mythical fears of survivalism and resistance in a data-rich world of virtual environments and posthuman bodies—which is where many of the SF- and technology-conscious youth cultures have been assembling in recent years.[17] Some would argue, however, that the ideas and values of the 1960s counterculture were only truly fulfilled in groups like the People's Computer Company, which ran Community Memory in Berkeley; or the

Homebrew Computer Club, which pioneered personal microcomputing.[18] So, too, the Yippies had seen the need to form YIPL, the Youth International Party Line, devoted to "anarcho-technological" projects, which put out a newsletter called *TAP* (alternately the *Technological American Party* and the *Technological Assistance Program*). In its depoliticized form, which eschewed the kind of destructive "dark-side" hacking advocated in an earlier incarnation, *TAP* was eventually the progenitor of *2600*. A significant turning point, for example, was *TAP*'s decision not to publish plans for the hydrogen bomb (the *Progressive* did so)—bombs that would destroy the phone system, which the *TAP* "phone phreaks" had an enthusiastic interest in maintaining.

There is no doubt that the hacking scene today makes counter-cultural activity more difficult to recognize and therefore to define as politically significant. It was much easier in the 1960s to *identify* the salient features and symbolic power of a romantic preindustrialist cultural politics in an advanced technological society, especially when the destructive evidence of America's supertechnological invasion of Vietnam was being screened daily. However, in a society whose technopolitical infrastructure depends increasingly upon greater surveillance, and where foreign wars are seen through the lens of laser-guided smart bombs, cybernetic activism necessarily relies on a much more covert politics of identity. Access to closed digital systems requires discretion and dissimulation, the authentication of a signature or pseudonym, not the identification of a real surveillable person, so there exists a crucial operative gap between authentication and identification. (As security systems move toward authenticating access through biological signatures—the biometric recording and measurement of physical characteristics such as palm or retinal prints, or vein patterns on the backs of hands—the hackers' staple method of systems entry through purloined passwords will be further challenged.) By the same token, cybernetic identity is never exhausted; it can be recreated, reassigned, and reconstructed with any number of different names and under different user accounts. In fact, most hacks or technocrimes go unnoticed or unreported for fear of publicizing the vulnerability of corporate security systems, especially when the hacks are performed by disgruntled employees taking their vengeance on management. So, too, authoritative identification of any individual hacker, whenever it occurs, is often the result of accidental leads rather than systematic detection. For example, Captain Midnight, the video pirate who commandeered a satellite a few years ago to interrupt broadcast television viewing, was traced only because a member of the public reported a suspicious conversation heard over a crossed telephone line.

Eschewing its core constituency among the white male preprofessional-managerial class, the hacker community may be expanding its parameters outward. Hacking, for example, has become a feature of young-adult novel genres for girls.[19] The elitist class profile of the hacker prodigy as that of an undersocialized college nerd has become democratized and customized in recent years; it is no longer exclusively associated with institutionally acquired college expertise, and increasingly it dresses streetwise. In a recent article that documents the spread of the computer underground from college whiz-kids to a broader youth subculture termed "cyberpunks," after the movement among SF novelists, the original hacker phone phreak Captain Crunch is described as lamenting the fact that the cyberculture is no longer an "elite" one, and that hacker-valid information is much easier to obtain these days.[20]

For the most part, however, the self-defined hacker underground, like many other proto-countercultural tendencies, has been restricted to a privileged social milieu, further magnetized by its members, understanding that they are the apprentice architects of a future dominated by knowledge, expertise, and "smartness," whether human or digital. Consequently, it is clear that the hacker cyberculture is not a dropout culture; its disaffiliation from a domestic parent culture is often manifest in activities that answer, directly or indirectly, to the legitimate needs of industrial R&D. For example, this hacker culture celebrates high productivity, maverick forms of creative work energy, and an obsessive identification with on-line endurance (and endorphin highs)—all qualities that are valorized by the entrepreneurial codes of silicon futurism. In a critique of the hacker-as-rebel myth, Dennis Hayes debunks the political romance woven around the teenage hacker:

> They are typically white, upper-middle-class adolescents who have taken over the home computer (bought, subsidized, or tolerated by parents in the hope of cultivating computer literacy). Few are politically motivated although many express contempt for the "bureaucracies" that hamper their electronic journeys. Nearly all demand unfettered access to intricate and intriguing computer networks. In this, teenage hackers resemble an alienated shopping culture deprived of purchasing opportunities more than a terrorist network.[21]

While welcoming the sobriety of Hayes's critique, I am less willing to accept its assumptions about the political implications of hacker activities. Studies of youth subcultures (including those of a privileged middle-class formation) have taught us that the political meaning of certain forms of cultural "resistance" is notoriously difficult to read. These meanings are either highly coded or expressed indirectly through media— private peer languages, customized consumer styles, unorthodox leisure patterns, categories of insider knowledge and behavior—that have no fixed or inherent political significance. If cultural studies of this sort have proved anything, it is that the often symbolic, not wholly articulate, expressivity of a youth culture can seldom be translated directly into an articulate political philosophy. The significance of these cultures lies in their embryonic or *protopolitical* languages and technologies of opposition to dominant or parent systems or rules. If hackers lack a "cause," then, they are certainly not the first youth culture to be characterized in this dismissive way: the Left in particular has suffered from the lack of a cultural politics capable of recognizing the power of cultural expressions that do not wear a mature political commitment on their sleeves.

The escalation of activism in the professions in the past two decades has shown that it is a mistake simply to condemn the hacker impulse for its class constituency. To cede the "ability to know" on the grounds that elite groups will enjoy unjustly privileged access to technocratic knowledge is to cede too much of the future. Is it of no political significance at all that hackers' primary fantasies often involve the official computer systems of the police, armed forces, and defense and intelligence agencies? And that the rationale for their fantasies is unfailingly presented as a defense of civil liberties against the threat of centralized intelligence and military activities? Or is all of this merely a symptom of an apprentice elite's fledgling will to masculine power? The activities of the Chinese student elite in the pro-democracy movement have shown that unforeseen shifts in the political climate can produce startling new configurations of power and resistance. After Tiananmen Square, Party leaders found it imprudent to purge those high-tech engineer and computer cadres who alone could guarantee the future of any planned modernization program. On the other hand, the authorities rested uneasy knowing that each cadre (among the most activist groups in the student movement) is a potential hacker who can have the run of the communications house if and when he or she wants.

On the other hand, I do agree with Hayes's perception that the media have pursued their romance with the hacker at the cost of underreporting the much greater challenge posed to corporate employers by their employees. Most high-tech "sabotage" takes place in the arena of conflicts between workers and management. In the ordinary, everyday life of office workers—mostly female—a widespread culture of unorganized sabotage accounts for infinitely more computer downtime and information loss every year than is caused by destructive "dark-side" hacking by celebrity cybernetic intruders. The sabotage, time theft, and strategic monkeywrenching deployed by office workers in their engineered electromagnetic attacks on data storage and operating systems might range from the planting of time or logic bombs to the discreet use of electromagnetic Tesla coils or simple bodily friction: "Good old static electricity discharged from the fingertips probably accounts for close to half the disks and computers wiped out or down every year."[22] More skilled operators, intent on evening a score with management, often utilize sophisticated hacking techniques. In many cases, a coherent networking culture exists among console operators, where, among other things, tips about strategies for slowing down the pace of the work regime are circulated. While these threats from below are fully recognized in boardrooms, corporations dependent upon digital business machines are obviously unwilling to advertise how acutely vulnerable they actually are to this kind of sabotage. It is easy to imagine how organized computer activism could hold such companies ransom. As Hayes points out, however, it is more difficult to mobilize any kind of labor movement organized upon such premises:

> Many are prepared to publicly oppose the countless dark legacies of the computer age: "electronic sweatshops," military technology, employee surveillance, genotoxic water, and zone depletion. Among those currently leading the opposition, however, it is apparently deemed "irresponsible" to recommend an active computerized resistance as a source of worker's power because it is perceived as a medium of employee crime and "terrorism."[23]

Processed World, the "magazine with a bad attitude," with which Hayes has been associated, is at the forefront of debating and circulating these questions among office workers, regularly tapping into the resentments borne out in on-the-job resistance.[24]

While only a small number of computer users would categorize themselves as "hackers," there are defensible reasons for extending the restricted definition of *hacking* down and across the caste hierarchy of systems analysts, designers, programmers, and operators to include all high-tech workers—no matter how inexpert—who can interrupt, upset, and redirect the smooth flow of structured communications that dictates their position in the social networks of exchange and determines the pace of their work schedules. To put it in these terms, however, is not to offer any universal definition of hacker agency. There are many social agents, for example, in job locations who are dependent upon the hope of technological *reskilling* and for whom sabotage or disruption of communicative rationality is of little use; for such people, definitions of hacking that are reconstructive, rather than deconstructive, are more appropriate. A good example is the crucial role of worker technoliteracy in the struggle of labor against automation and deskilling. When worker education classes in computer programming were discontinued by management at the Ford Rouge plant in Dearborn, Michigan, United Auto Workers members began to publish a newsletter called the *Amateur Computerist* to fill the gap.[25] Among the columnists and correspondents in the magazine have been veterans of the Flint sit-down strikes, who see a clear historical continuity between labor organization in the 1930s and automation and deskilling today. Workers' computer literacy is seen as essential, not only to the demystification of the computer and the reskilling of workers, but also to labor's capacity to intervene in decisions about new technologies that might result in shorter hours and thus in "work efficiency" rather than worker efficiency.

The three social locations I have mentioned above all express different class relations to technology: the location of an apprentice technical elite, conventionally associated with the term *hacking;* the location of the high-tech office worker, involved in "sabotage"; and the location of the shopfloor worker, whose future depends on technological reskilling. All therefore exhibit different ways of *claiming back* time dictated and appropriated by technological processes, and of establishing some form of independent control over the work relation so determined by the new technologies. All, then, fall under a broad understanding of the politics involved in any extended description of hacker activities.

The Culture and Technology Question

Faced with these proliferating practices in the workplace, on the teenage cult fringe, and increasingly in mainstream entertainment, where over the past five years the cyberpunk sensibility in popular fiction, film, and television has caught the romance of the outlaw technology of human/ machine interfaces, we are obliged, I think, to ask old questions about the new silicon order that the evangelists of information technology have been deliriously proclaiming for more than twenty years. The postindustrialists' picture of a world of freedom and abundance projects a bright millenarian future devoid of work drudgery and ecological degradation. This sunny social order, cybernetically wired up, is presented as an advanced evolutionary phase of society in accord with Enlightenment ideals of progress and rationality. By contrast, critics of this idealism see only a frightening advance in the technologies of social control—whose owners and sponsors are efficiently shaping a society, as Kevin Robins and Frank Webster put it, of "slaves without Athens" that is exactly the inverse of the "Athens without slaves" promised by the silicon positivists.[26] To counter the postindustrialists' millenarian picture of a postscarcity harmony in which citizens enjoy decentralized access to free-flowing information, it is necessary, then, to emphasize how and where actually existing cybernetic capitalism presents a gross caricature of such a postscarcity society.

   One of the stories told by the critical Left about new cultural technologies is that of monolithic, panoptical social control, effortlessly achieved through a smooth, endlessly interlocking system of surveillance networks. In this narrative, information technology is seen as the most despotic mode of domination yet, generating not just a revolution in capitalist production but also a revolution in living—"social Taylorism"— touching all cultural and social spheres in the home and the workplace.[27] Through gathering of information about transactions, consumer preferences, and creditworthiness, a harvest of information about any individual's whereabouts and movements, tastes, desires, contacts, friends, associates, and patterns of work and recreation becomes available in dossiers sold on the tradable information market, or is endlessly convertible into other forms of intelligence through computer-matching. Advanced pattern recognition technologies facilitate the process of surveillance, while data encryption protects it from public accountability.[28]

While the debate about privacy has triggered public consciousness about these excesses, the liberal discourse about ethics and damage control in which that debate has been conducted falls short of the more comprehensive analysis of social control and social management offered by Left political economists who see information, increasingly, as the major site of capital accumulation in the world economy. What happens in the process by which information, gathered up by data-scavenging in the transactional sphere, is systematically converted into intelligence? A surplus value is created for use elsewhere. This surplus information value is more than is needed for public surveillance; it is often information, or intelligence, culled from consumer polling or statistical analysis of transactional behavior, that has no immediate use in the process of routine public surveillance. This surplus bureaucratic capital is used to forecast social futures, and consequently is applied to the task of managing in advance the future behavior of mass populations. This surplus intelligence becomes the basis of a whole new industry of futures research that relies upon computer technology to simulate and forecast the shape, activity, and behavior of complex social systems. The result is a system of social management that far transcends the questions about surveillance that have been at the discursive center of the privacy debate.[29]

To challenge further the idealists' vision of postindustrial light and magic, we need only look inside the semiconductor workplace itself, home to the most toxic chemicals known to man (and woman, especially since women of color often make up the majority of the microelectronics labor force), where worker illness is measured not in quantities of blood spilled on the shopfloor but in the less visible forms of chromosome damage, miscarriages, premature deliveries, and severe birth defects. Semiconductor workers exhibit an occupational illness rate that, by the late 1970s, was already three times higher than that of manufacturing workers, at least until the federal rules for recognizing and defining injury levels were changed under the Reagan administration. Protection gear is designed to protect the product and the clean room from the workers, not vice versa. Recently, immunological health problems have begun to appear that can only be described as a kind of chemically induced AIDS, rendering the T-cells dysfunctional rather than depleting them like virally induced AIDS.[30] In corporate offices, where extraordinarily high stress patterns and illness rates are reported among VDT operators, the use of keystroke software to monitor and pace office workers has become a routine part of job performance evaluation programs. Some 70 percent of corporations use electronic surveillance or other forms of quantitative monitoring of their workers. Every bodily movement, especially trips to the toilet, can be checked and measured. Federal deregulation has meant that the limits of employee workspace have in some government offices shrunk below that required by law for a two-hundred-pound laboratory pig.[31] Critics of the labor process seem to have sound reasons to believe that rationalization and quantification are at last entering their most primitive phase.

What I have been describing are some of the features of that critical Left position—sometimes referred to as the "paranoid" position—on information technology which imagines or constructs a totalizing, monolithic picture of systematic domination. While this story is often characterized as conspiracy theory, its targets—technorationality, bureaucratic capitalism—are usually too abstract to fit the picture of a social order planned and shaped by a small, conspiring group of centralized power elites.

Although I believe that this story, when told inside and outside the classroom, for example, is an indispensable form of "consciousness-raising," it is not always and everywhere the best story to tell. While I am not comfortable with the "paranoid" labeling, I would argue that such narratives do little to discourage paranoia. The critical habit of finding unrelieved domination everywhere has certain consequences, one of which is to create a siege mentality, reinforcing the inertia, helplessness, and despair that such critiques set out to oppose in the first place. The result is a politics that can speak only from a victim's position. And when knowledge about surveillance is presented as systematic and infallible, self-censoring is sure to follow. In the psychosocial climate of fear and phobia aroused by the virus scare, there is a responsibility not to be alarmist or scared—especially when, as I have argued, such moments are profitably seized upon by the sponsors of control technology. In short, the picture of a seamlessly panoptical network of surveillance may be the result of a rather undemocratic, not to mention unsocialist, way of thinking, predicated upon the definition of people solely as victims. It echoes the old sociological models of mass society and mass culture, which cast the majority of society as passive and lobotomized in the face of modernization's cultural patterns. To emphasize, as Robins and Webster and others have done, the power of the new technologies to transform despotically the "rhythm, texture, and experience" of everyday life, and meet with no resistance in doing so, is not only to cleave, finally, to an epistemology of technological determinism, but also to dismiss the capacity of people to make their own use of new technologies, and to view technology as a contested site.[32]

The seamless "interlocking" of public and private information and intelligence networks is not as smooth and even as the critical school of hard domination would suggest. Compulsive gathering of information is no *guarantee* that any interpretive sense will be made of the files or dossiers. In any case, the centralized, "smart" supervision of an information gathering system would require, as Hans Magnus Enzensberger once argued, "a monitor that was bigger than the system itself"; " a linked series of communications . . . to the degree that it exceeds a certain critical size, can no longer be centrally controlled but only dealt with statistically."[33] Some would argue that the increasingly covert nature of surveillance indicates that the "campaign" for social control is not going well, and one of the most pervasive popular arguments against the panoptical intentions of technology's masters is that their systems do not work very well. Every successful hack or computer crime in some way reinforces the popular perception that information systems are not infallible. And the announcements of military-industrial spokespersons that the fully automated battlefield is on its way run up against an accumulated stock of popular skepticism about the operative capacity of weapons systems. These misgivings are born of decades of distrust for the plans and intentions of the military-industrial complex, and were quite evident in the widespread cynicism about the Strategic Defense Initiative. Just to take one example of unreliability, the military communications system worked so poorly and so farcically during the U.S. invasion of Grenada that commanders had to call each other on payphones: ever since, the command-and-control code of Arpanet technocrats has been C^5—Command, Control, Communication, Computers, and Confusion.[34] The Gulf War saw the most concerted effort on the part of the U.S. military-industrial-media complex to suppress evidence of such technical dysfunctions, which alone accounted, in the buildup to the war and its opening weeks, for much higher U.S. casualty figures than those sustained in actual combat. The absence of the ineffective B-1 bomber, the most sophisticated weapons system of the 1980s, went largely unnoticed. As weeks went by, the crowing of the military and the public media about U.S. supertechnology sounded more and more hollow. The Pentagon's vaunted information system proved no more—and often less—resourceful than the mental agility of its operators and analysts.

I am not suggesting that alternatives can be forged simply by encouraging disbelief in the infallibility of existing technologies. But technoskepticism, while not a *sufficient* condition for social change, is nonetheless a *necessary* condition. Stocks of popular technoskepticism are crucial to the task of eroding the legitimacy of those cultural values that prepare the way for new technological developments: values and principles such as the inevitability of material progress, the "emancipatory" domination of nature, the innovative autonomy of machines, the efficiency codes of pragmatism, and the linear juggernaut of liberal Enlightenment rationality—all increasingly under close critical scrutiny as a wave of ecological consciousness sweeps through the electorates of the West. Technologies do not shape or determine such values, which already pre-exist the technologies; the fact that they have become deeply embodied in the structure of popular needs and desires provides the green light for accepting certain kinds of technology. In fact, the principal rationale for introducing new technologies is that they answer to already existing intentions and demands that may be perceived as "subjective" but are never actually within the control of any single set of conspiring individuals. As Marike Finlay has argued, just as technology is possible only in given discursive situations (one of which is the desire of people to have it for reasons of empowerment) so capitalism is merely the site, and not the source, of the power that is often autonomously attributed to the owners and sponsors of technology.[35]

No frame of technological inevitability has not already interacted with popular needs and desires; no introduction of new machineries of control has not already been negotiated to some degree in the arena of popular consent. Thus the power to design architecture that incorporates different values must arise from the popular perception that existing technologies are not the only ones; nor are they the best when it comes to individual and collective empowerment. It was this perception—formed around the distrust of big, impersonal, "closed" hardware systems, and the desire for small, decentralized, interactive machines to facilitate interpersonal communication—that "built" the PC out of hacking expertise in the early 1970s. These desires and distrusts were as much the partial "intentions" behind the development of microcomputing technology as deskilling, worker monitoring, and information gathering are the intentions behind the corporate use of that technology today. The machinery of countersurveillance is now up and running. The explosive growth of public data networks, bulletin-board systems, and alternative information and media links, and the increasing cheapness of desktop publishing, satellite equipment, and international databases are as much the result of local political "intentions" as the fortified net of globally linked, restricted-access information systems is the intentional fantasy of those who seek to profit from centralized control. The picture that emerges from this mapping of intentions is not an inevitably technofascist one, but rather the uneven result of cultural struggles over values and meanings. These local advances are further assisted by the contradictions of capitalism itself, since market demand for ever cheaper and more resourceful technologies is putting video cameras and computing power into the hands of ordinary people who have traditionally experienced technology only as its object of surveillance, or, at best, its passive operator.

It is in the struggle over values and meanings that the work of cultural criticism takes on its special significance as a full participant in the debate about technology; a debate in which it is already fully implicated, if only because the culture and education industries are rapidly becoming integrated within the vast information service conglomerates. The media we study, the media we publish in, and the media we teach are increasingly part of the same tradable information sector. So, too, our common intellectual discourse has been significantly affected by the recent debates about postmodernism (or culture in a postindustrial world) in which the euphoric, addictive thrill of the technological sublime has figured quite prominently. The high-speed technological fascination that is characteristic of the postmodern condition can be read, on the one hand, as a celebratory capitulation by intellectuals to the new information technocultures. On the other hand, this celebratory strain attests to the persuasive effect associated with the new cultural technologies, to their capacity (more powerful than that of their sponsors and promoters) to generate pleasure and gratification and to win the contest for intellectual as well as popular consent.

Another reason for the involvement of cultural critics in the technology debates has to do with our special critical knowledge of the way cultural meanings are produced—our knowledge about the politics of consumption and what is often called the politics of representation. This knowledge demonstrates that there are limits to the capacity of productive forces to shape and determine consciousness, insisting on the ideological or interpretive dimension of technology as a culture that can and must be used and consumed in a variety of ways not reducible to the intentions of any single source or producer; technology's meanings cannot simply be read off as evidence of faultless social reproduction. It is a knowledge, in short, that refuses to add to the "hard domination" picture of disenfranchised individuals watched over by some scheming panoptical intelligence. Far from being understood solely as the concrete hardware of sophisticated electronic objects, technology must be seen as a lived, interpretive practice for people in their everyday lives. To redefine the shape and form of that practice is to help create the need for new kinds of hardware and software.

One of this chapter's aims has been to describe and suggest a wider set of activities and social locations than is normally associated with the practice of hacking. If there is a challenge here for cultural critics, it might be the commitment to making our knowledge about technoculture into something like a hacker's knowledge, capable of penetrating existing systems of rationality that might otherwise be seen as infallible; a hacker's knowledge, capable of reskilling, and therefore of rewriting, the cultural programs and reprogramming the social values that make room for new technologies; a hacker's knowledge, capable also of generating new popular romances around the alternative uses of human ingenuity. If we are to take up that challenge we cannot afford to give up what technoliteracy we have acquired in deference to the vulgar faith that tells us it is always acquired in complicity and is thus contaminated by the toxin of instrumental rationality; or because we hear, often from the same quarters, that acquired technological competence simply glorifies the inhuman work ethic. Technoliteracy, for us, is the challenge to make a historical opportunity out of a historical necessity.

NOTES

1. John Markoff, *New York Times,* May 30, 1989.

2. Bryan Kocher, "A Hygiene Lesson," *Communications of the ACM* 32, 1 (January 1989), p. 3.

3. Jon A. Rochlis and Mark W. Eichen, "With Microscope and Tweezers: The Worm from MIT's Perspective," *Communications of the ACM* 32, 6 (June 1989), p. 697.

4. Philip Elmer-DeWitt, "Invasion of the Body Snatchers," *Time* (September 26, 1988), pp. 62–67.

5. Judith Williamson, "Every Virus Tells a Story: The Meaning of HIV and AIDS," in *Taking Liberties: AIDS and Cultural Politics,* ed. Erica Carter and Simon Watney (London: Serpent's Tail/ICA, 1989), p. 69.

6. "Pulsing the system" is a well-known intelligence process in which, for example, planes deliberately fly over enemy radar installations in order to determine what frequencies they use and how they are arranged. It has been suggested that Morris Sr. and Morris Jr. worked in collusion as part of an NSA operation to pulse the internet system, and to generate public support for a legal clampdown on hacking. See Allan Lundell, *Virus! The Secret World of Computer Invaders That Breed and Destroy* (Chicago: Contemporary Books, 1989), pp. 12–18. As is the case with all such conspiracy theories, no actual conspiracy need have existed for the consequences—in this case, the benefits for the intelligence community—to have been more or less the same.

7. For details of these raids, see *2600: The Hacker's Quarterly* 7, 1 (Spring 1990).

8. "Hackers in Jail," *2600: The Hacker's Quarterly* 6, 1 (Spring 1989), pp. 22–23. The recent Secret Service action that threatened to shut down *Phrack,* an electronic newsletter operating out of St. Louis, confirms *2600*'s thesis: nonelectronic publication would not be censored in the same way.

9. This is not to say that the new laws cannot themselves be used to protect hacker institutions however. *2600* has advised operators of bulletin boards to declare them private property, thereby guaranteeing protection under the Electronic Privacy Act against unauthorized entry by the FBI.

10. Hugo Cornwall, *The Hacker's Handbook,* 3rd ed. (London: Century, 1988) pp. 2–6. In Britain, for the most part, hacking is still looked upon as a matter for the civil, rather than the criminal, courts.

11. Discussions about civil liberties and property rights, for example, tend to preoccupy most of the participants in the electronic forum published as "Is Computer Hacking a Crime?" in *Harper's* 280, 1678 (March 1990), pp. 45–58.

12. See Hugo Cornwall, *Data Theft* (London: Heinemann, 1987).

13. Bill Landreth, *Out of the Inner Circle: The True Story of a Computer Intruder Capable of Cracking the Nation's Most Secure Computer Systems* (Redmond, Wash.: Tempus, Microsoft, 1989), p. 10.

14. *The Computer Worm: A Report to the Provost of Cornell University on an Investigation Conducted by the Commission of Preliminary Enquiry* (Ithaca, N.Y.: Cornell University, 1989).

15. Ibid., p. 8.

16. A. K. Dewdney, the "computer recreations" columnist at *Scientific American,* was the first to publicize the details of this game of battle programs in an article in the magazine's May 1984 issue. In a follow-up article in March 1985, "A Core War Bestiary of Viruses, Worms, and Other Threats to Computer Memories," Dewdney described the wide range of "software creatures" that readers' responses had brought to light. A third column, in March 1989, was written in an exculpatory mode to refute any connection between his original advertisement of the Core War program and the spate of recent viruses.

17. Andrew Ross, *No Respect: Intellectuals and Popular Culture* (New York: Routledge, 1989), p. 212.

18. The definitive computer liberation book is Ted Nelson's *Computer Lib: Dream Machines* (Redmond, Wash.: Tempus, 1987, rev. ed.; original pub. 1974).

19. See Alice Bach's "Phreakers" series, which narrates the mystery-and-suspense adventures of two teenage girl hackers: *The Bully of Library Place* (New York: Dell, 1987), *Double Bucky Shanghai* (New York: Dell, 1987), *Parrot Woman* (New York: Dell, 1987), *Ragwars* (New York: Dell, 1987), and others. The hacker has also appeared recently as a demonized figure in other genres: *The Hacker*, a horror novel by Chet Day (Pocket Books, 1989); and as a new Batman comic series, drawn by Pepe Moreno.

20. John Markoff, "Cyberpunks Seek Thrills in Computerized Mischief," *New York Times* (November 26, 1988), pp. 1, 28.

21. Dennis Hayes, *Behind the Silicon Curtain: The Seductions of Work in a Lonely Era* (Boston: South End, 1989), p. 93.

One striking historical precedent for the hacking subculture, suggested to me by Carolyn Marvin, was the widespread activity of amateur or "ham" wireless operators in the first two decades of the twentieth century. Initially lionized in the press as boy-inventor heroes for their technical ingenuity and daring adventures with the ether, this white middle-class subculture was increasingly demonized by the U.S. Navy (whose signals the amateurs prankishly interfered with), which was crusading for complete military control of the airwaves in the name of national security. The amateurs lobbied with democratic rhetoric for the public's right to access the airwaves, and, although partially successful in their case against the Navy, lost out ultimately to big commercial interests when Congress approved the creation of a broadcasting monopoly after World War I in the form of RCA. See Susan J. Douglas, *Inventing American Broadcasting 1899–1922* (Baltimore: John Hopkins University Press, 1987), pp. 187–291.

22. "Sabotage," *Processed World* 11 (Summer 1984), pp. 37–38.

23. Hayes, *Behind the Silicon Curtain*, p. 98.

24. *Bad Attitude: The Processed World Anthology*, ed. Chris Carlsson with Mark Leger (London: Verso, 1990), contains highlights from the magazine's first eight years.

25. *The Amateur Computerist*, available from R. Hauben, PO Box 4344, Dearborn, MI 48126.

26. Kevin Robins and Frank Webster, "Athens without Slaves . . . or Slaves without Athens? The Neurosis of Technology," *Science as Culture* 3 (1988), pp. 7–53.

27. See, for example, the collection of essays edited by Vincent Mosco and Janet Wasko, *The Political Economy of Information* (Madison: University of Wisconsin Press, 1988); Kevin Robins and Frank Webster, *Information Technology: A Luddite Analysis* (Norwood, N.J.: Ablex, 1986).

28. Tom Athanasiou and Staff, "Encryption and the Dossier Society," *Processed World* 16 (1986), pp. 12–17.

29. See Kevin Wilson, *Technologies of Control: The New Interactive Media for the Home* (Madison: University of Wisconsin Press, 1988), pp. 121–25.

30. Hayes, *Behind the Silicon Curtain*, pp. 63–80.

31. "Our Friend the VDT," *Processed World* 22 (Summer 1988), pp. 24–25.

32. See Kevin Robins and Frank Webster, "Cybernetic Capitalism," in *The Political Economy of Information*, ed. Vincent Mosco and Janet Wasko (Madison: University of Wisconsin Press, 1988), pp. 44–75.

33. Hans Magnus Enzensberger, "Constituents of a Theory of the Media," *The Consciousness Industry*, tr. Stuart Hood (New York: Seabury, 1974).

34. Barbara Garson, *The Electronic Sweatshop: How Computers are Transforming the Office of the Future into the Factory of the Past* (New York: Simon & Schuster, 1988), pp. 244–45.

35. See Marike Finlay's Foucauldian analysis, *Powermatics: A Discursive Critique of New Technology* (London: Routledge & Kegan Paul, 1987). A more conventional culturalist argument can be found in Stephen Hill, *The Tragedy of Technology: Human Liberation versus Domination in the late Twentieth Century* (London: Pluto, 1988).

Ravi Sundaram

Beyond the Nationalist Panopticon: The Experience of Cyberpublics in India

In his now-classic text on postmodernism Frederic Jameson spoke of an "inverted millenarianism" which has come to characterize our time, where all anticipations of the future have been replaced by a sense of the end of various social imaginaries (1992, p. 1). Writing from a country that is located firmly within the periphery of late capitalism, there are many senses in which the old ideologies of nineteenth-century modernity are in deep crisis—the great *promesse de bonheur* of nationalism and Marxism has failed to materialize in South Asia. What we are instead witnessing is a dramatic and simultaneous process of both deterritorialization as well as territorialization where received notions of order, based on historical associations of citizenship, borders, time, and history, are being actively reworked.

At first glance, there seems cause for celebration. What Nietzsche called the "consuming historical fever" of modernity—the tendency to monumentalize history and to impose the burden of the millennium on all human practices—seems well behind us. But the world as we see it does not present a pretty sight—particularly at the margins of the metropolis. To take recourse to a Hegelianism, it is as if the World-Spirit, defeated at the final moment of self-consciousness, has enacted a terrible revenge for sacrificing its grand vision. In particular, for the Third World citizen, searching for identity among the ruins of the now-decaying artifacts of nationalism, it seems more and more clear that the storm of progress has passed on, with no promise of returning.

However, here lies the paradox. At the very moment that modernity could free itself from its nineteenth-century variant, the power of the West, which was *the* imaginative embodiment of the modern, seems more

This chapter is printed by permission of the author. © by Ravi Sundaram. Discussions with Ranjani Mazumdar, Shiv Visvanathan, Ashis Nandy, and Ravi Vasudevan helped clarify various issues in this paper. The responsibility for errors remains with the author.

fragile than ever. For the first time since the sixteenth century there seems to be a secular shift in the centers of wealth from the West to Asia's eastern frontier, and possibly China. The old state-system of modernity—based on secure borders and sovereignty—has collapsed, in the West itself, canonical notions of subjectivity, representation, and freedom have taken a battering from which one suspects they will never fully recover. The great Western millennium, beginning with the violence of the Crusades and culminating in European power, may end with the very idea of modernity seriously compromised.

What has this to do with the engagement with virtual spaces in a country like India? In the first place, the dynamics of India's movement into electronic spaces have occurred within the backdrop of the crisis of Western modernity and its product: the territorial state based on a particular concept of sovereignty. Further, it seems to me that it is the very fragility of the "West" that gives cyberspace a particular attractiveness for Third World users, at least in the case of India. This fragility of the Western imaginary in the real world contrasts with a certain efflorescence in virtual spaces. It is this disjunction that informs new modes of travel by Third World elites to the West, through virtual space. These are modes that need to be addressed as occupying a distinct space which depart from the old borders that defined the Third World's relationship to the West.

It is here that the old Third Worldist/classical Marxist critique of "cyberspace" seems limited. Such critiques have focused on the museumization of Third World cultures in the space of the Web, or the domination of multinational capital in the political economy of the information superhighway. There is a strong element of truth in both positions, but neither can explain the complex implication of virtual spaces in local/regional strategies for remapping *national* identity. In any event, while the relationship to an imaginary West is important to cyberpractices in India, this relationship by no means exhausts the complexity and local interconnectedness of such practices. What is needed when looking at cyberpractices in India is what Ernesto Laclau has called a "radical contexualisation," where the violent abstractions of "West," "capital," and "nation" do not erase the richness and contradictions of initiatives into virtual space.

India, Cyberspace, and the Publics

If one were to adopt a certain diffusionary model of the spread of cyberpractices in India, we would have to consider the following:

a) The simple fact of India being a peripheral society in the capitalist world economy: with one of the lowest saturation rates of telephones in the world; only a small minority of the population has electricity.

b) India has no tradition of cyberpunk, in fact, there is no indigenous science-fiction tradition. Most existing cultural communities have remained ambivalent about technology. Historically, representations of science and technology have been state-sponsored and social-realist in form.

Despite this, a significant number of people are linked to electronic networks in India and the number is fast growing. For a Third World country with inequalities like India this is quite remarkable. The reasons for this shall be examined in the course of this essay. What is significant is that "cyberspace" has emerged as a significant term in public discourse in India, becoming the focal point of much coverage and speculation in the media. Behind all of this is the growing community of users. Until now anonymous, and lacking the "heroic" qualities of the old nationalist scientist, the contemporary user lacks any visible representation of his or her agency.

I have tried to map the "user" into three, overlapping cyberpublics. "Public" is used here very loosely, indicating a cybercommunity in the making, where mutual rituals of initiation and excursion are only now being invented. The three cyberpublics are those of the nationalist state, the transnational elite, and the space between the market and the state occupied by various bulletin boards and social movement networks. While the boundaries of all the three publics are fuzzy,[1] they are also uneven in internal differentiation and modes of address. The cyberpublics are a relatively new phenomenon in India. What is attempted here is a very preliminary examination of these communities-in-formation, by mapping certain practices of the nationalist organization of space, and its consequences for agency and movement.

Nationalist policies employed a certain social cartography which attempted to organize space, representation, and identity. Maps generated 'borders' which sought to institutionalize identity, frame representations of citizenship, and mediate the relationship with the West and modernity (Krishna 1994). Mapping activities which were backed by the state's monopoly of legitimate violence were also implicated in a particular version of postwar modernism. The metaphor of the map is also useful to highlight different strategies which emerged in the postnational period, which sought to reorganize space by dislocating it from territory, and posit new forms of identity.

Cyberpublic 1: Nationalism

There is a general consensus among writers that the anticolonial struggle in India produced a rich constellation of overlapping, contested visions of nation and nationalism. Given the wide range of social mobilization, this was to some extent inevitable. For a long time competing discourses within the anticolonial movement on issues of identity, modernity, and "building the nation" remained dormant; it was only after the experience of development following independence that some of the older questions and dissenting views, notably those raised by Gandhi's practice, assumed greater significance.

The postcolonial period after 1947 saw a significant reconstitution of nationalism under the leadership of Jawaharlal Nehru, the first prime minister. The new turn consisted in affirming the need for an accelerated transition to modernity through the building of the rational institutions of state order, which could functionally reorganize national space for the purposes of accumulation and industrialization. In any event, the Gandhian cultural constellation was seen as dysfunctional for the needs of rational accumulation and state administration. What emerged is what Lefèbvre (1991) has called the construction of an 'abstract space'—not accessible through ordinary experience, and the preserve of purposive-rational modernizers—the Third World *aufklarer.*

Gandhi's discourse had included a reassertion of 'place' as a site of genuine experience and action—consider his symbolic evocation of the village as a site of anticolonial politics (Nandy 1996). This was in contrast to the abstract temporal cartography of colonialism which held out the developmental possibility of the railway imaginary as a means to overcome the 'village society'; colonial ideology further stressed the use of English as a passport to the cultural world-system and the virtues of colonial law as a sui generis means toward order and nonrevolutionary evolution. Gandhi's evocation of the everyday and a new aestheticised anticolonial strategy[2] (the innovative use of counter-commodities like *Khadi* or homespun cloth) was viewed with skepticism by the Nehruvian inheritors of postindependence India—as such, Gandhi was accorded a hagiographic status in official histories and marginalized.

While postindependent nationalists like Nehru excluded Gandhian economics from state-building strategies, they were quick to incorporate the growing discourse of *development* and make it an instrument of state policy. Much has been written on the postwar invention of 'development' as a necessary process to modernity held out by both Americanism and Sovietism alike—the attendant social and cultural disasters and the dislocation of millions from historic modes of living (Escobar 1995). What interests us here is how the developmentalist state in India carried out a particular spatial mapping that would play an important role in the development of the nationalist cyberpublic.

In the first place, right from the 1950s on the space of the 'global' underwent a certain bracketing. The conquest of the national space and its consolidation was seen as a necessary precondition to a thoroughgoing incorporation into the world economy. In addition, the 'national economy' became a shorthand representational device for *the nation itself.*[3] This deployment of the 'national economy' was, to be sure, the reaction to two hundred years of colonial exploitation and India's peripheral status in the world economy. What is important for our purposes is that the 'economy,' as in Lefèbvre's abstract space, was embedded in a matrix accessible only to a privileged and 'enlightened' class of modernizers. Further, the economy was conceived as a space clear of the cultural ambivalent inherent in the village, or "traditional community." The sociologist and thinker Zygmunt Bauman speaks of Western modernity's great fear of ambivalence which was inscribed into the project from the very beginning:

> The new, modern order took off as a desperate search for structure in a world suddenly denuded of structure. Utopias that served as beacons for the long march to reason visualized a world without margins, without left-overs, the unaccounted for. . . . The visualized world differed from the lost one by putting assignment where blind fate ruled. The jobs to be done were now gleaned from an overall plan, drafted by the spokesmen of reason; *in the world to come, design preceded orders.* (1992, p. xv, emphasis ours)

What obtained says Bauman, was a *legislative* modernity where *soi disant* intellectuals/modernizers saw 'society' as a tabula rasa—as an object of gardening and the elimination of ambivalence. While developmental planning in India was based on securing nationalist economic development, it was also firmly embedded in the discourses of the 'gardening' state—where development, through the reduction of poverty and inequality, was the movement toward 'order.' The modernist grid of the Plan (borrowed from Soviet experiences) was invested with phantasmagoric qualities: plan = development = order was part of the utopia of development. Here development/order went hand in hand with what David Harvey has called the logic of space-time compression under capitalist modernity (1989). Here the annihilation of space by time due to the expansion of global capital has led to the 'disembedding of social relations' and the homogenization of vast spaces of the world economy. Temporal acceleration was a significant part of the imaginary of developmentalism—this was inherent in the logic of 'catching up' with the core areas of the world economy by privileging a certain strategy of growth that actively delegitimatized local and 'traditional' practices.

What obtained was an imaginary that was strikingly common to Bentham's Panopticon. The original Panopticon was conceived as a prison where 'rational' methods of confinement were deployed to ensure the visibility of all the prisoners to the warden's gaze, while he himself remained out of their sight. Here the residents of the Panopticon live in an ordered, supervised environment committed to an abstract ideal of freedom. In the eyes of its innovators the political technology of the Panopticon had the great merit of imposing order while simultaneously preventing the oppressed from visualizing power. As Zizek points out,[4] it was the great "dark spot" as to who was at the center that gave the Panopticon its greatest use.

For it was this abstract center, that space of anonymity from the nation and the everyday, that gave the Nehruvian developmental bureaucracy its greatest relief. Meritocratic, upper-caste, and English-speaking, the state-managers of postindependence India cultivated an anonymity that was seen as necessary for a legislative modernity—an abstract vision that would transcend sectional, regional, and religious claims.

Temporal acceleration, development, and 'order' were, indeed the *focus imaginarius* of Nehruvian nationalism's struggle for modernity. In terms of historical practice such an imaginary had to be mediated through the claims of a republican democratic politics. The periodic remapping of political/social space by political actors/movements through the regime of political representation meant that the claims of panoptical political technology were continuously contested. As we shall see, the rise of new social movements of oppressed castes by the late 1980s seriously threatened the exclusivist vision of the Panopticon. This, along with the new globalism, compromised Nehruvianism's old 'map' of the national space.

Building the Network

Frederic Jameson calls architecture the privileged site of postmodern representation because it is able to speak best to the new spatiality of postmodernism. All twentieth-century movements have their iconography—Nehruvian nationalism included. As opposed to the Gandhian evocation of the Village, Nehruvian nationalism privileged the Dam. The Dam was the "temple of modernity," it evoked the power of secular labor over nature. It was Nehruvian nationalism's great dream of controlling and disciplining energy. In newsreels and in print, Indians were exhorted to 'visit' Bhakra Nangal—the first major postindependence dam site—through a previrtual tour.[5] As Deshpande (ibid.) points out, the identification of the Dam (along with sites that produced steel and electricity) as a site of postindependence nationalist 'journeys' was based on privileging the 'economy' and production as markers of patriotism and national development. This privileging of the 'economic' was by the 1970s grafted to a highly centralized and repressive state whose self-representation was dynastic rule by the Nehru-Gandhi family. 'Development' was paralleled by state-sponsored compulsory sterilization drives aimed at the poor.[6] This project ended in political defeat for Indira Gandhi and the Congress Party. In the 1980s the Congress was back in power—but the old nationalist architecture was in considerable crisis. A new approach was put into place in the early 1980s, actively encouraged by Rajiv Gandhi (Nehru's grandson) who later became prime minister in 1984. This new constellation had two main components.

The first was to ensure temporal acceleration while at the same time perform the task of emancipating the state-managers from the everyday, the interaction with *place*. In other words the annihilation of space through time would obtain without the messy political problems that spatiality and its associated politics produced.[7] What was needed was a solution that would shift from old-style nationalist policies, seen by the elite as restricting initiative and growth. This was resolved by an evacuation of the 'national' space ('globalization'), a process that would accelerate by the late 1980s and the early 1990s. Under pressure from the IMF and the World Bank, the old import substitution regime was gradually dismantled and controls on domestic industry and transnational companies lifted.

The end-result of all these moves was a decisive reconstruction of the old nationalist imaginary in ways that would dissolve it to the point of no recognition. 'Development' remained an issue but was reconstituted as a problem of *communication.* The way forward was computerization, networking, and a new visual regime based on a national television network. The computer soon became the iconic space around which almost all representation, both state and commercial cohered—the effect on nationalist discourse was incredible.[8] As opposed to the Nehruvian focus on nineteenth-century physical instruments of accumulation (steel, energy, coal), state discourse after 1984 posed a virtual space where issues of development would be resolved. Through public lectures, television programs, and press campaigns, state-managers simulated this new space, which though unseen was seen as transcending the lack inherent in Nehruvian controls.[9] This new image of the computer was akin to pure reification—as the old critical theorists like Lukács had described in *History and Class Consciousness.* Except this largely unseen object[10] was also a simulation machine, generating a new form of abstract space (the network) which would accelerate the transition to modernity and the 'West.' In the event, the old panoptics of Nehruvianism could not but undergo a subtle revision. The 'national' was reaffirmed but through a new discourse which complicated the notion of borders and sovereignty that were so central to the old visual regime. 'Development' was redefined, *pace* Virilio, as a problem of speed and information. The more accurate information you had, the better your chances in joining the West.

By the mid-1980s the state promoted the setting up of a national network which would connect all major district centers and state capitals of the country and process vast amounts of information relating to development and administration. This was the development of NIC—the National Informatics Centre. The NIC had been set up in the mid-1970s to promote computerization in administration, but really took off in the 1980s with the inauguration of a satellite-linked network: NICNET. NICNET is easily the largest network in the country today. It links up all district, state, and national centers, and runs large databases on social science, medicine, and law; it works all state research institutes in the country (1994). Apart from putting all these centers on the NICNET e-mail network, NIC provides users with Web and internet access. Internally, large research projects on artificial intelligence and CAD are being undertaken. Today NICNET has easily eclipsed India's other state network ERNET (Educational and Research Communication Network) to become a visible public presence.

NICNET was not just about more computers in administration and education—it intended to change the very technics of power. Hence its aggressive 'public' profile. Very simply, NICNET sought to mold a new state cyberpublic from the late 1980s on, through regular, well-publicized demonstrations on networking, e-mail, and international connectivity. It was unusual for a state organization in India to adopt such an aggressive public profile. This brought NICNET in conflict with other institutions of the state which argued for the older, more centralized bureaucratic forms of control.[11] NICNET was run by people who understood the need for a new panoptics of state power—the older methods of surveillance would just not work.[12] While privately, NICNET administrators pooh-pooh any efforts to censor the networks, they have campaigned publicly for opening up the airways to "ordinary citizens." NICNET opened some of its airways for public access to put forth its case for ending old-style state restrictions on communication.

The NICNET experiment attempted to rework the old modernist grid of Nehruvianism which was based on a representational realism, a production of identity based on Westphalia-style national borders and a model of development which privileged the 'economy' as a site of national renewal and subsequent transition to modernity. The networks that the NIC put into place did not *directly* confront the old model but instead sublated it, retaining elements of the old imaginary (development, a new 'nation,' temporal acceleration). The crucial innovation was to introduce a simulated space which would accelerate that which was lacking in the old.

NICNET's new grid for the network was in effect a simulation of the earlier panoptics: each district connected to the state capital, the latter connected to the national capital. District-level campaigns retained Nehruvianism's old crusading vision. Reporting on the effort to set up a district infomatics center in the Gurdaspur district in the state of Punjab in 1989 the NIC wrote:

> The NIC Gurdaspur Centre first became operational in 1989. In the beginning was the toughest part. A new technology, a new way of doing things, a new way was being advocated. The people of Gurdaspur had to be made aware of the utility of infomatics, government personnel had to be trained in computers. Gaining acceptance was no easy task, but help was always at hand. Thanks to the support and co-operation of the district administration and the unstinted efforts of the NIC officials the Gurdaspur District Centre got off to a flying start. . . .

Deemed to be an underdeveloped district, today Gurdaspur is well on its way to development. (1995, p. 8)

This language continues to play on 'development,' and deploys the old crusading language of Nehruvian modernity vis-à-vis tradition but adds the new simulated space of connectivity. This new panoptical technology is based on the allusion to a new space which would guarantee the unity of the national—the space of which is constantly threatened by internal strife and global incorporation. This phantasmic neo-national space is complicated by two factors. The first is the simple brutal fact of peripheralization—constant network breakdown which militates against a seamless web of communication.[13] The second is the multiplication of networks, which cancels the monopolistic legitimacy of panoptic power.

The Trope of the Journey

In many ways cybertime in a peripheral society like India can be seen as radically revising the idea of the Journey. Traveling was the basis of knowledge about others in precolonial civilizations—including that in South Asia. The great temporal community of the traveler was largely oral, and based on sustained interaction with other communities. In the world of the *Dar-ul-Islam*, which dominated the premodern world, traveling was the duty of every Muslim either on the Haj to Mecca at least once a lifetime or to learn about other Islamicate states in the realm. These were not the journeys of the Creole intelligentsia in the period of print-capitalism—a premium was placed on *oral* interaction with fellow Muslims. This was a journey without frontiers—the traveler was constrained only by physical limitations. Time operated through different registers—there was no temporal compulsion on the traveler's return: travelers came back after many years.

The nationalist journey was the great invention of Gandhi. The Gandhian journey began with a paradox: it used the railway network—a network that Gandhi had denounced as having destroyed the precolonial narrative and experiences of the Journey. For Gandhi:

> The railways disrupted the Indian body politic, which Gandhi saw as a huge body digesting and assimilating different cultural elements. He then showed how the two technologies, reflected in pilgrimage on foot and by rail, represented memory and erasure respectively. The pilgrim's progress was an act of faith. The arduous act of pilgrimage to different corners of India gave the pilgrim a sense of both neighborhood and nationhood, helping him to internalize both similarities and differences. The mechanical negotiation or 'ingestion' of territory through rail travel erased the sanctity of places and turned them into physical spaces. (Nandy 1995, pp. 181–82)

In fact, Gandhi's privileged journey was that of foot-travel. The practice of walking meant a different language of interaction, of knowledge. It was only through walking that humans came into interaction with different cultures and peoples—which the abstract motions of locomotion could not provide. The allusion to the premodern traveler/pilgrim could not be more direct.

However, Gandhi did use the railway extensively during anticolonial agitation. To counter his own ambivalence vis-à-vis the industrial capitalism that the railway represented, Gandhi carved out an area *within* the train—the Third Class compartment—a stimulated space where the people 'entered' the train as with the humility of the premodern pilgrim. Gandhi's innovation was to transform the Journey as a vehicle to politically remap the nation. This was also a journey-as-dialogue, with conventions of public interaction, anticolonial agitation, and spectacle. In a typically Gandhian way, the railway grid (colonialism's most ambitious disciplinary structure) was subjected to a series of inversions. This was despite Gandhi's own ambivalence vis-à-vis modern technology.[14] The Gandhian journey was also a *public* journey, with the attendant modes of representation constructing an imaginary community of nationalism.[15] Colonialism had introduced the political technology of the modern state with the attendant modes of political control. The Frontiers of old became Borders, and modern citizens in postcolonial India were subject to new mechanisms of political identity in the republican order. The Journey was now governed by clear notions of sovereignty, passports became a necessary precondition for travel abroad. Indian Nationalism was always suffused with a kind of cartographic anxiety (Krishna 1992)—a phenomenon made worse by the Partition of the subcontinent and the wars with Pakistan and China. Further, state-sponsored import-substitution industrialization drives regulated the import of foreign consumer items—what obtained was a kind of physical separation of the 'everyday' from the West for about twenty years. Filiality to the border became a marker for citizenship with state campaigns against smuggling intensifying during the authoritarian period of the 1970s. The state's claim was that of a monopoly over identity formation, citizenship, and national representation.

While the state-centered social space of the 'national' climaxed in the mid-1970s, the journey was being mapped out in different ways in the popular realm. From the 1950s on popular cinema generated a series of films where a large part of the narrative unfolded in the West—with Indian characters. While the melodramatic structure of the films was based on a final resolution at the point of origin (India), what is interesting is the visual mapping of the 'West' simulating a virtual tour of that geographic space for millions who had been denied that chance. This substitute travelogue, though mediated through the foundational myth of nationalism (the return to the originary soil in the end), nevertheless speculated on the space 'outside' the border, which remained mysterious to the vast majority of citizens.[16] Nationalism operated within the framework of a representational realism: an affirmation of corporeal identity through a framing of the 'real,' which in turn was referred to a particular place: the Nation. The Journey of late nationalism has begun to be eclipsed by the implications of a journey through virtual space, raising profound implications for the existing forms of identity.

James Clifford has written that "a journey makes sense as a 'coming to consciousness,' its story hardens around your identity. (Tell us about your trip)" (1988, p. 167). The journeys into virtual space have allowed for the destabilizing of this 'hardening' of identity Clifford talks about—it has generated new 'boundary stories' about virtual cultures in India. For one, the idea of the Border, so central to state-nationalism, is transcended. The new experience of cyberspace that allows the citizen to "arrive without having set out" disrupts the old realist confines of the Journey: for the first time in the history of the visual regimes set in motion by the West, the Third World is allowed entry into hitherto prohibited spaces.

This journey is nevertheless embedded in new networks of power which seek to emancipate the self from the space of the nation/periphery, from the political sphere, and remake the lost community in cyberspace through the deployment of new landscapes—the ethnoscapes of postnationalism (Appadurai 1993). In effect, the old nationalist landscape went through a double process. The first was a certain deterritorialization of the old nationalist space, confined by the Border. At the same time a process of transnational territorialization was under way, where India was being invented in virtual space by sections of nonresident Indians. This was a process that was actively promoted by the Indian state.

The invention of the 'NRI' (nonresident Indian) served to rework the old cartography of nationalism, where identity formation was possible only within the Border. With the NRI, the border was extended to those spaces outside national sovereignty where the diaspora lived. In the old nationalist imaginary people left the home of the nation only to return— sometime. This was not to happen with the vast masses of Indians who migrated to the West from the 1950s on. The discourse of the 1980s restated the idea of the emigrants 'return'—only to make it temporally indeterminate. The 'natural' affinity to place in the old nationalist discourse was reworked in the 1980s with growing globalization and the emerging space-time compression. The NRI was now seen as forming part of a 'national' community—yet emancipated from the 'place' of the nation.

In the 1980s the figure of the NRI was evoked as someone who would help in the new-style development process. As the NRI had a natural affinity for the Home, any capital he or she brought in to the country would be motivated by patriotic feelings in contrast to say, multinational capital. In the 1990s, however, as national development receded the NRI began to play a prominent role in the new virtual spaces. The idea of the Return now became less relevant—it was achieved in pure space.

Cyberpublic II: The New Elites and the Diaspora

By 1995, both state and private networks had spread to connect around 120,000 users in India.[17] Although this number may seem small in comparison with the West, it represents one of the largest figures for a Third World country. There is also every possibility that this number will rapidly accelerate in the near future. ERNET, the main state network after NIC-NET, plans to network around 8,000 colleges—in addition to the 6,000 institutions it already connects.[18] Various problems dog the existing networks. While the networking market has been seeing a steady 100 percent yearly growth, the area of Wide Area Networking (WAN) suffers from poor infrastructure which is still inadequate for high-speed connectivity.

A quick profile of users in India will help us understand the kind of cultural practices centered around the Net. While individual users who use dial-up connections from home are growing with the expansion of private networks, large numbers of network users still work from offices, research institutions, and public terminals. As expected, most users are male and come from the middle and upper strata in both economic and social terms of community and caste. A large proportion of them are from research institutions and universities—although private networks may soon equal the share of the state networks once the telecommunication policy is liberalized.

What has the rapid spread of virtual space through the networks meant in terms of public discourse? The coming of the Web and discussions on the information superhighway in the United States has reached India with such a force that it is difficult to pick up a newspaper without reading an average of two stories a day on 'cyberspace.' The term 'cyberspace,' coming in the context of increased globalization and the dismantling of the old nationalist regime, speaks to a variety of discourses and practices—which posit a new virtual landscape. This new landscape is being actively created by a complex intersection of various practices which seek to map new borders of identity formation. Here the old landscape of the nation, the citizen, and the political are being rewritten through new Journeys.

In the first instance, discourses on the Web seem to reproduce old nationalist tropes. Most commercial providers have sought to pose their Web strategy as providing a window for 'Indian' capital to the 'world.' In this reading, Indian company sites on the Web have the best possible chance for "national" development. It is in virtual space, where emancipated from the territorial limits of old-style capitalism (which was inherently biased in favor of the West), Indian capital could take its message. The Web's great advantage for Indian companies, says one of the commercial providers, is "that (here) even smaller players get world-wide publicity."[19] The discourse here is very different from the allusions to the Frontier myth that dominated early Western representations of the Web. In the Indian case the commercial message is clear: to be a genuine 'national' capitalist, you must transcend the Border and enter virtual space. For it is here that the peripheral status in real time will be transcended.

Behind the impetus toward virtual space is the retreat of large sections of the old upper-caste, Anglicized elite from the political sphere. It is important not to be crude in speaking about this phenomenon, but since the 1980s the cartography of the political public is being actively rewritten, with the emergence of new social movements of oppressed castes. The rise of these movements has had the effect of challenging the old panoptics of the state which was predicated on a homogenizing legislative modernity, led by an enlightened elite of modernizers. The symbolic space of this elite lay in the abstractions of national development and the 'economy.' The rise of the new movements has initiated a process that may effectively unseat the old elites from the political sphere; the certainties of the old nationalist journey no longer obtain.

For the old elite the state is no longer the secure kingdom of cultural hegemony and identitarian certainty.[20] The social landscape has undergone an effective Haussmannization marked by upper-caste retreats from the old grid of politics and abstract nationalist identity. The large metropolitan centers are being reconfigured to accommodate a reinvented suburbia free from subaltern intrusions: these are simulated landscapes of designer villages, transplanted American postmodern designs, and private security. But what of the nationalist journey? My argument is that for this elite-in-retreat, the old journey is being transcended by the encounter in virtual space. In its place new practices have emerged which have sought to both overcome the corporeality of the nationalist border, while at the same time attempting to create new "nationalist" electronic communities conforming to a 'Hindu' imaginary. (Since the mid-1980s an aggressive Hindu nationalist movement has grown in India, seeking to rebuild nationalism around themes of a threatened religious identity, majoritarianism [where the state acts in interests of the majority community], and violent attacks on the Muslim minority.)

The second cyberpublic is therefore doubly coded by practices that seek to map out a new conception of space beyond the nation, while at the same time attempting to inscribe a new Hindu nationalist community in virtual space.

At the center of this new landscape has been the growth of new technologies of representation which have had the effect of disrupting the old tropes of anticolonialism and Nehruvian nationalism. If the village and the 'economy' functioned as a representational shorthand for the Gandhian and Nehruvian imaginary respectively, the new cultural landscapes in the 1990s saw a complex of initiatives centered around new narratives of consumption and desire which resist easy historicist classification. These new practices have entered around the rapid growth of television, video, music, and one of the world's largest film industries. Foreign satellites now beam images to India, effectively breaking the state's monopoly over television; India now produces one of the world's largest video- and audiocassette industries, largely centered around the film industry.

The new cultural space is crisscrossed by a fluidity of national/regional/global cultural styles mediated by the recognition of a new agent—the "consumer-subject"; the old moral codes regulating desire are being reconstituted and a new Hindu nationalist imaginary attempts to cannibalize all these new practices for its political project. There is no doubt that for the time being at least, the claims of a legislative modernity are suspect. The panoptic vision of a regulated cultural practice, while voiced periodically, lacks the authority and legitimacy as in the past.[21] In this liminal space mediated by various cross-practices, the elite cyberpublic occupies a hybrid space which attempts to emancipate itself from the nation, its Border, and its political public. The modes of representation allude to a fluid space, where the nation is present yet thoroughly displaced, informed by a hybrid language, styles, and volatile mixture of both presence and absence.

In this context, the Web offers the phantasmic possibility of playing with an identity that recognizes displacement. The Journey into virtual space is the journey beyond the nation. For the Web traveler, a typical member of the displaced elite public in India, the West is re-created/simulated as a *simultaneous* presence. There is a certain experience of Web travel when logging on from the Third World, that almost evokes Benjamin's analysis of Baudelaire's *flaneur*, or the stroller in Second Empire Paris. The Web traveler in the elite cyberpublic seeks out the virtual space of the Web to experience the "shock of the new," which Benjamin calls the distinctive feature of modernity. The images of the Web, like the city in Haussmannized Paris, are shot through with a phantasmic space where exist dream worlds of desire and consumption—the arcades in Benjamin's story and the Web sites for our traveler. The city for the *flaneur* has a labyrinthine character, with secret passages, a web of experiences and unknown dreams which is sought out by the stroller.

This is where Baudelaire's *flaneur*—the mythic hero of modernism—and the late-twentieth-century Indian Web traveler part. For the *flaneur*, the crowd was the great veil between himself and the phantasmagoria of the city. For the Web traveler of the elite cyberpublic, the journeys into virtual space perform the opposite function—of an emancipation from the "crowd" of real time.

Web strolling from India is an entry into a space whose virtuality enhances the feeling of being in the "West." In the context of other experiences of space-time acceleration brought about by the television revolution of the 1990s, this feeling is magnified. This is an entirely new geography of desire, almost exclusively centered around sites in the West.[22] This is quite distinct from the new ethnographies of travel in the West.

Writing from a Western setting, James Clifford (1988, p. 14) points out:

> An older topography of experience and travel is exploded. One no longer leaves home confident of finding something new, another time or space. Difference is encountered in the adjoining neighborhood, the familiar turns up at the ends of the earth.

For the Indian Web traveler, the incursions assume a search for a mythic space of modernity, where "newness" is emancipated from territory. Like Baudelaire's *flaneur* who sought out the crowd in his search for 'ever-new,' the Web traveler journeys on the highway to look for the new. The websites constitute a simulated exhibition ('places of pilgrimage to the fetish commodity'—Benjamin) where the traveler-consumer, like the visitor to the nineteenth-century site, is asked, "look at everything, touch nothing" (Benjamin in Frisby 1986, p. 254). These pilgrimages have the effect of an experience of modernity (shock, ecstasy, entry into power-knowledge spaces) hitherto unknown in the periphery—even for the elite. Yet they are fleeting experiences—burdened by real-time constraints.[23] For the Web traveler the fleeting experience of transcending the Border, rather than long-term immersion into virtual space is the norm.

As mentioned before, Web journeys are informed by a double-coding, one side of which is the elite cyberpublic's emancipation from the old nationalist grid. The other side is the creation of a naturalized space of "India" on the Web—initiated largely by Indians in the Diaspora. Dominated by expatriate Indians sympathetic to Hindu nationalism, these websites[24] pose Hindu identity as isomorphic with India: a space purged of ambivalence.[25] It is almost as if the old legislative modernity of the Nehru period has been transplanted to virtual space, purged of its democratic political sphere.[26] In the virtual space of 'India' on these websites, Hindu identity becomes an artifact—a contestable process is replaced by a reified boundary. For the NRI, the virtual space of India finally replaces the actual pressure of the Return. The Journey is now a sanitized one no longer fraught with tension—the shock and complaints of peripheral poverty, the perplexities of cultural self-questioning. Here the websites act as markers of homogenized spiritual space, with rigid cultural borders, where "India" functions as a virtual museum for those for whom Hinduism can fulfill the great unfulfilled dream of legislative reason—a world without ambivalence.

Both practices discussed in this cyberpublic are, in the old Marcusean sense, affirmative. In the first instance journeys into virtual space function as either a new postnational hybridity emerging from the new elite enclosures of India. Here "hybridity" seems to have a very different function from the heroic status accorded to that term by postcolonial intellectuals in the West. In the case in point hybridity emerges from the new liminal cultural landscape of 1990s India but also performs an act of closure vis-à-vis the popular-political. The aporetic position of this elite cyberpublic vis-à-vis the national through its new cyberjourneys is compromised by its complicity with the power of both local and multinational capital. On the other hand, the Hindu nationalist attempt to territorialize virtual space with its rhetoric of origins, of contamination, and of naturalization remains the most reactionary attempt to invert the old Journey.

Cyberpublic III: Bulletin Boards, Activists, and the Search for Alternatives

The third cyberpublic remains the most ambiguous of the three domains of cyberdiscourses in India. Existing in the fluid space between the state cyberpublic and the elite domains of the Web, this cyberpublic contains within it a wide range of actors seemingly unconstrained by either the state or the transnational market. The map of this cyberpublic is typically rhizomic, a constantly shifting zone of activist networks, small bulletin boards, and dissident scientists. The borders of this cyberpublic are, once again fuzzy, sometimes operating within the grid of the national state network, sometimes playfully intruding into the more privileged Web space dominated by private/multinational capital. Less hybrid than experimental, this public speaks to the possibility of radical reconstitution of electronic space as one which touches real time through its myriad surfaces.

In the opening up of electronic space beyond the frontiers of the state/market dichotomy, bulletin boards (BBS) have played a crucial role. Numbering just a few score until the last year BBS have mushroomed not only in all the major metros but also in small towns all over India. Led by a combination of small business persons, computer/telecom graduates and those with skills acquired in the trade, the BBS cater to a sector of the population who find both the Web and the state networks inaccessible due either to prohibitive costs or lack of an imaginative space.

Initial BBS discussions concentrated largely on the computer trade—reflecting the users' immediate concerns. Recent discussions have been broader, concentrating on politics and sexuality. The latter is a topic that most system operators tend to be wary of (most discussions are not yet on-line), but this has not prevented frank discussions of issues that have hitherto remained invisible from the public sphere. The BBS have to walk a fine line, with recent media stories about sites offering explicit pictures for free downloading. State regulation is also a threat, with recent legislation threatening to tax the BBS heavily—this threat seems to have receded, for the moment.

The BBS are a recent phenomenon in India, and one is yet to see even the kind of experiments that the CommuniTree group initiated in the United States in the 1970s. However, any comparison, or even the suggestion of 'fitting' India in an evolutionary schema based on the development of virtual systems in the West would be wrong. For a very, very long time to come electronic space will be out of the reach of the majority of the urban population, let alone those who live in the villages. The importance of the BBS remains elsewhere.

Existing between the space of state control and the power of global capital, BBS offer a novel form of agency within the discourse on virtuality. To sections of the urban population disembedded by globalization and subject to the shock-like experience of the new Haussmannized city, the BBS offer an important zone of engagement and the possibility of a new performative space. I use the latter with some economy. It is the very novelty of the BBS, along with its semiunderground status, which opens up the possibility of experimentation—a process which has just begun. As sites proliferate, so will the variety of experiences, the inventiveness and technosocial practices. Women, hitherto a marginal voice in the BBS community, are slowly making their appearance. Strategies of self-representation remain dominantly realist, with users only using different 'faces' when discussing explicit issues. The historical identity of realism and scientific practices was hammered into the popular space by nationalism—hence the tendency to stick to a realist mode during the initial moment of initiation into technoculture by most novices. It seems to me that the time for a genuine *aufhebung* to a new nonrealist mode of representation is already on the agenda of the BBS community.

Social Movements, Nationalism, and Technoculture

Most social movements in India have as their point of departure the cartography of the postindependence nationalist state. As pointed out earlier, the state privileged a model of development, iconized the dam and the steel mill as the imaginative reference points of development. The official discourse on science and technology remained within the framework of the developmental modernism imposed on the periphery. Here science and technology were opposed to culture, abstracted from notions of play, creative tradition, and aesthetic experiment.

What is important is that the sites of nationalist science were symbolized by the magnified products of developmental modernism—the dam and the steel mill. The new social movements that emerged after the 1970s generated a critique of the technological/developmental imaginary of nationalism, stressing a range of alternative practices. The various movements (those of women, untouchables, antidam) did not pose a cohesive alternative; their opposition to state-sponsored technological practices was generally uniform.

Thus when the computer was initially introduced in the 1980s as the neo-modernist successor to the dam, the hostility of both the new social movements and the old Left was total. They parodied the utility of the computer in a peripheral society like India, a critique that generally echoed then prevalent notions of utility and sustainability and concerns about workforce cutbacks. The fact that the computer was introduced with the old-style developmental rhetoric made the movements even more suspicious.[27] Today in the 1990s, the movements have come to not only accept the computer but also the creative possibility of networking. This is a dramatic change, for which a number of factors have been cited. In the first place the old movements are in crisis—many have disintegrated and joined the NGO sector. The crisis of old-style nationalism and Marxism has reshaped the old reference points for the movements. The fast-growing NGO sector is linked to global donors, the sector's incorporation into global electronic space is only a matter of time. It seems to me that these factors are a necessary but not sufficient explanation for the widespread acceptance of electronic networks among the movement-community. At the heart of the transition are hidden issues of desire and identity which have been brought into play.

The old sites of the large-dam and the steel mill were enlarged symbols of the nationalist will-to-power, generating the violence of displacement and the destruction of local communities. As violent symbols these 'sites' are still the focus of large movements. On the other hand, the world of virtual space that exists 'behind' the computer lacks any corporeal violence associated with developmentalism. It seems to me that virtual spaces began to evoke a world of pleasure and initiation for individual activists, without the violence of developmental modernism. A certain aesthetics of experimentation had already been experienced by activists in their search for alternatives to developmentalist disasters. With the coming of e-mail, the internet, and later bulletin boards a liminal space emerged, where utopian desires for modernity, the possibility of experimentation "without destruction," overlapped with the pleasures of initiation rituals into technoculture. Further there is the possibility of a dialogue with the self: the more rounded forms of identity in the nationalist period mistrusted ambivalence. In every sense new boundaries of imagination and agency have been created.

To be sure, only a very small minority of activists are still connected. Those who are either urban, relatively affluent, or have access to global funds. But what is remarkable is the widespread legitimacy of electronic space among dissenters and activists, who would be equally critical of the technological monuments of nationalism. It could be argued that the entry of virtual spaces posed technology as a cultural practice in a way that the developmental modernism[28] of the Nehruvian period (with the singular emphasis on monuments) could never do. It anticipates a new situated technosocial space, perhaps a 'cyborgness' for the periphery. I use these terms with considerable hesitation, for reasons that will be spelled out later. But following Donna Haraway's call for "situated knowledges" we can argue that new sets of practices could emerge in India which may mark the transition from the binary spaces of developmental modernism.

Developmental modernism operated within a Third World version of what Foucault has called the blackmail of the Enlightenment. Foucault uses this formulation in his famous essay on Kant's "What is Enlightenment" to refer to the violence of the philosophical choices presented by the Enlightenment: "you either accept the Enlightenment and remain within the tradition of rationalism . . . ; or else you criticize the Enlightenment and then try to escape from its principles of rationality" (1984, p. 43). There is, says Foucault, simply no other choice: *no tertiium datur.* In the Indian case, the canvas was less Olympian: it operated within the rather simplistic oppositions of development/science/progress versus tradition/reaction/stasis. For many decades the first triad was overwhelmingly hegemonic, based on the state's monopoly of power and violence and even extending to old-Left oppositional movements. It is only in the recent decades with the rise of the new social movements, that elements of a genuine *aufhebung* have emerged. The old oppositions do not hold securely anymore—a discursive space of questioning has emerged, energizing new exchanges on technology, tradition, and popular experimentation.[29] The current generation of activists has been reared on this new diet, where opposition to large dams and displacement could go hand in hand (wherever possible) with ventures into virtual space. This transition is so significant that it is remarkable that it has gone unnoticed.[30] Nevertheless, despite the richness and potentials of the third cyberpublic in negotiating a space between the market and the state, the access to virtual space still remains a privilege. The plans of the state network ERNET to connect eight thousand colleges and schools will undoubtedly expand this public; however, there is an urgent need to fight for cheap, publicly accessible networks. The current neo-liberal mood of the ruling elite is hostile to any public space in the electronic media, a long battle is ahead for activists.[31] Conclusion: Artificiality, modernity, and alternative futures in the periphery.

We can now go back to Jameson's characterization of our time as an invented millenarianism. It cannot be denied that for the spectacular nineteenth-century ideologies, Marxism, liberalism, and nationalism, the logic of disillusionment is complete. The great *aufhebung* has, in fact, not obtained—rather the idea of the end confronts all. The grand social subjects of the nineteenth century (the proletariat, the middle class, the nation) have been confronted by a landscape of death. We are faced with the death of the social, the death of the subject, the death of the author, the death of the real, the death of the nation. It is as if Adorno and Horkheimer's gloomy prognosis in the *Dialectic of Enlightenment* has come true: modernity has become a giant graveyard.

Yet it is important to caution against premature burials. The punctuality of death is not a useful metaphor in social theory. It may be important to listen to Nietzsche's savage warning, "The destruction of an illusion does not produce truth but only one more piece of ignorance, an extension of our 'empty space,' an increase of our 'desert'" (1967, p. 327). It may not be necessary to go all the way with Nietzsche, but living in a country at the periphery of the great centers of capital accumulation, it may be useful to point out that an unmediated celebration of the "new" is not just feasible, nor very interesting. Every historical appearance of "newness" in South Asia, right from colonialism to developmental modernism, has been embedded in violence and domination. No amount of dialectical sophistry on modernity can erase that history.

At the same time it is not necessary to take adopt the position of a nineteenth-century *Kulturkritik* in denouncing the cultural social landscape of our time and deny that fundamental reorganizations in the terms of representation, subjectivity, and the old boundaries of Western modernity are taking place. In this sense Benjamin's classically modernist notions of the present as "eternal recurrence" or the "present as new" have little use in a world where old-style homogenization of consumption spaces under capitalism have little for new strategies of capital accumulation (Huyssen 1995, p. 26).

For India, cyberspace and the cultural experiences that it evoked arrived as a representation of the "new," in ways that were complexly coded.[32] A number of loose cyberpublics emerged, reflecting in diverse ways state, neo-elite, and popular strategies to negotiate the crisis of the old Panoptical grid of nationalism and the supremacy of the Border.

Cyberculture also came to India within the framework of a new package of globalization which contained within it a potent mixture of pleasure and danger. If on the one hand globalization has unleashed a new discourse of consumption which unsettled the old nationalist/Marxist denial of the consuming public, the 'global' also comes on the backs of the power of transnational corporations with their tremendous power and contempt for popular participation. As Virilio points out, "the new communications technologies will only further democracy if, and only if, we oppose from the beginning the caricature of global society being hatched for us by big multinational corporations throwing themselves at a breakneck pace down the information superhighways" (1995).

To be sure, the inherent feature of "newness" is contingency which may contain the possibilities of its demise. Says Huyssen, "our fascination with the new is always already muted, for we know that the new tends to include its own vanishing, the foreknowledge of its obsolescence at its very moment of appearance. The time span of the new shrinks and moves towards vanishing point" (1995, p. 26). The problem is, as I have mentioned above, the "new" does not reproduce itself globally in fundamentally similar ways. Despite all the fantasies of virtuality, "newness" is inscribed in a different set of overlapping imaginaries when it reaches a "Third World" country like India. *Contra* Virilio speed has many faces, and will continue to do so for some time to come.

What is needed is an approach that is sensitive to the situated character of knowledge-formations and cyberpractices in the Third World, where narratives of critique and fluidity in the West may have entirely different consequences when re-presented in a country like India. Consider, for example, the emergence of new technopractices in the West which have led to the blurring of the historical distinctions between nature and culture, between the body and the machine. As a number of writers (Haraway 1991; Rabinow 1992; Stone 1992) have pointed out, the overlap of technology, biology, and culture have led to a situation where the old distinctions between organicity and technology, between culture, experience, and science (where one or the other was privileged), have lost their old categorical fixity. This involves a crucial transition from the old Enlightenment oppositions of nature and culture from which flowed the representations of human praxis and subject formation. As Rabinow argues, while nature is being remodeled on the model of culture, the latter is reconstructed on the basis of nature (ibid).

What happens to this transition when it reaches the periphery of world-capitalism? The new biosociality is surely implicated in the plunder of plant species from the Third World and their relocation to the West; technoculture is a clear part of the new expansion strategies of multinational companies in the periphery. Technosociality in India speaks to elite strategies of withdrawal and reoccupation of the national, it is also implicated in a culture of distinction and distance from the lives of the underprivileged.

My point is not to score old-Left polemical points against the new cultural constellations emerging in the West. The issue is that these constellations are part of a new flexible system of accumulation on a world-scale where "a dizzying sense of bodily freedom" for one may mean displacement, and loss in another part of the world we live in. To neglect this would not just be unreflexive on the part of all of us who 'live' and enjoy the pleasures of virtual space—it would be connivance with the new grammar of power emerging on a world scale.

However, boundary stories do matter, even on the periphery. The stories have been less about the individual body-space of the citizen and more about rewriting new maps of the national, where cyberjourneys have punctured the old Panoptic Border. From the point of view of situating knowledge we can rewrite Haraway's brilliant statement on the idea of 'nature': "The certainty of what constitutes the *nation* . . . (that is, as) a source of insight, a subject for knowledge, and a promise for innocence—is undermined, perhaps fatally" (1985, p. 70).

NOTES

1. The fuzziness of the publics militate against any attempt to "mirror" a particular public with a social/cultural group. The old logics of certainty no longer obtain here.

2. One does not have to read Michel de Certeau to understand Gandhi but the former's *The Practice of Everyday Life* (1984) throws interesting light on "sly" practices of resistance.

3. As Deshpande points out, "The construction of the national economy very soon becomes synonymous with the nation itself. This enshrining of the nation as the synechdonic representation of the nation was perhaps one of Nehru's distinctive personal contributions to the nationalist cause" (1993, p. 24).

4. This is what Zizek says: "In Bentham's Panopticon one finds virtuality at its purest. You never know if someone is at the centre. If you knew someone was there, it would have been less horrifying. Now it's just an 'utterly dark spot,' as Bentham calls it. If someone is following you and you are not sure, it is more horrible than if you know there is somebody. A radical uncertainty" (1995).

5. Architecture never played an iconic role in postindependence India. Nationalism, unlike colonialism, never sponsored major historicist buildings. It was content in importing some of the worst versions of construction. The only concession to the avant-garde was the invitation to Le Corbuseir to design the city of Chandigarh—surely one of the most soulless urban sites in postindependence India. For an interesting analysis of Corbuseir, see Tafuri (1976).

6. A national state of emergency was declared in 1975, and all civil and political rights were suspended until early 1977. The emergency was in many ways the denouement of the Panopticon, the beginning of the end of the old centralized state of developmental nationalism.

7. During this period India was going through considerable political turmoil, with challenges from ethnic/regional movements leading to a cycle of political violence.

8. Political leaders, led by the then prime Minister Rajiv Gandhi, would insist on posing beside a desktop; most advertisements after the early 1980s were sure to include images of computers as part of their sales pitch.

9. Early efforts were focused on developing a local telecommunications production base. These efforts were centered around the Centre for Development and Communication (C-DOT). Led by the charismatic Sam Pitroda, C-DOT's efforts were later subverted by a section of the bureaucracy sympathetic to transnational capital.

10. Even though the computer campaign had started in the early 1980s, this writer saw his first computer in 1988, when he had to submit his first assignment as a graduate student in a U.S. University.

11. There were ugly public slanging matches between NICNET and the state monopoly which acts as India's international gateway—the Videsh Sanchar Nigam Limited (VSNL).

12. In an interview with this writer the director of the NIC maintained that controls on the Net were meaningless—and the state would do nothing to check the dissemination of pornographic and explicit material.

13. Even in Delhi, the average period that this writer has managed to stay on-line is twenty minutes! The simple facts of Third World connectivity: phones break down, as does electricity.

14. This was in contrast with Marx's optimistic view. Marx tended to see the role of the railway in India very much as Baudelaire saw Baron von Haussmann's reconstruction of Pans in the mid-nineteenth century—the process of destruction being creative and opening up imaginative modes of social action.

15. A comparative journey, though more focused in time and space, is the Long March undertaken by the Chinese Red Army in 1934–35.

16. The exhibition mode had a limited presence in India. The great nationalist exhibition was Expo 72 at New Delhi in 1972. In a space especially built for the purpose various

Indian states and, more important, foreign (mainly Western) countries put up halls for visitors. This was the Third World and Nehruvian version of Crystal Palace—and was a huge success at the time. Since then exhibitions have tended to be more focused, concentrating on commercial and business interests rather than recreating the spectacle like quality of Expo 72.

17. The state networks (NICNET, ERNET, VSNL) have around one hundred thousand users and the private providers (Axess, RPG-SPRINT, Dartmail, UUNET) have around twenty thousand among them. This is an approximate calculation. We must also bear in mind that accounts, particularly in research institutions, tend to be shared by a group of users.

18. See, "Boom Time for Computer Networks," *Pioneer,* April 5, 1995.

19. "What's on at http://www.indianic.com" by Sudha Nagaraj, *Pioneer,* April 26, 1996.

20. The political wagers of the upper-caste elite have shifted from the state to the Hindu nationalist movement; yet other sections support the Congress Party.

21. This refers to periodic calls by groups for censorship of films with explicit sequences in the name of moral order. Most state attempts at widening censorship guidelines have been ineffective. In fact a recent Supreme Court judgment struck down the verdict of a lower court preventing the screening of the controversial film *Bandit Queen.*

22. Sites located in India have also no popularity with travelers who live in the country. Almost all the "hits" for Indian sites have been from abroad. See *Times of India,* May 22, 1996.

23. Once again very few telephone connections manage to stay on-line for more than twenty minutes.

24. To be sure alternative sites set up by dissenting members of the expatriate community do exist. However these sites cannot compete with those of the right wing.

25. The website of the right-wing Hindu nationalist Bharatiya Janata party, which puts out an aggressive brand of nationalism based on majoritarianism and anti-Muslim rhetoric, is, in fact, operated from the United States! The irony of this was entirely lost on the leadership of the BJP.

26. It is important to remember that despite all the authoritarian and disciplinary initiatives of Nehruvian panoptics, the relationship to the democratic political sphere prevented the realization of a "totally administered society."

27. In addition the computer introduced was the old IBM PC with a rather unattractive modernist architecture. Macs were too expensive since clones did not exist.

28. Developmental modernism does not entertain any dialectical doubt that was possible in Europe. Here the emphasis was more on functional mapping of space based on the triad development/science/nation. The pioneers of developmental modernism were neither artists nor philosophers but what Rabinow in his work on the French modern calls "technicians of general ideas" (1989); anonymous state-managers who painstakingly built the model.

29. The demise of the aura of technology has also its flip side: the emergence of a technotopia of global consumption which has as its main address the elite of the second cyberpublic.

30. The rapid rise of a mass consumption culture in the urban areas has removed the older aura around "technical" articles of consumption. Articles which the urban mass public previously saw only as exhibitionary items (where, as Benjamin pointed out, "you could see, but not touch"), are now available for daily use. These include (public and private) telephones, televisions, and, for the more affluent, VCRs and refrigerators. Though many activists who are into cyberspace may be ambivalent toward the new consumption space, this is a world where they work and struggle.

31. Given the situation of widespread income inequalities of a country like India, some form of state funding is vital for the development of alternative networks. Private and multinational capital has shown little interest in developing alternative spaces. There is every likelihood this attitude will continue in the near future. The struggle must be, however, to create a genuine public space where access to networks is made as open as possible, without state controls.

32. Even as perceptive an observer like Paul Virilio tends to miss the mediated character of this transmission. Arguing that cyberspace heralds a new global time which is going to lead to the end of local times Virilio writes, "For the first time history is going to unfold within a one-time system: global time. . . . [I]n the very near future, our history will happen in universal time, itself the outcome of instantaneity—there only" (1995). Virilio counters this bleak scenario with possibility of the "accidents of accidents"—the cyberaccident, when networks crash and capitalism 'collapses.'

REFERENCES

Appadurai, Arjun. (1993). "Patriotism and Its Futures." *Public Culture*, 5:411–29.

Bauman, Zygmunt. (1992). *Intimations of Postmodernity*. London: Routledge.

Benjamin, Walter. (1968). *Illuminations*. New York: Schocken Books.

Berman, Marshall. (1982). *All that is Solid Melts into Air: The Experience of Modernity*. London: Verso.

Clifford, James. (1988). *The Predicament of Culture*. Cambridge, Mass.: Harvard University Press.

Davis, Mike. (1990). *City of Quartz: Excavating the Future in Los Angeles*. London: Verso.

de Certeau, Michel. (1984). *The Practice of Everyday Life*. Berkeley: University of California Press.

Deshpande, Satish. (1993). "Imagined Economies." *Journal of Arts and Ideas*, nos. 25–26, December, pp. 5–36.

Escobar, Arturo. (1994). *Encountering Development: The Making and Unmaking of the Third World 1945–1992*. Princeton: Princeton University Press.

Frisby, David. (1985). *Fragments of Modernity*. Cambridge: Polity.

Gandhi, Mohandas Karamchand. (1956). *Hind Swaraj*. In *Collected Works*. New Delhi Publications Division.

Haraway, Donna. (1985). "A Manifesto for Cyborgs." *Socialist Review* no. 80, 65–105.

———. (1991). *Simians, Cyborgs and Women: The Reinvention of Nature*. New York: Routledge.

Harvey, D. (1989). *The Condition of Postmodernity: An Enquiry into the Origins of Cultural Change*. Oxford: Blackwell.

Huyssen, Andreas. (1995). *Twilight Memories*. New York: Routledge.

INFOMATICS. Quarterly Newsletter of the National Informatics Centre, New Delhi.

Jameson, Fredric. (1992). *Postmodernism, or the Cultural Logic of Late Capitalism*. Durham, N.C.: Duke University Press.

Knox, Paul. (1992). *The Restless Urban Landscape*. Englewood Cliffs, N.J.: Prentice Hall.

Krishna, Sankaran. (1994). "Cartographic Anxiety: Mapping the Body Politic in India." *Alternatives* 4.

Lefèbvre, Henri. (1991). *The Production of Space*. Oxford: Blackwell.

Nandy, Ashis. (1995). *The Savage Freud*. New Delhi: Oxford.

———. (1996). "The Decline in the Imagination of the Village." Unpublished ms.

Nehru, Jawaharlal. (1946). *The Discovery of India*. New York: John Day.

NICNET. (1994). *Ten Years of the National Infomatics Centre*. New Delhi: NIC.

Nietzsche, Friedrich. (1967). *The Will to Power*. Ed. Walter Kaufman. New York: Vintage.

Rabinow, Paul. (1989). *French Modern: Norms and Forms of the Social Environment*. Cambridge, Mass.: MIT Press.

———. (1992). "Artificiality and Enlightenment: From Sociobiology to Biosociality." In J. Crary and S. K. Winter (eds.), *Incorporations*. New York: Zone Books.

Rossi, A. (1982). *Architecture in the City*. Cambridge, Mass.: MIT Press.

Simmel, G. (1971). "The Metropolis and Mental Life." In D. Levine (ed.), *On Individuality and Social Form*. Chicago, Ill.: University of Chicago Press.

Soja, Edward. (1989). *Postmodern Geographies: The Reassertion of Space in Critical Social Theory.* London: Verso.

Stam, Robert, and Ella Shohat. (1994). *Unthinking Eurocentrism.* London: Routledge.

Stone, Alucquère Rosanne. (1991). "Will the Real Body Stand Up: Boundary Stories about Virtual Cultures." In Bendedikt (ed.), *Cyberspace: The First Steps.* Cambridge, Mass.: MIT Press.

———. (1992). "Virtual Systems." In J. Crary (ed.), *Inscriptions.* New York: Zone Books.

Tafuri, Manfredo. (1976). *Architecture and Utopia.* Cambridge, Mass.: MIT Press.

Turkle, Sherry. (1995). *Life on the Screen: Identity in the Age of the Internet.* New York: Simon & Schuster.

Virilio, Paul. (1995). "Speed and Information." *CTheory.*

Zizek, Slavoj. (1995). "Civil Society, Fanaticism and Digital Reality: A Conversation with Slavoj Zizek." *CTheory.*

Guillermo Gómez-Peña

The Virtual Barrio
@ The Other Frontier
(or the Chicano *interneta*)[1]

> "*(Mexicans) are simple people. They are happy with
> the little they got. . . . They are not ambitious and com-
> plex like us. They don't need all this technology to com-
> municate. Sometimes I just feel like going down there
> & living among them.*"
>
> *—Anonymous confession on the Web*

Fighting My Own Endemic Technofobia[2]

I venture into the terra ignota of cyberlandia, without documents, a map, or an invitation at hand. In doing so, I become a sort of virus, the

This version of "The Virtual Barrio" is reprinted by permission of the author. An earlier iteration of this essay originally appeared in the book *Clicking In: Hot Links to a Digital Culture,* edited by artist Lynn Hershman, Bay Press, 1996. This version is considerably different. Besides rewriting certain sections and elaborating specific points, I have asked several theorists and artists, including Pedro Meyer, Joanna Frueh, Richard Scheckner, Francis Pisani, and others to "respond" to my ideas. Their opinions, in print, will appear as footnotes in the pages that follow. In the internet version, their responses will appear as "hypercards." Responses from readers and internet users to this version of the text will find their way into future versions (NAFTAZTC@aol.com). *Comenzamos!*—GP

1. My colleagues have pointed out various contradictions in this work: Mexican artist Pedro Meyer and *Le Mond* journalist Francis Pisani pointed out to me that I criticize the role of "victims" that Latinos assume regarding high-technology and yet I myself often assume the "tone" and positionality of a victim. Meyer also noted that it was strange that I chose to write the text in English since I criticize the use of English as lingua franca on the Net. I have chosen not to "correct" my "contradictions." Instead, I have incorporated the objections within the internal debate of the text.

2. I wish to express to the reader that this text, like most of my theoretical texts, suffers from an acute crisis of literary identity; partly, because it reflects my ever-shifting positionalities as a Mexicano/Chicano interdisciplinary artist and writer living and working in between two countries and many communities; but also because the text attempts to describe fast-changing realities and fluctuating cultural attitudes. As of now, I am still not sure

cyberversion of the Mexican fly: irritating, inescapable, and, hopefully, highly contagious.[3]

My "lowrider" laptop is decorated with a 3-D decal of the Virgin of Guadeloupe, the spiritual queen of Spanish-speaking America. It's like a traveling altar, an office, and a literary bank, all in one. Since I spend 70 percent of the year on the road, it is (besides my phone card of course), my main means to keep in touch with my agent, editors, and performance collaborators spread throughout many cities in the United States and Mexico. The month before a major performance project, most of the technical preparations, last-minute negotiations, and calendar changes take place in the mysterious territory of cyberspace. Unwillingly, I have become a technoartist and an information superhighway bandido.

I use the term "unwillingly" because, like most Mexican artists, my relationship with digital technology and personal computers is defined by paradoxes and contradictions. I don't quite understand them, yet I am seduced by them. I don't want to know how they work, but I love how they look and what they do. I criticize my colleagues who are acritically immersed in *las nuevas tecnologías,* yet I silently envy them. I resent the fact that I am constantly told that as a "Latino," I am supposedly "culturally handicapped" or somehow unfit to handle high-technology; yet once I have the apparatus right in front of me, I am tempted and uncontrollably propelled to work against it; to question it, expose it, subvert it, or imbue it with humor, radical politics, and *linguas polutas* such as Spanglish and Franglè.

Contradiction prevails. Two years ago, my collaborator Cybervato Roberto Sifuentes and I bullied ourselves into the hegemonic "space" of the Net. Once we were generously adopted by various communities (Arts Wire, Chicle, and Latino net, among others) we suddenly started to lose interest in maintaining ongoing conversations with phantasmagoric beings we had never met in person (and that I must say is a Mexican cultural prejudice: if I don't know you in person, I don't really care to converse with you). Then we started sending a series of poetic/activist "technoplacas" in Spanglish. In these short communiqués we raised some tough questions regarding access, identity, politics, and language. Since at the time we didn't

---

which might be the ideal format to articulate these ideas: a "personal" chronicle (as in the first two sections); a theoretical essay capable of containing contradictory voices (an anathema in academia); or an activist manifesto-like narrative (as in the last part).

3. I constantly shift from "I" to "we," and the "we" means at different times "my main collaborator Roberto Sifuentes and I"; "my (techno-art) colleagues and I"; "all Chicanos in the net" or "all outsiders/insiders in the net." The "we" is clearly contextual and temporary. I am fully aware of the risks of the use of "we," yet I cannot escape the following predicament: "We" all criticize the problems of a "master narrative" in the 1990s and yet, "we" all wish to belong to a community larger than our immediate tribe of collaborators. How to solve this, I still don't know.

quite know where to post them in order to get the maximum response, and the responses were sporadic and unfocused, our interest began to dim. And it was only through the gracious persistence of our technocolleagues that we decided to remain seated at the virtual table, so to speak.

Today, despite the fact that Roberto and I spend a lot of time in front of our laptops (when we are not touring, he is in New York, and I'm in San Francisco or Mexico City)—conceptualizing performance projects which incorporate new technologies, or redesigning our website—every time we are invited to participate in a public discussion around art and technology, we tend to emphasize its shortcomings and overstate our skepticism. Why? I can only speak for myself. Perhaps I have some computer traumas, or suffer from endemic digital fibrosis.

Confieso: I've been utilizing computers since 1988; however, during the first five years, I used my old Mac as a glorified typewriter. During those years I probably deleted accidentally here and there over three hundred pages of original texts which I hadn't backed up on disks, and thus was forced to rewrite them by memory. (Some of these "reconstructed texts" appeared in my first book *Warrior for Gringostroika*, Greywolf Press, 1994). The thick and confusing "user-friendly" manuals fell many times from my impatient hands. As a result of this, I spent many desperate nights cursing the mischievous gods of cyberspace, and dialing promising "hot lines" which rarely answered, or if they did, provided me with complicated instructions in computer Esperanto that I was unable to follow.

My bittersweet relationship to technology dates back to my formative years in the highly politicized ambiance of Mexico City in the 1970s. As a young self-proclaimed "radical artist," I was full of ideological dogmas: for me, high-technology was intrinsically dehumanizing (*enajenante* in Spanish), and it was mostly used as a means to control "us" little technoilliterate people politically. My critique of technology overlapped with my critique of capitalism. To me, "capitalists" were rootless (and faceless) corporate men who utilized mass media to advertise their useless electronic gadgets. They sold us unnecessary apparati which kept us both eternally in debt (as a country and as individuals) and conveniently distracted from "the truly important matters of life." Of course, these "important matters" included sex, music, spirituality, and "revolution" California-style (meaning, *en abstracto y bien fashionable*). As a child of contradiction, besides being a rabid "antitechnology artist," I owned a little Datsun and listened to my favorite U.S. and British rock groups in my Panasonic *importado,* often while meditating or making love as a means to "liberate myself" from capitalist socialization. My favorite clothes, books, posters, and albums had all been made by "capitalists" with the help of technology; but for some obscure reason, that seemed perfectly logical and acceptable to me.

Luckily, my family never lost their magical thinking and sense of humor around technology. My parents were easily seduced by refurbished and slightly dated American and Japanese electronic goods. We bought them as *fayuca* (contraband) in our Tepito neighborhood, and they occupied an important place in the decoration of our "modern" middle-class home. Our huge color television set, for example, was decorated as to perform the double function of entertainment unit and involuntary postmodern altar—with nostalgic photos of relatives, paper flowers, and assorted figurines all around it. So was the humongous *equipo de sonido*, next to it, with an amp, an 8-track recorder, two record players, and at least fifteen speakers which played all day long a syncretic array of music, including Mexican composer Agustin Lara, Los Panchos (of course with Eddie Gorme), Sinatra, Esquivel, and Eartha Kit. Cumbias, followed by Italian operas and rock 'n' roll alternated with *rancheras*. (In this sense, my father was my first involuntary instructor in postmodern thought.) Although I was sure that with the scary arrival of the first microwave oven in our traditional kitchen, our delicious daily meals were going to turn overnight into sleazy fast food, soon my mother realized that *el microondas* was only good to reheat cold coffee and soups. The point was to own it, and to display it prominently as yet another sign of *modernidad*. (At the time, in Mexico, modernity was perceived as synonymous with U.S. technology and pop culture.) When I moved north to California (and therefore into the future), I would often buy cheesy electronic trinkets for my family (I didn't qualify them as "cheesy" back then). During vacations, to go back to visit my Mexico City family with such presents ipso facto turned me into an emissary of both prosperity and modernity. Once I bought an electric *ionizador* for Grandma. She put it in the middle of her bedroom altar, and kept it there—unplugged, of course, for months. When I next saw her, she told me: "Mijito, since you gave me that thing [still unplugged], I truly can breathe much better." And she probably did. Things like televisions, short-wave radios, and microwave ovens; and later on ionizers, walkmans, crappie *calculadoras*, digital watches, and video cameras, were seen by my family and friends as *alta tecnologia* (high technology), and their function was as much pragmatic as it was social, ritual, sentimental, symbolic, and aesthetic.

It is no coincidence, then, that in my early performance work (1979–90), *chafa* (cheap) technology performed ritual and aesthetic functions as well. *Verbigratia*: For years, I used television monitors as centerpieces for my "video-altars" on stage. Fog machines, strobe lights and gobos, megaphones and voice filters have remained since then, trademark elements in my "low-tech/high-tech" performances. By the early 1990s, I sarcastically baptized my aesthetic practice, "Aztec high-tech art," and when I teamed with "Cyber Vato" Roberto Sifuentes, we decided what we were doing was "technorazcuache art." In a glossary of *"borderismos"* which dates back to 1994, we defined it as "a new aesthetic that fuses performance art, epic rap poetry, interactive television, experimental radio and computer art; but with a Chicanoccentric perspective and an sleazoide bent."

Mythical Differences

The mythology goes like this. Mexicans (and by extension other Latinos) can't handle high-technology. Caught between a preindustrial past and an imposed modernity, we continue to be manual beings; *Homo Faber* par excellence; imaginative artisans (not technicians); and our understanding of the world is strictly political, poetical, or metaphysical at best, but certainly not scientific or technological. Furthermore, we are perceived as sentimentalist and passionate creatures (meaning irrational). When we decide to step out of our anthropological realm, and utilize high-technology in our art (most of the time we are not even interested), we are meant to naively repeat what others—mainly Anglos and Europeans—have already done much better.

We, Latinos, often feed this mythology, by overstating our "romantic nature" and humanistic stances; or by assuming the romantic role of colonial victims of technology. We are always ready to point out the fact that social and personal relations in the United States, the strange land of the future, are totally mediated/distorted by faxes, phones, computers, and other technologies we are not even aware of; that the overabundance of information technology in everyday life is responsible for social handicaps, sexual neurosis, and cultural crisis in the United States.

Is it precisely our lack of access to these goods that makes us overstate our differences? "We," in the contrary, socialize profusely, negotiate information ritually and sensually, and remain in touch with our (still intact?) primeval selves. The mythology continues to unfold: since our families and communities are not exposed to the "daily dehumanizing effects of high-technology," we are somehow untouched by postmodern "illnesses" such as despair, fragmentation, or nihilism. "Our" problems are mainly political, not personal or psychological. This simplistic and extremely problematic binary worldview portrays Mexico and Mexicans as technologically underdeveloped, yet culturally and spiritually superior, and the United States as exactly the opposite.

Reality is much more complicated and ridden with contradictions. The average Anglo-American does not understand new technologies either; people of color and women in the United States don't have "equal access" to cyberspace. Furthermore, American culture has always led the most radical (and often childish) movements against its own technological development and back to nature. (In the 1990s, American Luddites tend to be much more purist and intolerant than their Mexican counterparts.) Meanwhile, the average urban Mexican (more than 70 percent of all Mexicans live in large cities) exposed to world transculture on a daily basis is already afflicted in varying degrees with the same "First World" existential malaises allegedly produced by high-technology and advanced capitalism. In fact, the new generations of Mexicans, including my hip *generación*-Mex nephews and my eight-year-old fully bicultural son, are completely immersed in and defined by MTV, personal computers, Nintendo, video games, and virtual reality (even if they don't own a computer). In fact, I would go as far as to say that in contemporary Mexico, generational borders can be determined by the degree of familiarity with high-technology and by cyberliteracy. Far from being the romantic preindustrial paradise of the American imagination, the Mexico of the 1990s is already a virtual nation whose cohesiveness and fluctuating boundaries are largely provided by transnational pop culture, television, tourism, free market (a dysfunctional version, of course), and yes, the internet.

But life in the ranchero global village is ridden with epic contradictions. Despite all this, still very few people south of the border are on-line, and those who are wired, tend to belong to the upper and upper middle classes, and are mostly related to corporate or managerial metiers. The Zapatista phenomenon is a famous exception to the rule. Technoperformance artist extraordinaire Subcomandante Marcos has been communicating with the "outside world" through extremely popular web pages sponsored and designed by U.S. and Canadian radical scholars. (It is still a mystery to me how his communiqués get from the jungle village of "La Realidad" in northern Chiapas, which still [as of 1997] has no electricity, to his web pages literally overnight.) However, these web pages are more known outside Mexico for a simple reason: Telmex, the Mexican Telephone company, makes it practically impossible for anyone living outside the main Mexican cities to use the Net, arguing that there are simply not enough lines to handle both telephone and internet users.

Every time my colleagues and I have attempted to create some kind of binational dialogue via digital technologies (i.e., link Los Angeles to Mexico City through satellite video-telephone), we are faced with a myriad of complications. In Mexico, the few artists with ongoing "access" to high-technologies who are interested in this kind of transnational technodialogue, with a few exceptions, tend to be socially privileged, politically uninformed, and aesthetically uninteresting.[4] And the funding

---

4. Pedro Meyer: I find several things wrong with this sentence. It is not only in Mexico that this is the case. It is the world over that high technologies are a matter reserved for the privileged. This is true since time began. Only that today the high technologies are electronic, and belong to the digital world. In the past it was other materials that were the subject of scarcity and thus privilege. For example, at the time of the Mexican muralist movement in the forties, it meant access to certain walls on which to paint. It's obvious that there were more painters than there were walls available on which to paint. So not everyone who wanted to, could paint on such walls, at least not on "important" walls of public buildings. Going back farther in time, the Aztecs gave the artists in their society a very special place in which to develop their skills, and again that was a matter of privilege. So it's disingenuous to consider that art when associated with privilege is necessarily diminished in its stature. One can find countless instances throughout the history of art, where privilege was never very far from the great masterpieces that were created. . . . The observation that the

sources down there willing to fund this type of project are clearly interested in controlling who is part of the experiment.

> "Rebecca (Solnit) thinks America on-line is like K-Mart and keeps getting lost in the aisles somewhere between press-on-nails and flash-sessions. This morning AOL fell asleep while I was forwarding your text to my brother (the Anglo-Sandinista one) and it disappeared. Maybe it's like a combination of K-Mart and the Argentinean military. What with all this disappearing, loco?"
>
> —Excerpt from an e-mail

Cyber-migras and "Webbacks"

Roberto and I arrived late to the debate, along with a dozen other Chicano experimental artists.

At the time, we were shocked by the benign or quiet (not naive) ethnocentrism permeating the debates around art and digital technology, especially in California. The master narrative was either the utopian and dated language of Western democratic values or a bizarre form of New Age anticorporate/corporate jargon. The unquestioned lingua franca was of course English, "the official language of science, information and international communications"; and the theoretical vocabulary utilized by both the critics and apologists of cyberspace was hyperspecialized (a combination of esperantic "software" talk; revamped poststructuralist theory— hadn't we already overcome poststructuralism in the early 1990s?—and

---

(Mexican) artists associated with new technologies are politically uninformed or aesthetically uninteresting, responds more than anything to a populist interpretation by a number of people in the world of culture who have taken it upon themselves to dismiss new technologies as if it were a symbol of status to be a technophobe. By taking on such a negative attitude, in reality what they are doing is covering up their own shortcomings and ignorance of what new technologies truly have to offer. What better way to deal with such limitations than to dismiss them outright. One would also have to factor in how reactionary and lacking in courage, such attacks against any changes really are. A true artist thrives in the exploration of all that there is new. To live on the cutting edge and try to make sense of it all, is no task for the weak of heart. It's always easier to stay within the comfort of the known, than to risk failure in the process of trying what is not yet under control. To say that the outcome is for the most part 'uninteresting' is to assume that the aesthetic choices already had an opportunity to work themselves out. New forms require a period, sometimes quite prolonged, of incubation. If we observe a beautiful person, do we acknowledge that that same person at one time was only a fetus with little or no resemblance to that future beautiful presence? And in the end, who does the judging? Are we conscious that such "aesthetic choices" of beauty are if nothing else just part of a set of cultural expressions of value and not universal truths?

nouvelle psychoanalysis), and largely depoliticized (i.e., postcolonial theory and the border paradigm were conveniently overlooked). If Chicanos, Mexicans, and other "people of color" didn't participate enough on the Net, it was solely because of lack of information or interest (not money or "access"). The unspoken assumption was that our true interests were "grassroots" (and by grassroots I mean, the streets in the barrio and our ethnic-based community institutions), representational, or oral (as if these concerns couldn't exist in virtual space). In other words, we were to remain painting murals, tagging, plotting revolutions in rowdy cafés, reciting oral poetry, and dancing *salsa* or *quebradita.* (Some colleagues believe that the mere fact that Roberto and I and a handful of other Chicanos are now temporarily sitting at the cybertable is already a huge political victory. Others more cynical believe that the reason why we get invited to the great rave of consciousness is to bring some Tex-Mex galore and tequila to an otherwise fairly puritan fiesta. Hopefully not.)

When we began to dialogue with U.S. artists working with new technologies, we were also perplexed by the fact that when referring to "cyberspace" or "the Net," they spoke of a politically neutral/race-less/gender-less and classless "territory"—a territory which provided us all with "equal access" and unlimited possibilities for participation, interaction, and belonging, especially "belonging" (in a time in which no one feels that they "belong" anywhere).[5] Yet there was never any mention of the phys-

---

5. Joana Frueh: I do not want to exchange my sex, and my genders are many. I am abundant in identities in my daily, intimate, and professional lives. My identities merge in variant combinations, then supersede each other as they operate according to the integrity of a single human being who is not a fragment or a figment of her own or someone else's imagination. People make up stories about themselves and others all the time. Narration-as-imagination can be fun, and dangerous, in real space.

I am not cyber-efficient, not a cyber-afficionado—I want soul-inseparable-from-the-body in front of me, arm in mine, in my bed. I feel affection for people in the flesh, for facing the source of a voice as it eases or lurches out of a body and touches me all over mine; I want to hear whether a voice resonates deeply, from a diaphragm, or thinly, from chest or throat. I want to participate in emotional upheavals, withdrawals, magnetics when I am part of a group's or individual's physical energy. I want Eros—rich and joyous connectedness—as close as I can get it, and that demands touching, sensibly and sensually. I see no other way to build the body of love.

Over time the summer I read a lot of Sade. In *The 12 Days of Sodom,* one of his funniest yet most somber novels, I met the expert whore and storyteller Mme. Duclos. She is radiant and brilliant, and she confidently displays her exceptionally beautiful buttocks. She is one of the few characters who leaves the isolated, impenetrable, and lugubrious Chateau de Silling alive. Silling is the gothic castle par excellence. It is the site of fascinating and grisly debauchery. The four protagonist debauchees, all men, do not victimize Duclos, for to their eyes, ears, minds, and bodies, she is splendid.

Duclos is forty-eight and so was I when I read *120 Days.* I identify with her, despite my problems with Sade and the ones I imagine Duclos to have with him, too, in a chapter I'll soon write for my new book *Monster/Beauty: Paradox of Pleasure.* There I will inhabit her. As I walk home from my neighborhood park, I AM her. This is the identity exchange of a litical and social loneliness, or the fear of the "real world" which propels so many people to get on-line, stay "there," and pretend they are having "meaningful" experiences of "communication" or "discovery" (two very American obsessions). To many of them, the thought of exchanging identities on the Net and impersonating other genders, races, or ages, without real (social or physical) consequences seemed extremely appealing and liberating, and by no means superficial or escapist.[6]

The utopian rhetoric around digital technologies, especially the one coming out of California, reminded Roberto and I of a sanitized version of the pioneer and frontier mentalities of the Old West; and also of the early century futurist cult to the speed, size, and beauty of epic technology (airplanes, trains, factories, etc.). Given the existing compassion fatigue regarding political art and art dealing with matters of race and gender, it was hard not to see this feel-good philosophy (or rather theosophy) as an attractive exit from the acute social and racial crisis afflicting the United States.

---

erary and art fetishist, me, of a Sade fan who adores his and Duclos's unity of body and mind. Mind-incomplete-without-body, Duclos, a pro-porn feminist like me, is one of my erotic identities, virtual as literature, real as a clarification of my needs and pleasures in erotic and aesthetic self-creation.

I like bodies to be high-content and high-context like art and the erotic. In my experience and to my knowledge, "low-context messages" (John Simmons, "Sade and Cyberspace," in James Brook and Lain A. Boal, eds., *Resisting the Virtual Life: The Culture and Politics of Information* [San Francisco: City Lights, 1995], 146–47) pervade cyberspace. There the human body radically distorts, reduces, and romanticizes into information, through which it may assume a nauseating omnipotence—the authority of disembodiment.

In 1989 Marvin Minsky wrote, "We are entering a new century in which you are connected to the world, to the virtual world. And much more intimately than you are connected to the real world. . . . Our connection with the real world is very thin, and our connection with the artificial world is going to be more intimate and more satisfying than anything that's come before" (Minsky, quoted in *SF Camerawork* [Spring–Summer 1993]: 4). Eros is very thick and very real, and I do not belong to Minsky's "our"; yet I know how thinness belongs to people's disaffection from daily and seismic figurative slaps-in-the-face and kicks-in-the-butt. Writing in 1989 from his position as Toshiba Professor of Media Arts and Sciences at the MIT Media Lab, Minsky says, "If it was possible, I would have myself downloaded" (Minsky, in "Is the Body Obsolete? A Forum," *Whole Earth Review* [Summer 1989]: 37). This is no way to court intimacy with aging or racialized bodies, for many people would choose to be an easy body, attractive and unproblematic in terms of normative social standards.

I do imagine that some people would choose "bodies" of terror, but one can be a cultural terrorist monster/beauty—in her body of origin. This requires creativity and courage.

Disembodiment inhibits Eros, which is both material and incorporeal. Eros propels people into physicality and sociality; Eros has physical and social consequences. Interactivity is not erotic connectedness, whose profusion occurs in the high-context plenitude of prosaic reality.

—MONSTER/BEAUTY/PERFORMING LIKE A FLORIBUNDA

6. Many feminist colleagues have expressed to me the fact that for women "exchanging genders" on the Net can be both "liberating" and transgressive.

Like the pre-multi-culti art world of the early 1980s, the new high-tech art world assumed an unquestionable "center," and drew a dramatic digital border. And "on the other side," there lived all the technoilliterate artists, along with most women, Chicanos, Afro-Americans, and Native Americans in the United States and Canada, not to mention the artists living in so-called Third World countries. Given the nature of this hegemonic cartography, those of us "illegal aliens" living south of the digital border were forced to assume once again the unpleasant but necessary roles of webbacks, cyberaliens, digital viruses, technopirates, and virtual coyotes (smugglers).

> "In the barrios of resistance, contemporary versions of the old kilombos, every block has a secret community center. There, the runaway youths called Robo-Raza II or 'floating greasers' publish anarchist laser-Xerox magazines, edit experimental home videos on police brutality (yes, police brutality still exists) and broadcast pirate radio and TV interventions like this one over the most popular programs. . . .
>
> These clandestine centers are constantly raided, but Robo-Raza II just moves the action to the garage next door. Those who get 'white-listed' can no longer get jobs in the 'Mall of Oblivion.' And those who get caught in *fraganti*, are sent to rehabilitation clinics where they are subjected to instant socialization through em-pedagogic videos (from the Spanish verb *emperor*, meaning to force someone to drink, and the Mayan noun *agogic*, *o sea*, a man without a self, like many of you)."
> —From "The New World Border," City Lights, 1996

Chicano Virtual Reality

Perhaps our first truly "high-tech" project which contributed to the politicization of the debates around new technologies was a performance designed for cable television with the obscure title of "Naftaztec: Pirate Cyber-TV for A.D. 2000."

On Thanksgiving Day 1994, the evening news broadcast to over 3.5 million American households was interrupted by two "post-Nafta cyber-Aztec TV pirates," transmitting their bizarre views on American culture "directly from their underground vato-bunker, somewhere between New York, Miami and Los Angeles."

In actuality, what the viewers were witnessing was an experiment of interactive multilingual television via satellite. Roberto and I had teamed with filmmakers Adrienne Jenik, Philip Dwja, and Branda Miller (from I-Ear Studio at Rennselear Polytechnic) to broadcast a simulacrum of a pirate television intervention to hundreds of cable television stations across the country. The stations' program directors had agreed to "play along" (which is unthinkable on PBS), and advertised the time slot under a fictional title.

For an hour and a half, the "information superhighway banditos" encouraged the perplexed viewers to call in and "respond" to the broadcast, which was a strange blend of radical politics, autobiographical material, and parodies of traditional television formats gone bananas. The style was very much like MTV, with five hand-held cameras in constant motion. During the broadcast, we demonstrated a "Chicano virtual reality machine" which "could turn personal and collective memories into video footage ipso facto" and "a virtual reality bandanna," which could "allow Anglos to experience first hand the psychological sensation of racism." We also received "live reports" via Picture Tel (video telephone) from the Electronic Cafe in Los Angeles. We spoke in English, Spanglish, Franglé, French, and a robo-language invented by us; and encouraged our viewers "to be intelligent, poetical, and performative" in their response. During the live broadcast, sometimes it would appear that the television station was struggling to regain the airwaves, but we managed to maintain control.

The performance was transmitted over computer networks as well, via "M-Bone," and those watching in cyberspace could interact with us, and with one another, by posting written and visual comments. We received dozens of phone calls and computer messages. One constant thread in most responses we received had to do with the fact that the viewers (and the M-Bone users) were amazed by how technically and visually sophisticated and "un-folksy" the program was, given the fact (although not always overtly stated) that it had been created by "Mexicans." Many others, clearly pissed by our arrogance ("2 Mexicans live on national television broadcasting uncensored material"), made reference to the fact that we should leave immediately the high-tech simulated space we created "illegally" and return to our "pyramid-infested past." The total cost of the project (including the rental of satellite time) I believe was under $7,000, and to our surprise, it received the prize of "best experimental video" at the San Antonio Cine Festival.

Ethno-Cyborgs and "Artificial Savages"

The current debates about the body and its relation to the new technologies have polarized tremendously the experimental arts community and particularly the performance art milieu. There are those in the "machine art" movement who advocate the total disappearance of the human body and its replacement with computer or mechanical robotics. Others believe that the body, though obsolete, can still remain in the center of the art event, but that new technology can equip it with prosthetic (perceptual and physical) extensions. A visceral reaction to these proposals can be found in the artists of "apocalypse culture" who have adopted a radical Luddite stance: to reclaim the body primitive as a site for pleasure and pain, and "return" (so they claim) to a sort of neotribal paganism, very much in the Western tradition of anarchist "dropout" culture. What Roberto and I are trying to do is explore other options: to politicize technology, to imbue it with humor and *linguas polutas;* to use it as a means to enhance the interactivity between performers and live audiences who unknowingly become voyeurs/tourists; and to gather cultural and political information of a very unique confessional nature, which will then be reinterpreted by and expressed through our "primitive," political, and erotic bodies. What the live audience ends up experiencing is a sort of visualization of their own postcolonial demons and racist mirages.

Our most recent "technodiorama" project, titled "El Mexterminator I," first premiered in Mexico City in March 1995 under the working title of "The Museum of Frozen Identity." Since then, different versions in progress have been performed in the United States, Spain, Italy, Austria, Canada, England, and Wales. In this project, Roberto and I utilize the visitors' (both physical and virtual visitors, that is) responses and "confessions" to design visual and performative representations of "the new mythical Mexican and Chicano of the 90's." In other words, the actual internet responses become the basis for the creation of a series of "ethno-cyborgs," co-created (or rather "co-imagined") collaboratively with thousands of anonymous Net-users. Unlike our previous diorama projects, the idea now is to cede our will to the internet users (and to the gallery visitors when we are able to have the necessary technology available at the performance site) in determining the nature and content of the "living dioramas," including how we should dress; what music we must listen to; and, most important, what kind of ritualized actions we should engage in and what type of interaction we are to have with the audience. What we do as performance artists then is to "embody" this information, reinterpret it, and stylize it. In this sense, the "ethno-cyborgs" and "artificial savages" incarnate profound fears and desires of contemporary Americans regarding the Latino "other," immigrants, and people of color, and function as mirrors for the (real and virtual) visitors to see the reflections of their own psychological and cultural monsters.

These performances always involve some form of physical interactivity with the audience. Visitors to the galley space are encouraged "to interact with the live specimens" in various modes: they can feed us, touch us, attempt to engage us in a conversation, handle our props ("at their own risk"), point replicas of weapons at us ("to experience the feeling of shooting at a live Mexican"), and occasionally, they are invited to "alter our identity" by changing our makeup and costumes, and even "replace us for a short period of time." Lately, we are setting up "identity makeover booths," where audience members undergo "instant" identity changes through special effects makeup and costumes. Whenever we can, we try to set up a bar inside the space to "carnivalize" the experience even more. When this happens, the behavior of the audience changes dramatically as they become less inhibited through the ingestion of tropical cocktails.

The complete version of "Mexterminator I" premiered in early 1998 during the "commemoration" of the Guadalupe-Hidalgo Treaty. It parodied an end-of-the-century "Museum of Experimental Ethnography," incorporating several "ethno-cyborgs" reflective of America's problematic relationship with cultural otherness in the 1990s (i.e., the Chicano as an "endangered species," the "Mad Mex" super criminal, the exotic "Cultural transvestite," La Zapatista stripper, etc.).

First Draft of Manifesto: Remapping Cyberspace

In the past years, many theoreticians of color, feminists, and activist artists have finally crossed the digital border without documents. This recent diaspora has forced the debates to become more complex and interesting. But since "we" (as of now, the "we" is still blurry, unspecific, and ever-changing) don't wish to reproduce the unpleasant mistakes of the cultural wars (1988–93)—nor do we wish to harass the brokers, impresarios, and curators of cyberspace as to elicit a new backlash—our strategies and priorities are now quite different.

"We" are no longer trying to persuade anyone that we are worthy of inclusion (we now know very well that we are either temporary insiders or insiders/outsiders at the same time). Nor are "we" fighting for the same funding (since serious funding no longer exists, especially for politicized experimental art), and the computer tycoons we all thought would eventually become progressive philanthropists are just oversized teenagers with no political understanding of culture whatsoever.

For the moment, what we (cyber-immigrants) desire is:

- to "politicize" the debate;
- to remap the hegemonic cartography of cyberspace;
- to develop a multicentric theoretical understanding of the (cultural, political, and aesthetic) possibilities of new technologies;
- to exchange a different sort of information (mythopoetical, activist, performative, imagistic); and
- to hopefully do all this with humor, inventiveness, and intelligence.

Chicano artists in particular wish to "brownify" virtual space, to "Spanglishize the Net," and to "infect" the linguas francas.

These concerns seem to have echoes throughout Latin America, Asia, Africa, and many so-called Third World communities within the so-called First World.

With the increasing availability of new technologies in our communities, the notion of "community art" and "political" or politicized art is changing dramatically. Now the goals, as defined by activist artists and theoreticians, are to find innovative grassroots applications to new technologies (i.e., to help the Latino youth literally exchange their weapons for computers and video cameras), and to link all community centers and artist collectives through the internet. Artist-made CD-ROMs and web pages can perform an extremely vital educational function: they can function as community "memory banks" (*encyclopedias chicanicas* so to speak), sites for encounter, dialogue, complicity, and exchange; as well as virtual bases of operation and action for trans/border grassroots projects.

To attain all this, the many virtual communities must get used to a new cultural presence—the webback (*el nuevo virus virtual*); a new sensibility; and many new languages spoken on the Net. As for myself, hopefully one day I won't have to write in English in order to have a voice in the new centers of international power.

Annotated Bibliography

Agre, Phil, and Marc Rotenberg, eds. 1998. *Technology and Privacy: The New Landscape.* Cambridge, Mass.: MIT Press. Responding to recent transformations in culture and information technologies—including computer networking, cryptography, privacy activism, and data-protection law—these essays provide a conceptual framework for the analysis and debate of privacy policy and for the design and development of information systems.

Aronowitz, Stanley, Barbara Martinsons, and Michael Menser, eds. 1996. *Technoscience and Cyberculture.* London: Routledge. A "cultural studies" approach to science and technology. Rejects the view that technology and science and culture are discrete objects and/or fields, and demonstrates that all are so intertwined that to consider one—genetic engineering, smart weapons, global technologies, or science studies, for example—is to implicate the other frameworks as well.

Baudrillard, Jean. 1983. *Simulations.* New York: Semiotext(e). A key theoretical tract in the postmodern project. Assumes the disintegration of panoptic/perspectival space and subject-object distinctions in late capitalism, along with the emergence of a global culture defined by hyperrealism rather than representation; simulation rather than the real. Disneyland posed as master-paradigm in a postideological consumer world after politics.

Baudry, Jean-Louis. 1974–75, Winter. "Ideological Effects of the Basic Cinematographic Apparatus." *Film Quarterly* 28, no. 2. A classic of 1970s "screen" theory. Argues that the technical—and psychoanalytic—apparatus of cinema positions the spectator in ideologically suspect ways. The technology emerges from, and includes, the viewing subject's innate desire to see itself (falsely) as whole and unified.

Bellamy, Robert V., and James R. Walker. 1996. *Television and the Remote Control.* New York: Guilford. An extensive study of how recent technological changes in television and video—cable, the VCR, the remote—have altered viewing activities in television's interactive age. Particularly useful in describing how programming and promotion are tied to new electronic technologies.

Benedict, Michael, ed. 1991. *Cyberspace: First Steps.* Cambridge, Mass.: MIT Press. A collected work that examines philosophical, cultural, graphic, and architectural issues involved in innovations in virtual reality, and in the emergence of a navigable, human-computer interface termed cyberspace.

Benjamin, Walter. 1969. "The Work of Art in the Age of Mechanical Reproduction." In *Illuminations,* tr. Harry Zorn. New York: Schocken Books, pp. 217–51. Mechanical reproduction in the arts of printing, photography, and cinema stripped twentieth-century art of its institutional and conceptual aura of uniqueness.

Technologies fragment performance, alter spectatorship, and animate the media arts in political as well as aesthetic ways.

Bordwell, David, Janet Staiger, and Kristin Thompson. 1985. *Classical Hollywood Cinema*. New York: Columbia University Press. A comprehensive, integrated, historical account of how the Hollywood mode of production came to dominate cinema practice in the twentieth century. Revisionist work that rejects technological determinism, psychoanalytic textual analysis, and ideological screen theory in favor of historical analysis this is simultaneously technical and formal; institutional and archival; industrial and aesthetic.

Brand, Stewart. 1987. *The Media Lab: Inventing the Future at MIT.* New York: Viking/Penguin. A look at the ideas, technical breakthroughs, community, and culture that surrounded one of the most influential research and development sites in the digital era. The Media Lab proved as important in conceptualizing new digital technologies as it did in developing hardware. Its legacy continues to influence both the corporate world of technology development and the critical orientation of computer publications.

Brinkley, Joel. 1997. *Defining Vision: The Battle for the Future of Television.* New York: Harcourt, Brace. An accessible, detailed account of the competition over, and history of, the development of HDTV and digital television in the United States.

Brook, James, and Iain A. Boal, eds. 1995. *Resisting the Virtual Life: The Culture and Politics of Information.* San Francisco: City Lights Books. An extensive critique of the "pernicious" ways that digital technologies and the mediated and simulated relationships they cultivate are deployed across society as substitutes for "face-to-face interactions." Criticism and activism directed at the inherent passivity of "programmed interactivity."

Browning, John, and Spencer Reiss. 1998, March, April, May. "Encyclopedia of the New Economy, Part I, II, and III." *Wired.* A provocative lexicon of digital buzzwords. Particularly valuable in demonstrating how global economics and corporate culture actually fuel what might otherwise be posed as the smart slang of cutting-edge and techno-hip.

Caldwell, John T. 1995. *Televisuality: Style, Crisis, and Authority in American Television.* New Brunswick: Rutgers University Press. Examines American televisual practice—programming, technology, corporate and production culture—since 1980. Demonstrates how electronic imaging technologies—videographic, cinematic, and digital—cannot be usefully understood without reference to the institutional, economic, aesthetic, and cultural practices that animate and perpetuate them. Critiques the autonomy typically afforded new production technologies and digital media, by examining their logics within both the culture of production and culture at large.

Comolli, Jean-Louis. 1986. "Technique and Ideology: Camera, Perspective, Depth of Field." In *Narrative, Apparatus, Ideology.* Ed. Philip Rosen. Tr. the British Film Institute. New York: Columbia University Press, 421–43. Provides a technical archaeology of the dominant ideology engineered into the classical Hollywood cinema apparatus. Explicates both iconic and conceptual determinations, and the steady accumulation of technical processes that worked to deny difference in the cinematic image, all in order to artificially create the dominant visual regime known as realism.

Conley, Verena Andermatt, ed. 1993. *Re-Thinking Technologies.* Minneapolis: University of Minnesota Press. A collection of philosophical essays that assess the problematic position of new technologies within the humanities. Builds on the work of Heidegger, Virilio, and Guattari, and applies philosophical and ideological meditations from continental and cybernetic theory within the Anglo-American context.

Crary, Jonathan. 1992. *Techniques of the Observer: On Vision and Modernity in the Nineteenth Century.* Cambridge, Mass.: MIT Press. Counters the traditional view that modern art initiated a radical break in visual representation, by demonstrating that the rupture with classical forms of vision occurred earlier in the nineteenth century, through the massive reorganization of knowledge, social practices, and techniques of observation.

Cubitt, Sean. 1991. *Timeshift: On Video Culture.* London: Routledge. Argues that video, rather than film or television, has the potential to become a uniquely democratic medium. Applies semiotic and postmodernist theory to various electronic media practices in order to describe the dynamics of contemporary video culture.

Dowmunt, Tony, ed. 1993. *Channels of Resistance: Global Television and Local Empowerment.* London: BFI. Challenges the problematic notion that global electronic media necessarily involves cultural homogenization. Numerous case studies show how local initiatives around the world are creating opportunities for national, regional, and ethnic identities to find expression in electronic media.

Druckerey, Timothy, ed. 1996. *Electronic Culture: Technology and Visual Representation.* New York: Aperture Foundation. A diverse collection of essays, by philosophers, critics, and theorists, that explore the problematic status of the image and visual experience in a networked, digital culture.

Ellul, Jacques. 1964. *The Technological Society.* New York: Alfred A. Knopf. A philosophical, historical critique of what the dominance of technique means for the present and future of society. The irreversible rule of technique makes spontaneous behaviors deliberate and rationalized in all areas of life, subordinating humanity itself to the internal logic of systematic technological development.

Enzensberger, Hans Magnus. 1974. *The Consciousness Industry.* New York: Seabury. A wide-ranging critique that examines capitalism, popular culture, and the "industrialization of the mind." The chapter entitled "Constituents of a Theory of Media" catalogues the "emancipatory" potential of new, portable, electronic media in the context of new Left activism.

Foucault, Michel. 1979. *Discipline and Punish: The Birth of the Prison System.* New York: Vintage Books. An influential, analytical model for how technologies—in this case the penal system—function in practice as social discourses animated by the play of power. Explicates the centrality of surveillance and the "panopticon" as psychosocial forces that order the body and discipline the subject in the modern era.

Gilder, George. *Life After Television: The Coming Transformation of Media and American Life.* New York: Norton, 1990, 1992. A celebratory view of computer culture by a futurist who bases his apologetic on two levels: the technical microcosms and possibilities of the silicon chip and fiber optics; and the extensive dysfunctions leveled on society by America's monolithic television industry and its minions. A democratic future will depend on the failure of HDTV, and ability of computer networks to shatter all forms of centralized power and control.

Hall, Doug, and Sally Jo Fifer, eds. 1990. *Illuminating Video: An Essential Guide to Video Art.* New York: Aperture Foundation/Bay Area Video Coalition, San Francisco. A large and eclectic collection of artists' writings on video art, from installation work to interactive CDs, intended both as a "map" of various independent practices and as a handbook that can elicit more challenging forms of electronic art criticism.

Hanhardt, John, ed. 1986. *Video Culture: A Critical Investigation.* Rochester: Visual Studies Workshop. A collected work that combines essays on ideology by Brecht, Benjamin, and Althusser with critical apologetics by video artists, art-world curators, and critics, including Douglas Davis, David Antin, David Ross, and Nam June Paik. An attempt from a modernist perspective to consider the "distinctive features" of video art and the forces acting on its history.

Haraway, Donna. 1985. "Manifesto for Cyborgs: Science, Technology, and Socialist Feminism in the 1980s." *Socialist Review* 80. An influential critical study of the mutating boundaries between nature and science, the body and technology, from the interdisciplinary perspective of feminist theory and cultural studies.

Hayward, Philip, and Tana Wollen, eds. 1993. *Future Visions: New Technologies of the Screen.* London: BFI. Examines an array of new technologies—widescreen cinema, IMAX, CGI, HDTV, and virtual reality systems—all to better explain how they work, and what their implications are for conventional film and television.

Heim, Michael. 1993. The *Metaphysics of Virtual Reality.* New York: Oxford University Press. Takes a philosophical approach (from Plato and Leibniz to Heidegger) to consider how computer technologies, searches, and virtual interfaces have altered everything from our grounding ontological frameworks to our use of logic, intuition, and modes of conceptualization.

Katz, Jon. 1997. *Media Rants: Postpolitics in the Digital Nation.* San Francisco: Hardwired. An animated, critical intervention into struggles between "netizens" and their alarmist government and corporate critics. The Net offers a revolutionary alternative to traditional politics and notions of the public sphere, in a generation that views politics, culture, and media as the same thing.

Kroker, Arthur, and Michael Weinstein. 1994. *Data Trash: The Theory of the Virtual Class.* New York: St. Martin's. Critiques political and economic interests involved in the technical class's "will toward virtuality." Written in response to the widely celebrated coming of the "information superhighway," the authors pose questions about labor and social relations, globalism, and digital "road kill" in the context of "cyberculture."

Levy, Pierre. 1997. *Collective Intelligence: Mankind's Emerging World in Cyberspace.* New York: Plenum Trade. Counters popular dystopian views of new technologies by proposing that, far from alienating us, the technology of cyberspace will have a humanizing influence on us, and will foster the emergence of what Levy terms a "collective intelligence." Describes a cooperative electronic-neural process that will actually validate the contributions of the individual and the insights of the human subject.

Lipsitz, George. 1990. *Time Passages: Collective Memory, and American Popular Culture.* Minneapolis: University of Minnesota Press. Key text in American studies that documents how television was used in the post–World War II social formation. Shows how technologies and their entertainment content worked to invoke older ethnic identities even as they remade them as part of a consensus more suitable for the U.S. project of suburbanization and consumerism. Suggests the extent to which media technologies have functioned instrumentally within broader ideologies of ethnicity, class, and nationalism.

Lyon, David. 1994. *The Electronic Eye: The Rise of Surveillance Society.* Minneapolis: University of Minnesota Press. Provides a detailed account of the role and extent of surveillance in contemporary society. Grounds Foucault's vision of modernity with examples from computerized and information-saturated society, and attempts to counter the pessimism and paranoia of postmodernism.

Mattleart, Armand. 1994. *Mapping World Communication: War, Progress, Culture.* Tr. Susan Emanuel and James A. Cohen. Minneapolis: University of Minnesota Press. Excavates the "technical networks" and genealogies through which global communications and "globalism" have come to take center stage. A skeptical, critical history of mass communications as a cultural force.

McLuhan, Marshall. 1964, 1994. *Understanding Media: The Extensions of Man.* Cambridge, Mass.: MIT Press. Key work in popularizing and dramatizing the central role of electronic media in contemporary society, and in suggesting how technologies are far more than material tools. An influential, speculative lexicon of the comprehensive ways that changes in media affect changes in human consciousness and perception.

Morse, Margaret. 1998. *Virtualities: Television, Media Art, and Cyberculture.* Bloomington: Indiana University Press. Synthesizes analyses and critical accounts of diverse contemporary practices—the performance of news, television graphics, mediated space, cybernetic orality, and video installation—within a broader discussion of virtual reality's dependence on fictions of presence, immersion, and an aesthetics of transcendence.

Mumford, Lewis. 1963. *Technics and Civilization,* New York: Harcourt, Brace, and World. An account of history based on distinctions between technologies and the psychic and cultural dispositions of eras and peoples, including explanations of the relationship between the development of the mechanical clock and capitalism, the notion of epic construction projects as sociological "megamachines," and the general warning that technologies threaten humane values in contemporary society.

Negroponte, Nicholas. 1995. *Being Digital.* New York: Alfred A. Knopf, 1995. A collection of short essays by the founding director of MIT's Media Lab, that describe the extent to which modern culture—from business practices to entertainment to social interaction—has been transformed by computer technology. Suggests future changes not simply in the manner that we interface with information technologies, but in the fundamental ways that we learn, work, and live in the digital age.

Penley, Constance, and Andrew Ross, eds. 1991. *Technoculture.* Minneapolis: University of Minnesota Press. A collection of critical cultural studies on fan subcultures, AIDS activists, hackers, technoporn, rap groups, and rock stars. All examine and target the management of repressive technocultures and the political potential of creative appropriation.

Plant, Sadie. 1997. *Zeros + Ones: Digital Women + The New Technoculture.* New York: Doubleday. An intertextual manifesto on the relationship between women and machines, from the Industrial Revolution to the passions of the internet. Envisions both a feminist legacy and postpatriarchal future in the networks, connections, and nonlinear experiences of Net and computational technologies.

Postman, Neil. 1992. *Technopoly: The Surrender of Culture to Technology.* New York: Alfred A. Knopf. Challenges the "scientistic" basis of contemporary society by tracing an historical and ideological progression from cultures based on tools, technologies, technocracies, and finally technopolies—the deleterious "submission of all forms of cultural life to the sovereignty of technique and technology."

Ritzer, John. 1996. *The McDonaldlization of Society: An Investigation into the Changing Character of Contemporary Social Life,* Thousand Oaks, Calif.: Pine Forge Press/Sage. A sociological critique of the process by which the principles of the fast-food restaurant are coming to dominate more and more sectors of both American and international cultures. The social effects of a corporate-managerial ethos based on efficiency, calculability, predictability, and control.

Ross, Andrew. 1998. *Real Love: In Pursuit of Cultural Justice,* New York: NYU Press. Interrogates the cultural forms through which economic forces take their daily toll upon our labor, communities, and environment. Includes a particularly good account, entitled "Jobs in Cyberspace" (pp. 7–34), of the function and contradictions of Silicon Alley "sweatshops" that have emerged in the post-information superhighway era.

Roszak, Theodore. 1986, 1994. *The Cult of Information: A Neo-Luddite Treatise on High-Tech, Artificial Intelligence, and the True Art of Thinking.* Berkeley: University of California Press. Criticizes the place of computer technologies in the culture of the 1990s. Argues the "data glut" obscures basic questions of justice and purpose, subverting productivity along with sound thinking.

Rushkoff, Douglas. 1994. *Cyberia: Life in the Trenches of Cyberspace.* San Francisco: HarperCollins. An engaging, participant observation description of the underground Net culture that preceded the mainstreaming of the Net in the form of America Online and the information superhighway. Rushkoff describes the personalities and predilections of hackers, cyberpunks, digital ravers, and computer artists as electronic communities formed, morphed, and dissolved. A churning world of on-line seekers, geeks, and anarchist visionaries.

Schiller, Daniel. 1999. *Digital Capitalism: Networking the Global Market System.* Cambridge, Mass.: MIT Press. The best account of the internet from a global political economic perspective. Deconstructs the commercial mythologies that sell and legitimize digital media, details the relationships between the goals of multinational media corporations and their investors, and describes how the selective adaptations of new technologies drive capitalist expansion.

Sconce, Jeffrey. 2000. *Television Ghosts: An Occult History of Electronic Media.* Durham: Duke University Press. Presents a cultural history of "liveness" and "telepresence" in electronic media, examining the prevalence of supernatural, paranormal, and otherwise occult discourses in the popular imagination of telegraphy, wireless, radio, television, and new media technologies. Case studies chart a historical transition in occult media metaphors, arguing that initial fascination with the seemingly fantastic properties of media technology in and of itself has gradually developed into a supernatural belief in the sovereignty of the "worlds" produced by electronic media.

Segaller, Stephen. 1998. *Nerds 2.01: A Brief History of the Internet.* New York: TV Books, L.L.C./Oregon Public Broadcasting. The first comprehensive account of the personalities, start-ups, corporations, financial events, and technical breakthroughs that comprise the emerging history and global cultural phenomenon of the internet. One of an increasing number of accounts in the late 1990s that see the internet, rather than the PC, as the key trigger technology of the digital age. Based on interviews and oral histories from a litany of Silicon Valley CEOs.

Seiter, Ellen. 1999. *Television and New Media Audiences.* London: Oxford University Press. Case studies of media consumption suggest how audience research can account for the complex uses of technologies in the domestic sphere and the classroom, the relationship between gender and genre, and the varied interpretation of media technologies and media forms. Reviews the most important research on television audiences and recommends the use of ethnographic, longitudinal methods for the study of media consumption and computer use at home as well as in the workplace.

Silverstone, Roger. 1994. "The Tele-technological System." In *Television and Everyday Life,* ed. Roger Silverstone and Eric Hirsch. London: Routledge, pp. 78–103. The "origins" of any new media technology can only be understood if one considers how a "socio-technical system" is also "tactically" manufactured alongside any new hardware development. Silverstone argues that technologies successfully emerge "as a result of the potential space created within a network for the actions of individuals." The industrially sanctioned role of user/buyer/viewer "agents" within the "consumption junction" of the home is a key to understanding how any new technologies succeed.

Slack, Jennifer D. 1994. *Communication Technologies and Society: Conceptions of Causality and the Politics of Technological Intervention.* Norwood, N.J.: Ablex. A good critical survey of theories of technology-and-society, and of the fields of technology assessment, and appropriate technology.

Spigel, Lynn. 1992. *Make Room for Television: Television and the Family Ideal in Post-war America.* Chicago: University of Chicago Press. A social and historical study of how suburban, middle-class domesticity was engineered along with television viewing technologies in the 1940s and 1950s. Examines how the institutional and commercial convergence of postwar public discourses—popular magazines, advertising, architecture, and programming practices—animated and resolved class and gender anxieties through the popularization of television.

Springer, Claudia. 1996. *Electronic Eros: Bodies and Desire in the Postindustrial Age.* Austin: University of Texas Press. Explores and analyzes technoerotic imagery in recent films, cyberpunk literature, comic books, and television, and shows how futuristic images encode current debates concerning gender roles and sexuality.

Turkle, Sherry. 1995. *Life on the Screen: Identity in the Age of the Internet.* New York: Simon & Schuster. One of the best and most influential accounts of how people navigate simulated worlds, create virtual personalities, and forge on-line relationships. Explores how computers impact subjectivity, and provoke new ways to think about politics, community, sexual identity, and basic conceptions of self.

Virilio, Paul. 1995. *Open Sky.* New York: Verso. A speculative lament over the perceived loss of real geographical spaces, distances, intimacy, the nuclear family, marriage, and democracy, by one of digital culture's most influential theorizers. An enigmatic proposal for a new ethics and ecology of perception able to counter the "infantilism" of cyberhype, for a culture struggling to achieve what he terms "escape velocity."

Virilio, Paul. 1994. *The Vision Machine.* London: BFI. Follows McLuhan, Baudry, and Crary, in attempting to provide an overarching history of seeing and social existence. A survey of technologies of perception, production, and the current logic of public representations.

Wasko, Janet. 1994. *Hollywood and the Information Age.* Austin: University of Texas Press. Investigation of the effects of new technologies on production, distribution, exhibition, and marketing of the filmed entertainment industry. Considers ancillary markets, globalization, theme parks, and merchandising as Hollywood positions itself for the twenty-first century.

Wiener, Norbert. 1967. *Cybernetics: Or Control and Communications in the Animal and Machine.* Cambridge, Mass.: MIT Press. Originally published in 1948. One of the foundational texts in cybernetics and information science. Based on experiments with radar, antiaircraft tracking, and "servo-"governed machines in the 1940s, cybernetics developed into a technical science based on the intelligent use and processing of feedback by computational technologies. The prefix "cyber" was eventually liberated from information technology engineering and popularized as a term for all things digital (e.g., cyberspace, cyberglove).

Williams, Mark. 2000. *Remote Possibilities: A History of Early Television in Los Angeles.* Berkeley: University of California Press. A provocative, historical case study of pre-Net electronic media. Examines the emergence of local and regional television broadcasting, precommercial initiatives, the geographic struggle between locals and national networks, and the ideological ways that nationalism, networking, and regulation negotiated local interests.

Williams, Raymond. 1974. *Television: Technology and Cultural Form.* New York: Schocken Books. Influential book that conjoined in a single study analysis of the institutional and industrial logic of broadcasting, along with considerations of programs and the ideological possibilities of new technologies. Elaboration of key notion of "flow" in electronic media, and alternatives to technological determinism.

Winston, Brian. 1986. *Misunderstanding Media.* Cambridge, Mass.: Harvard University Press. Revised and expanded as: Winston, Brian. 1998. *Media, Technology and Society—A History: From the Telegraph to the Internet.* London: Routledge. A comprehensive and detailed history of various nineteenth- and twentieth-century electronic media. Offers a useful, systematic model for explaining technological developments—based on the interplay of social necessity and suppression—and challenges popular and academic mythologies about information "revolutions."

Contributors

IEN ANG is professor of cultural studies and director of the Research Centre in Intercommunal Studies at the University of Western Sydney, Nepean, Australia. She received her Ph.D. at the University of Amsterdam and has published widely in the fields of audience research, media, and cultural studies. Her books include *Watching Dallas, Desperately Seeking the Audience,* and *Living Room Wars: Rethinking Media Audiences for a Postmodern World.*

JOHN THORNTON CALDWELL is chair of the department of film and television at the University of California at Los Angeles. He is the recipient of awards from the National Endowment for the Arts and Regional Fellowships; his films and videos have been screened in festivals in Amsterdam, Berlin, Paris, and New York and broadcast in the United States and abroad. He is the author of *Televisuality: Style, Crisis, and Authority in American Television* and is currently completing a film on migrant farmworker housing, *Rancho California (por favor),* and a book entitled *Theory Machines.*

CYNTHIA COCKBURN is a researcher and writer in the Development of Sociology at the City University, London. Her books include *Brothers: Male Dominance and Technological Changes, Machinery and Dominance: Women, Men and Technical Know-how,* and (with Susan Ormrod) *Gender and Technology in the Making.* She is currently researching gender and identity in the contexts of war, following her book *The Space Between Us: Negotiating Gender and National Identities in Conflict.*

HELEN CUNNINGHAM is a senior lecturer in communication, culture, and media studies at Middlesex University, United Kingdom, where she specializes in teaching about youth culture, new media technologies, and media audiences. She contributed to *Digital Diversions, Youth Culture in the Age of Multimedia* (Taylor and Francis, 1988), and *Young People, Creativity and New Technologies: The Challenge of Digital Arts* (Routledge, 1999).

HANS MAGNUS ENZENZBERGER is a prominent contemporary German poet and essayist. His book *poems for people who don't read poems* was published in 1968, and his influential collected works *The Consciousness Industry: On Literature, Politics, and the Media* was published in 1974. His essays have appeared in *Partisan Review, Encounter, The New York Review of Books, Modern Occasions,* and the *New Left Review.*

GUILLERMO GÓMEZ-PEÑA was awarded a multiyear MacArthur Foundation Fellowship in 1991 for his critically acclaimed work as a writer and multimedia performance artist. From 1984 to 1990 he founded and participated in the "Border

Arts Workshop" and contributed to the national radio program "Crossroads." He is the author of the book *Warrior for Gringostroika,* and his work on digital media and cultural and racial identity examines the possibilities of Chicano virtual reality, technopirating, cyber-coyotes, and Mexican hypertexts. In 1997 his book *The New World Border* received the American Book Award.

ARTHUR KROKER is critic, theorist, and author of *Spasm: Virtual Reality, Android Music, Electric Flesh, The Possessed Individual, The Postmodern Scene,* and *Data Trash: The Theory of the Virtual Class.* A contributing editor of *Mondo 2000,* contributor to *Technoscience and Cyberculture,* and co-editor of the influential electronic journal *CTHEORY,* Kroker is professor of political science at Concordia University, Montreal.

BILL NICHOLS is director of the graduate program in cinema studies at San Francisco State University and has taught at the University of California at Santa Cruz and Los Angeles. He is the editor of the agenda-setting two-volume anthology, *Movies and Methods,* and author of several books on documentary film, ethnography, and issues of cultural representation, including the widely cited *Ideology and Image, Representing Reality,* and *Blurred Boundaries.*

ANDREW ROSS is professor and director of the graduate program in American Studies at New York University. His research examines media and culture, intellectual history, science, social theory, and ecology and technology. His books include *No Respect: Intellectuals in Popular Culture, The Failure of Modernism, Strange Weather: Culture, Science, and Technology in the Age of Limits,* and *Real Love: In Pursuit of Cultural Justice.* He co-edited the anthology *Technoculture* and edited *Universal Abandon?: The Politics of Postmodernism.*

ELLEN SEITER is professor of communication at the University of California at San Diego, where she teaches media and cultural studies, media production, and women's studies. Her books include *Remote Control: Television Audiences and Cultural Power* (1989), which she co-edited, *Sold Separately: Parents and Children in Consumer Culture* (1993), and *Television and New Media Audiences* (1999). A new media consultant and producer, she recently completed an interactive CD-ROM project for children on superheroes and consumerism entitled *HeroTV* (www.southmoon.com).

VIVIAN SOBCHAK is associate dean and professor of film and television studies at the UCLA School of Theater, Film, and Television. Her work focuses on media theory and its intersections with philosophy, perceptual studies, and historiography. Her books include *Screening Space: The American Science Fiction Film, The Address of the Eye: A Phenomenology of Experience,* and two anthologies, *The Persistence of History: Cinema, Television and the Modern Event* and *Meta-Morphing: Visual Transformation in the Culture of Quick Change.*

ALLUCQUÈRE ROSANNE STONE is associate professor and director of the Advanced Communications Laboratory (ACTLab) at the University of Texas at Austin, senior artist at the Banff Center for the Arts, and former resident fellow at the Humanities Research Institute, University of California at Irvine. She has been a filmmaker, rock 'n' roll music engineer, neurologist, social scientist, cultural theorist, and performer. Her numerous publications include *The Empire Strikes Back:*

A Posttranssexual Manifesto and the *War of Desire and Technology at the Close of the Mechanical Age*.

RAVI SUNDARAM is Fellow in the Center for the Study of Developing Societies, Delhi, India. His areas of interest focus on alternative futures and transnational cultural studies, and his publications have addressed issues of temporality, nationalism, and modernity. His books include *Time, Modernity, and the Modern World System*, and his current research examines globalization and the crisis of the national imaginary.

MICHAEL A. WEINSTEIN is professor of political philosophy at Purdue University, photography critic for *New City* in Chicago, and a rap poet. He has published nineteen books, ranging from culture theory to metaphysics, including *Data Trash: The Theory of the Virtual Class*.

RAYMOND WILLIAMS was professor of political science at Stanford University and then University Reader in Drama at Cambridge University. His many books on communication—based on critical, sociological examinations of literature, drama, and the media—include *Communications, The Sociology of Culture*, and *Television: Technology and Cultural Form*.

BRIAN WINSTON is head of the school of communication design and media at the University of Westminster. He has previously headed departments at the National Film and Television School of Great Britain, Glasgow University, New York University, Penn State, and Cardiff University of Wales. He began his career at Granada Television in 1963 on *World in Action*, worked for the BBC's current affairs department, has worked as an independent documentary filmmaker, and, in 1985, won a prime-time Emmy for documentary script writing (WNET, New York). His ninth and latest book is *Media Technology and Society: A History, From the Telegraph to the Internet*.

Index